AF411370

Advances on Smart Cities and Smart Buildings

Advances on Smart Cities and Smart Buildings

Editors

Michele Roccotelli
Agostino Marcello Mangini

MDPI • Basel • Beijing • Wuhan • Barcelona • Belgrade • Manchester • Tokyo • Cluj • Tianjin

Editors
Michele Roccotelli
Department of Electrical and
Information Engineering
(DEI), Politecnico di Bari
Italy

Agostino Marcello Mangini
Department of Electrical and
Information Engineering
(DEI), Politecnico di Bari
Italy

Editorial Office
MDPI
St. Alban-Anlage 66
4052 Basel, Switzerland

This is a reprint of articles from the Special Issue published online in the open access journal *Applied Sciences* (ISSN 2076-3417) (available at: https://www.mdpi.com/journal/applsci/special_issues/Smart_Cities_Buildings).

For citation purposes, cite each article independently as indicated on the article page online and as indicated below:

LastName, A.A.; LastName, B.B.; LastName, C.C. Article Title. *Journal Name* **Year**, *Volume Number*, Page Range.

ISBN 978-3-0365-4015-3 (Hbk)
ISBN 978-3-0365-4016-0 (PDF)

Contents

About the Editors

Michele Roccotelli received an MScEng degree in Automation Engineering from the Polytechnic of Bari in 2010. He received his Ph.D. in Electrical and Information Engineering from the Polytechnic of Bari, Italy (Apr. 2016). He has been a visiting scholar (Jan. 2015-Jul. 2015) at the École Normale Supérieure de Cachan, Paris, France. He is currently an assistant professor at Polytechnic University of Bari, at the Department of Electrical and Information Engineering. His research interests include energy-efficiency systems, models and strategies for smart grids and smart buildings, and electric mobility systems and services. He has been a member of the program committees of several international conferences, such as the organizing committee of the IEEE SMC 2019. He is co-author of +40 articles published in international conferences and journals.

He has been/is involved in the following H2020 EU-funded research and innovation projects in the electromobility sector: (1) NeMo (https://nemo-emobility.eu/); (2) ELVITEN (https://www.elviten-project.eu/en/); (3) eCharge4Drivers (https://echarge4drivers.eu/).

Agostino Marcello Mangini received a Laurea degree in electronics engineering and Ph.D. degree in electrical engineering from the Polytechnic of Bari, Bari, Italy, in 2003 and 2008, respectively. He has been a Visiting Scholar with the University of Zaragoza, Zaragoza, Spain. He is currently an Assistant Professor with the Department of Electrical and Information Engineering, Polytechnic of Bari. He has authored or co-authored over 90 printed publications. His current research interests include the modeling, simulation, and control of discrete-event systems, Petri nets, supply chains and urban traffic networks, distribution and internal logistics, management of hazardous materials, management of drug distribution systems, and healthcare systems. Dr. Mangini was on the program committees of the 2007–2015 IEEE International SMC Conference on Systems, Man, and Cybernetics and the 2009 IFAC Workshop on Dependable Control of Discrete Systems. He was on the Editorial Board of the 2017 IEEE Conference on Automation Science and Engineering.

applied sciences

Editorial

Advances on Smart Cities and Smart Buildings

Michele Roccotelli * and Agostino Marcello Mangini

Department of Electrical and Information Engineering, Politecnico di Bari, 70126 Bari, Italy;
agostinomarcello.mangini@poliba.it
* Correspondence: michele.roccotelli@poliba.it

Citation: Roccotelli, M.;
Mangini, A.M. Advances on Smart
Cities and Smart Buildings. *Appl. Sci.*
2022, *12*, 631. https://doi.org/
10.3390/app12020631

Received: 2 January 2022
Accepted: 6 January 2022
Published: 10 January 2022

Publisher's Note: MDPI stays neutral
with regard to jurisdictional claims in
published maps and institutional affiliations.

Modern cities are facing the challenge of combining competitiveness on a global city scale and sustainable urban development to become smart cities. A smart city is a high-tech intensive and advanced city that connects people, information, and city elements using new technologies in order to create a sustainable, greener city; competitive and innovative commerce; and an increased quality of life. This Special Issue collects the recent advancements on smart cities and covers different topics and aspects.

Technological innovations have revolutionized the lifestyle of society and have led to the development of advanced and intelligent cities. The term smart city has recently become synonymous for a city that is characterized by an intelligent and extensive use of Information and Communications Technologies (ICTs) in order to allow the efficient use of information. In this context, new solutions and tools are offered to tourists to optimize and customize itinerary planning. In particular, graph theory and optimization algorithms are used to find the optimal touristic itinerary paths and, a multi-algorithms strategy is used to maximize the number of attractions (PoIs) to be visited on these paths [1].

A review of the sensors deployed in a smart city is conducted in [2], both from a technological and functional point of view. Sensors play an important role, as they gather relevant information from the city, citizens, and the corresponding communication networks and transfer the information in real-time. Although the use of these sensors is diverse, their application can be categorized in six different groups: energy, health, mobility, security, water, and waste management. Based on these groups, this review presents an analysis of different sensors that are typically used in efforts toward creating smart cities. Insights about different applications and communication systems are provided as well as the main opportunities and challenges faced when making a transition to a smart city. Ultimately, this process is not only about smart urban infrastructure, but more importantly, it is about how these new sensing capabilities and digitalization developments improve the quality of life of the citizens who live in these cities.

In addition, the global decarbonization and electrification of the world's energy demands has led to the quick adoption of Electric Vehicle (EV) technology. There is an emerging need to provide a wide network of fast Vehicle-to-Grid (V2G) charging stations to satisfy the energy demand and to guarantee the sufficient autonomy of such vehicles. Accordingly, V2G charging stations must be prepared to work properly with every manufacturer and operator and to provide reliable designs and validation processes. To support such processes, the development of power electric vehicle emulators with V2G capability is essential [3]. In [3], the design and development details of an electric vehicle emulator for testing V2G chargers with power factor grid correction functionality are provided, and the ability to emulate a real V2G EV to handle fast charge/discharge with an EV charger is validated experimentally. Another solution to support the attractiveness of EVs by enhancing the autonomy of batteries and by limiting the range anxiety is provided by the analysis in [4]. The integration of an auxiliary power unit (APU) can extend the range of a vehicle, making them more attractive to consumers. Recently, many extended-range electric vehicle systems and configurations have been proposed to recover energy. However, an extensive analysis of the most relevant technologies that recover energy, the current topologies and configurations of such devices, and the state-of-the-art of control methods used to manage energy is necessary to identify the best solution. The analysis presented mainly focuses on finding maximum fuel economy, reducing emissions, minimizing the system's costs, and providing optimal driving performance. The evaluation of range extenders for electric vehicles aims to guide researchers and car manufactures to generate new topologies and configurations for EVs with optimized range, improved functionality, and low emissions.

The problem of forecasting the electricity demand is not only linked to EVs. Today, buildings are still the main contributors of energy consumption within a smart city. In particular, the long-term electricity demand forecast is essential for the energy provider to analyze the future demand and for the accurate management of the demand response. Forecasting the consumer electricity demand with efficient and accurate strategies will help the energy provider to optimally plan generation points, such as solar and wind, and to produce energy accordingly to reduce the rate of depletion. An efficient and accurate forecasting model is provided by [5] to study the daily consumption of the consumers from their historical data and to forecast the necessary energy demand from the consumer's side. The proposed recurrent neural network gradient boosting regression tree (RNN-GBRT) forecasting technique allows the demand for electricity to be reduced by studying the daily usage patterns of consumers. The efficiency of the proposed forecasting model is compared with various conventional models.

The prediction methods are also adopted in vehicle sharing systems. In this framework, the work in [6] deals with the long-term prediction of bike rental/drop-off demands at given bike station locations in the expansion areas. The real-world bike stations are mainly built-in batches for expansion areas. To address the problem, they propose LDA (Long-Term Demand Advisor), a framework that can be used to estimate the long-term characteristics of newly established stations. In LDA, several engineering strategies are proposed to extract discriminative and representative features for long-term demands. Moreover, for the original and newly established stations, several feature extraction methods and an algorithm are proposed to model the correlations between urban dynamics and long-term demands. Real-world data from New York City's bike-sharing system are evaluated to show that the LDA framework outperforms baseline approaches.

Furthermore, ref. [7] proposes an analysis of European cycle logistics projects, studying how the corresponding supporting policies have an impact on their economic performance in terms of profit and profitability. First, they identify project success factors by geographic area and project-specific characteristics; then, they statistically test possible dependence relationships with supporting policies and economic results. Moreover, they provide a value-based identification of those characteristics and policies that more commonly lead to better economic results. The work can serve as a basis for the prioritization and contextualization of those project functionalities and public policies to be implemented in a European context. It was determined that cycle logistics projects in Europe achieve high profit and profitability levels and that the current policies are generally working well and the support of them as well as their profit and profitability vary across the bike model utilized: mixing cargo bikes and tricycles generates the highest profit and profitability, whilst a trailer–tricycle–cargo bike mix paves the way for high volumes and market shares.

Some real-use smart city case studies are presented in [8,9]. For instance, in [8], the Campus City project is presented. In the Campus City initiative, the Challenge Living Lab platform is used to promote research, innovation, and entrepreneurship, with the intention of creating urban infrastructure and creative talent (human resources) that solves different community, industrial, and government Pain Points within a Smart City ecosystem. The main contribution is the presentation of a working model and the open innovation ecosystem used in Tecnologico de Monterrey that could be used as both a learning mechanism as well as a base model for scaling it up into a Smart Campus and Smart City. A discussion on the findings of the model and challenge implementation is provided, showing that the Campus City initiative and the Challenge Living Lab allow the identification of highly relevant and meaningful challenges while providing a pedagogic framework in which students are highly motivated, engaged, and prepared to tackle different problems that involve government, community, industry, and academia.

Scaling up the analysis, another practical example is given by the City of Monterrey, Mexico, which is presented at two planning scales [9]: at the metropolitan and local levels. Both scales of analysis measure accessibility to main destinations using walking and cycling as the main transport modes. The results demonstrate that the levels of accessibility at the metropolitan level are divergent, depending on the desired destination as well as on the planning processes (both formal and informal) from different areas of the city. At the local level, the Distrito Tec Area is diagnosed in terms of accessibility to assess to what extent it can be considered a part of a 15 min city. The results show that Distrito Tec lacks the desired parameters of accessibility to all destinations to be a 15 min city. Nevertheless, there is a considerable increase in accessibility levels when cycling is used as the main mode of transportation. The current research project serves as an initial approach to understand the accessibility challenges of the city at different planning levels by generating useful and disaggregated data. Finally, it concludes by providing general recommendations that should be considered during planning processes that are aimed at improving accessibility and sustainability.

Funding: This research received no external funding.

Conflicts of Interest: The authors declare no conflict of interest.

References

1. Mangini, A.M.; Roccotelli, M.; Rinaldi, A. A Novel Application Based on a Heuristic Approach for Planning Itineraries of One-Day Tourist. *Appl. Sci.* **2021**, *11*, 8989. [CrossRef]
2. Ramírez-Moreno, M.A.; Keshtkar, S.; Padilla-Reyes, D.A.; Ramos-López, E.; García-Martínez, M.; Hernández-Luna, M.C.; Mogro, A.E.; Mahlknecht, J.; Huertas, J.I.; Peimbert-García, R.E.; et al. Sensors for Sustainable Smart Cities: A Review. *Appl. Sci.* **2021**, *11*, 8198. [CrossRef]
3. García-Martínez, E.; Muñoz-Cruzado-Alba, J.; Sanz-Osorio, J.F.; Perié, J.M. Design and Experimental Validation of Power Electric Vehicle Emulator for Testing Electric Vehicle Supply Equipment (EVSE) with Vehicle-to-Grid (V2G) Capability. *Appl. Sci.* **2021**, *11*, 11496. [CrossRef]
4. Puma-Benavides, D.S.; Izquierdo-Reyes, J.; Calderon-Najera, J.D.D.; Ramirez-Mendoza, R.A. A Systematic Review of Technologies, Control Methods, and Optimization for Extended-Range Electric Vehicles. *Appl. Sci.* **2021**, *11*, 7095. [CrossRef]
5. Dash, S.K.; Roccotelli, M.; Khansama, R.R.; Fanti, M.P.; Mangini, A.M. Long Term Household Electricity Demand Forecasting Based on RNN-GBRT Model and a Novel Energy Theft Detection Method. *Appl. Sci.* **2021**, *11*, 8612. [CrossRef]
6. Hsieh, H.-P.; Lin, F.; Jiang, J.; Kuo, T.-Y.; Chang, Y.-E. Inferring Long-Term Demand of Newly Established Stations for Expansion Areas in Bike Sharing System. *Appl. Sci.* **2021**, *11*, 6748. [CrossRef]
7. Giglio, C.; Musmanno, R.; Palmieri, R. Cycle Logistics Projects in Europe: Intertwining Bike-Related Success Factors and Region-Specific Public Policies with Economic Results. *Appl. Sci.* **2021**, *11*, 1578. [CrossRef]
8. Huertas, J.I.; Mahlknecht, J.; Lozoya-Santos, J.D.J.; Uribe, S.; López-Guajardo, E.A.; Ramirez-Mendoza, R.A. Campus City Project: Challenge Living Lab for Smart Cities. *Appl. Sci.* **2021**, *11*, 11085. [CrossRef]
9. Gaxiola-Beltrán, A.L.; Narezo-Balzaretti, J.; Ramírez-Moreno, M.A.; Pérez-Henríquez, B.L.; Ramírez-Mendoza, R.A.; Krajzewicz, D.; Lozoya-Santos, J.D.-J. Assessing Urban Accessibility in Monterrey, Mexico: A Transferable Approach to Evaluate Access to Main Destinations at the Metropolitan and Local Levels. *Appl. Sci.* **2021**, *11*, 7519. [CrossRef]

applied sciences

Review

Sensors for Sustainable Smart Cities: A Review

Mauricio A. Ramírez-Moreno [1], Sajjad Keshtkar [2], Diego A. Padilla-Reyes [1], Edrick Ramos-López [1], Moisés García-Martínez [3], Mónica C. Hernández-Luna [4], Antonio E. Mogro [1], Jurgen Mahlknecht [1], José Ignacio Huertas [1], Rodrigo E. Peimbert-García [1,5,*], Ricardo A. Ramírez-Mendoza [1], Agostino M. Mangini [6], Michele Roccotelli [6], Blas L. Pérez-Henríquez [7], Subhas C. Mukhopadhyay [5] and Jorge de Jesús Lozoya-Santos [1]

[1] School of Engineering and Sciences, Tecnologico de Monterrey, Eugenio Garza Sada 2501 Sur, Monterrey 64849, Mexico; mauricio.ramirezm@tec.mx (M.A.R.-M.); diego.padilla@tec.mx (D.A.P.-R.); edramos@tec.mx (E.R.-L.); a01366355@itesm.mx (A.E.M.); jurgen@tec.mx (J.M.); jhuertas@tec.mx (J.I.H.); ricardo.ramirez@tec.mx (R.A.R.-M.); jorge.lozoya@tec.mx (J.d.J.L.-S.)

[2] ESIME Unidad Ticoman, Instituto Politecnico Nacional, Mexico City 07340, Mexico; posgradoesimet@ipn.mx

[3] School of Engineering and Sciences, Tecnologico de Monterrey, San Luis Potosi 78211, Mexico; moises.garcia@tec.mx

[4] Faculty of Economy, Universidad Autonoma de Nuevo Leon, San Nicolas de los Garza 66451, Mexico; monica.hernandezlu@uanl.edu.mx

[5] School of Engineering, Macquarie University, Sydney, NSW 2109, Australia; subhas.mukhopadhyay@mq.edu.au

[6] Department of Electrical and Information Engineering, Politecnico di Bari, 70126 Bari, Italy; agostinomarcello.mangini@poliba.it (A.M.M.); michele.roccotelli@poliba.it (M.R.)

[7] Precourt Institute for Energy, Stanford University, Stanford, CA 94305, USA; blph@stanford.edu

* Correspondence: rodrigo.peimbert@tec.mx

Citation: Ramírez-Moreno, M.A.; Keshtkar, S; Padilla-Reyes, D.A.; Ramos-López, E.; García-Martínez, M.; Hernández-Luna, M.C.; Mogro, A.E.; Mahlknecht, J.; Huertas, J.I; Peimbert-García, R.E.; et al. Sensors for Sustainable Smart Cities: A Review. *Appl. Sci.* **2021**, *11*, 8198. https://doi.org/10.3390/app11178198

Academic Editor: Juan-Carlos Cano

Received: 29 June 2021
Accepted: 31 August 2021
Published: 3 September 2021

Publisher's Note: MDPI stays neutral with regard to jurisdictional claims in published maps and institutional affiliations.

Abstract: Experts confirm that 85% of the world's population is expected to live in cities by 2050. Therefore, cities should be prepared to satisfy the needs of their citizens and provide the best services. The idea of a city of the future is commonly represented by the smart city, which is a more efficient system that optimizes its resources and services, through the use of monitoring and communication technology. Thus, one of the steps towards sustainability for cities around the world is to make a transition into smart cities. Here, sensors play an important role in the system, as they gather relevant information from the city, citizens, and the corresponding communication networks that transfer the information in real-time. Although the use of these sensors is diverse, their application can be categorized in six different groups: energy, health, mobility, security, water, and waste management. Based on these groups, this review presents an analysis of different sensors that are typically used in efforts toward creating smart cities. Insights about different applications and communication systems are provided, as well as the main opportunities and challenges faced when making a transition to a smart city. Ultimately, this process is not only about smart urban infrastructure, but more importantly about how these new sensing capabilities and digitization developments improve quality of life. Smarter communities are those that socialize, adapt, and invest through transparent and inclusive community engagement in these technologies based on local and regional societal needs and values. Cyber security disruptions and privacy remain chief vulnerabilities.

Keywords: smart networks; digitization; smart transit; cyber security; smart city; smarter communities; smart sensors; smart meters; internet of things (IoT); facial recognition; cyber privacy

1. Introduction

"Smart Cities" or "intelligent cities" are the cities of the future that offer innovative solutions to improve the quality of life of urban communities in a sustainable and equitable manner. The idea of a "Smart City" represents more efficient cities that better manage their resources, services and technologies and, above all, put them at service of the citizens. People-centric planning will allow a better management and efficiency of the city through the deployment of smart infrastructure. By 2050, 85% of the world's population is expected to live in cities [1–3]. This means that in the following decades, urban centers will face a growing number of problems such as: (1) energy supply, (2) CO_2 emissions, (3) mobility systems planning, (4) raw materials and goods provision, and (5) the provision of health and security services to all the residents in these rapidly growing population centers [4].

To better respond to the increasing volatility driven by climate change, pandemics like COVID-19, and connected political and economic fluctuations, cities must be redesigned to increase their adaptive capacity and resilience [5]. Schemes and models of more liveable cities need to be created, where digital technologies are the key elements for the sustainable development and organic growth of cities [6]. Vulnerabilities of the shared economy in smart cities, such as transport services and their enabling technologies are being tested by the global pandemic and cyber security risks. However, new apps and internet businesses flourished, such as food delivery and online shopping [7–9].

Skepticism towards pervasive digitization, sensing, monitoring, and visualization capacities deployment, and other smart communication technologies used by both the private sector and government, arise, among other factors, from the concerns of citizens regarding the processing of their data, and, therefore, their privacy [10]. The use of new technologies such as artificial intelligence, where personal data play a critical role, are facing increasing challenges [11]. Support for the establishment of codes of conduct are being promoted by the industry itself, and use cases will be key so that its development does not suffer [1]. In the years to come, problems such as intentional disinformation, the evolution of so-called digital rights, or the consolidation of digital identity, must be addressed. The solutions oriented to solve these problems require a long-term perspective, which will steadily become more and more evident as Smart City implementations have become a norm across the globe [12].

With the incoming necessity of communities to become smarter, many applications have started to arise on different countries; and a literature revision that discusses such applications will be useful as a reference for future smart city implementations. In this review, the literature covering several applications of sensors for smart cities is summarized and discussed, in order to fulfill three main objectives: (1) To provide a revision of the most important Smart Cities implementations across the world; (2) To describe the main applications of sensors for smart cities across six main topics (health, security, mobility, water and waste management, and energy efficiency); and (3) To identify common challenges and opportunities of smart city deployments in the proposed six topics.

2. Literature Review

A systematic search of relevant scientific literature was conducted in this study through databases such as Scopus, Google Scholar, and IEEE Xplore. The literature search on sensors for smart cities was divided into six main sectors: health, security, water, waste, energy, and mobility. The criteria for selecting and revising relevant papers from each section, followed the PRISMA methodology [13], as well as these principles:

1. Recent (2010–2020) literature was reviewed to ensure a revision of the current state of the art of the technologies applied to smart cities environment; prioritizing papers published during the 2015–2020 period. Figure 1a shows the distribution of the years of publications of the revised papers for this review;
2. Among the selected literature for each section, the most cited papers, including journal articles and conference proceedings were revised extensively. Figure 1b presents the distribution of the number of references with an increasing number of citations of the revised literature. Papers from Q1 and Q2 journals were given priority over Q3 and Q4 journals; as well as journals with impact factors higher than 1.0. Figure 1 shows (c) the quartiles, and (d) journal impact factor of the revised papers in this this review. Additionally, Figure 2a shows the type of references (journal, conference proceedings, books, webpages, and theses) selected and its percentage;
3. In the six main sectors, preference was given to articles where the main topic was the use of sensors exclusively for the subject evaluated. A wide range of studies

from the exploration of theoretical aspects up to practical applications were included. Figure 2b shows the percentage of revised papers under categories "Health", "Security", "Mobility", "Water", "Waste", "Energy", and "Smart Cities";

4. For each of the six sectors, different keywords were used to find relevant literature across each field. A list of keywords used for each section is presented as follows:

   (a) Health: Key terms were searched in the publication title, abstract, and keywords, and include: "smart city", "smart cities", "sensors", "wearable sensors", "body sensors", "smart health", "smart healthcare", "healthcare sensors", "healthcare applications", and "internet of things";

   (b) Water: The selection of the articles for the survey was carried out using the keywords "water" AND "sensors" AND "smart cities";

   (c) Waste: The selection of the articles for the survey was carried out using the keywords "waste" AND "sensors" AND "smart cities";

   (d) Mobility: Among the main keywords and combination of keywords used for this search were "mobility" AND "sensors" AND "smart cities". Other keywords such as "traffic", "vehicle", "pedestrian" AND "sensors" AND "smart cities" were also included;

   (e) Energy: The following keywords were included in the search: "energy consumption", "thermal comfort", "energy-consuming systems", "greenhouse gas emissions", "HVAC system", "lighting systems", "buildings energy consumption", "urban space energy consumption", "Key Performance Indicators", "Light Power Density (LPD)", "alternative energy source", "smart buildings", "smart lighting", "smart citizens", "ecological buildings", "virtual sensors", "BIM modeling", "energy consumption sensors";

   (f) Security: Keywords from this topic include: "Cybersecurity", "Sustainable Development", "Environment Security", "Society Security", "Human Security" AND "sensors" AND "smart cities".

Following the aforementioned principles, a total of 193 references were reviewed in detail; of which 129 come from journals, 46 from conference proceedings, 9 from books, 8 from web pages, and 1 from a Ph.D. thesis.

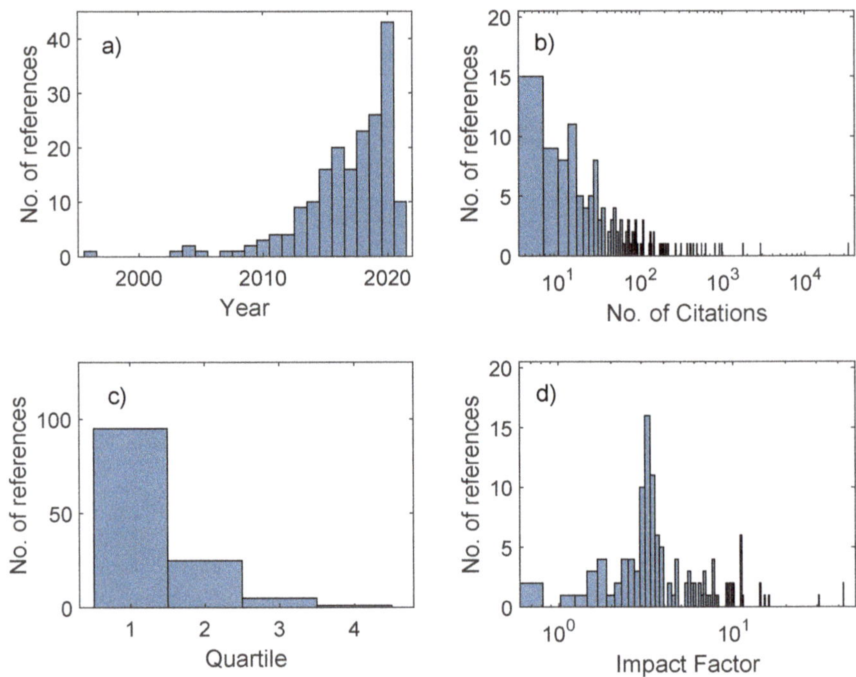

Figure 1. Histograms showing the distributions of different features of the papers selected for this review: (**a**) Year, (**b**) Number of citations, (**c**) Journal Quartile and (**d**) Journal Impact Factor.

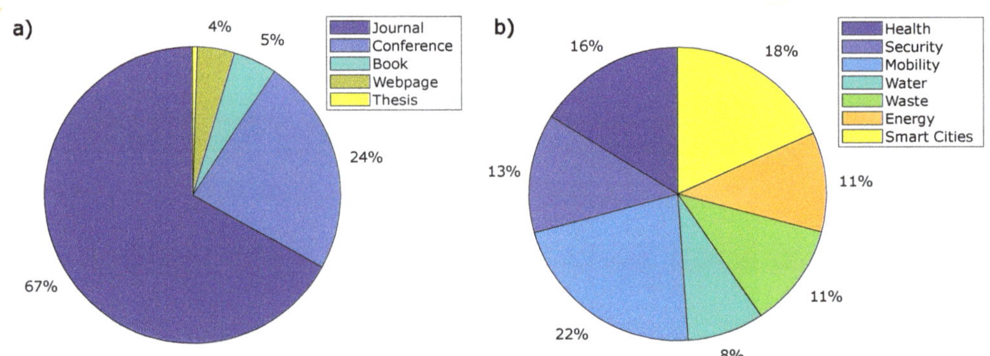

Figure 2. Pie charts representing the distribution of the (**a**) Type of reference and (**b**) Topic of the selected references for this review.

3. Results

3.1. Smart City

The goals of smart cities initiatives are to develop economically, socially, and environmentally sustainable cities [5,6,12]. Generally, the ideal model of a smart city is based on the incorporation of the following subsystems and technologies: distributed energy generation (micro-generation) [14], smart grids (interconnected and bidirectional smart networks) [15], smart metering (intelligent measurement of energy consumption data) [16], smart buildings (eco-efficient buildings with integrated energy production systems) [17], smart sensors (intelligent sensors to collect data and keep the city connected) [18], eMobility (implementation of electric vehicles) [19], information and communication technologies (ICT) [20] and smart citizens (key piece of a smart city) [21].

Cities constitute a complex socio-technical system [22,23]. In order to design the best solutions for cities, their inhabitants, social entities and governments need to be considered [2,3].

More pleasant spaces and places are required to live, boost competitiveness and productivity. To achieve this, the development of communication technology, such as 5G, is imperative. There also exists a growing need for initiatives of Industry 4.0 to permeate cities, since small or medium-sized enterprises (SMEs) represent the largest business fabric in developing countries [5]. In this sense, it is necessary to increase operators' access to reliable 5G infrastructure, allowing optimal deployment and economic rationality of the networks. In addition, governments are expected to provide access to resources and tools that facilitate the deployment speed and offer the necessary infrastructure for the latest generation networks [24].

3.2. Smart Cities in the World

In this section, some smart cities deployments across the world are presented. In Europe: Tampere, Helsinki, Amsterdam, Vienna, Copenhagen, Stockholm, Milton Keynes, London, Malaga, Barcelona, Santander, Paris, and Geneva [4,25–27]. In Asia: Singapore, Hong Kong, Shanghai, Beijing, Songdo, Seoul, and also smart cities in Taiwan, Indonesia, Thailand, and India [4,25,27–29]. In North America: Toronto, Vancouver, New York, Washington, and Seattle [4,25,27]. In South America: Medellin and Rio de Janeiro [27,30,31]. In Oceania: Melbourne, Perth, Sydney, Brisbane, and Adelaide [27,32].

The most common smart city project implementations across these cities include: development of collaborative business districts in Barcelona [21,25] and Hong Kong [4,27]; citizen security by traffic monitoring in Rio de Janeiro [4,31], and natural disaster monitoring in Singapore and Indonesia [29,33]; public service, smart government, and communication transformation in cities of China (Wuhan, Shanghai, Beijing, Dalian, Tianjin, Hangzhou, Wuxi, Shenzhen, Chengdu, and Guangzhou) [28]; adaptation of cultural spaces in Medellin [30]; citizen engagement and data enhancement in New York and Washington [27]; deployment of experimental testbeds and living labs in Santander [26] and London [27]; integration of local and foreign universities in Tampere [27] and Songdo [25]; deployment of smart green projects and policies in Seoul [27] and Toronto [4]; fiber optic and smart grids in Geneva [27]; energy efficiency and innovation enhancement in Vienna [27]; improved water consumption in Copenhagen, Hong Kong, and Barcelona [4]; deployment of electric charging stations in Malaga and Paris and Amsterdam [4]; Big data integration and analysis in India and Thailand [29]; wired communities, pedestrian mobility, and mass transit solutions in Sydney, Brisbane, Adelaide, Melbourne, and Perth [32]; carbon emission reduction in Seattle [25]; smart waste collection systems in Helsinki [25], Songdo [25], Barcelona [25]; smart parking in Milton Keynes [25]. Table 1 shows the main Smart City implementations reported in the literature for health, security, mobility, water, waste management, and energy efficiency, for the countries reviewed in this section. A more extensive search on smart city deployments around the world was performed using the results from the literature review, as well as information gathered from related websites [34–36]. The results of this search are presented in Figure 3.

Table 1. Smart cities implementations around the world.

City	Health	Security	Mobility	Water	Waste	Energy
Tampere [27]			Smart transportation			
Helsinki [25]			Car charging facilities		Automated waste collection	Smart grids
Amsterdam [4]	ICT in health, Health Lab					Clean energy generation
Vienna [27]			Smart parking, car sharing			Energy efficiency
Copenhagen [4]			Bike lane network	Water quality monitoring	Optimized waste disposal	Energy efficiency
Stockholm [4]				Water management policies	Waste management system	
Milton Keynes [25]			Smart parking, MotionMap app		Sensors in recycling centers	Smart metering app
London [27]			App for public transport		Smart Waste collection	
Malaga [4]			Electric vehicles, charging stations			Smart grids, clean energies, smart lighting
Barcelona [4]	Remote healthcare	Incident detectors at home	Traffic and public transport management		Smart Containers	Centralized heating/cooling
Santander [26]			Smart Parking, GPS monitoring	Smart park irrigation		Smart public lighting
Paris [4]	eHealth, smart medical records		Bike sharing, charging stations			
Geneva [27]			Smart transportation			Fiber-optic, smart grid networks
Singapore [33]		Siren alerts for natural disasters	Traffic maps, public transport apps	Apps for water consumption tracking		Apps for energy consumption tracking
Hong Kong [4,27]		Smart card IDs for citizens	Open, real-time traffic data		Smart waste management	
Shanghai [28,37]			Pedestrian movement analysis (Big data)			
Beijing [28,38]			V2E solutions, smart cards for transportation			
Songdo [25,27]	Remote medical equipment and checkups		Self-charging electric vehicle technology		Underground waste suction system	Smart buildings
Seoul [27]			Bus service based on data analytics			
Taiwan [39]		Smart defense system for law enforcement				
Indonesia [33]		Flood monitoring and report app				
Thailand [33]		Tsunami and flood monitoring		Water management app		
India [4]			Smart transport systems			Clean energy, green buildings
Toronto [28]			Smart urban zone growth			

Table 1. *Cont.*

City	Health	Security	Mobility	Water	Waste	Energy
New York [27]		Sensors deployment after 9/11 attacks				Energy efficiency using LEDs
Washington DC [27]			Bike sharing, smart stations			Sensor-based LED streetlights
Seattle [25]		Flood monitoring, law-enforcement cameras, gunshots GPS tracking	Smart trafic lights	Real-time precipitation monitoring		Reduction of CO_2 emissions
Medellin [30]			Outdoor electric stairs and air wagons			
Rio de Janeiro [31]		GPS/video monitoring installation in police cars	Traffic monitoring using cameras			
Melbourne [27]			Smart parking, open urban planning, metro Wi-Fi			Energy efficiency, smart grid, smart lighting
Perth [40]		Cyber-security and digital forensics				
Sydney [32]			ICTs in daily urban transport			
Brisbane [32]			Pedestrian spines			
Adelaide [32]			Wired communities			

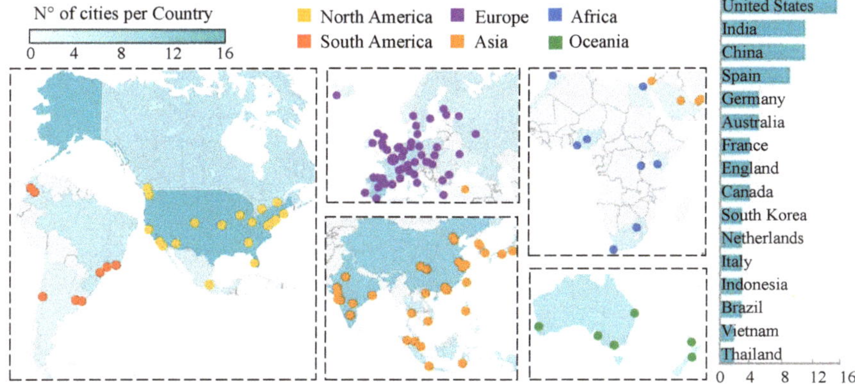

Figure 3. Map of smart cities across the world. The shade of color represents different number of smart cities per country. Cities are represented with colored points in the map, identified in six main regions: Africa, Asia, Europe, North America, Oceania, and South America. The bars on the right side show the countries with highest number of smart cities.

3.3. Sensors

The general architecture of an intelligent management system consists of readings (sensors), gateways (communication), and workstations (instructions, analytics, software, and user interface) [41–43].

3.3.1. Sensors for Health Monitoring

Healthcare has become a prolific area for research in recent years, given that new sensor technology allows real-time monitoring of the patients' state. Smart healthcare provides healthcare services through smart gadgets (e.g., smartphones, smartwatches, wireless smart glucometer, etc.) and networks (e.g., body area and wireless local area network), offering different stakeholders (e.g., doctors, nurses, patient caretakers, family members, and patients) timely access to patient information and the ability to deploy the right procedures and solutions, which reduces medical errors and costs [44].

Biosensors are fundamental when monitoring health, and different applications can be identified in medical diagnoses [45], and antigen detection [46], among others. Inorganic flexible electronics have witnessed relevant results, including E-skin [47], epidermal electronics (see Figure 4) [48], and eye cameras [49]. Some common materials used to create these sensors are carbon-based or conductive organic polymers, which present poor linearity [50,51]. However, more reliable, and flexible sensors have been created at lower cost, better linearity, and shorter response time, such as piezoresistive sensors integrating nano-porous polymer substrates [52].

New sensor developments are creating relevant opportunities in the health industry. Current procedures for sensing proteins are commonly based on noisy wet-sensing methods. A more robust procedure is carried out by means of graphene sensors that avoid the drifting of electrical signals, resulting in more stable and reliable signals. These sensors also reduce detection times [53]. Nonetheless, when having these robust sensors connected to the human body, one critical challenge is their communication through the wireless sensor network, since the IEEE 802.15.4 standard can hardly be adapted to multi-user interfaces [54]. Still, some solutions have been proposed, such as replacing the ultra-wideband (UWB) with a gateway, so sensor nodes stop and switch to 'sleep mode' until new information transmission is needed again. This allows lowering the energy consumption and collisions and increase the speed and number of users [55].

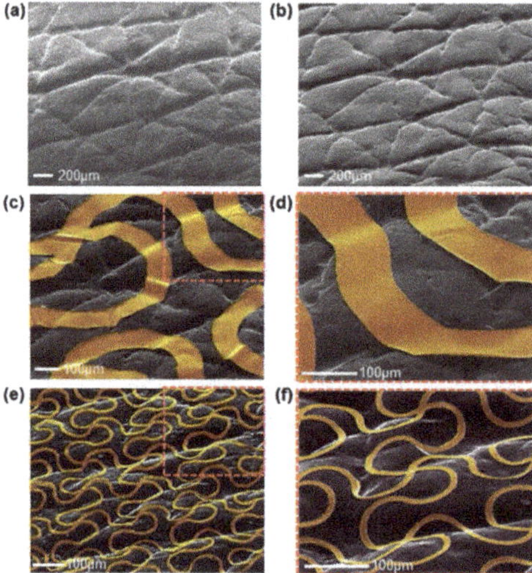

Figure 4. (**a**) Epidermal electronic system (EES) made of a skin replica created from the forearm, before and; (**b**) after application of a spray-on-bandage; (**c**) Colourized microscopy image of the EES with conductive gold films of 100 μm and; (**e**) its magnified view. (**d**) Microscopy image of EES with gold films of 10 μm and; (**f**) its magnified view [48].

3.3.2. Sensors for Mobility Applications

Three main systems of urban mobility: vehicles, pedestrians, and traffic, are considered in this section. Due to the increasing number of vehicles every year in urban settlements, traffic jams, pollution, and road accidents tend to increase as well [56]. These problems suggest that there is an emergent need for intelligent mobility solutions. One such solution is intelligent traffic control, oriented to avoid traffic jams and optimize traffic flow [18].

Due to repetitive starts and stops, fuel consumption and carbon emissions increase during traffic jams [18]. Therefore, providing solutions for traffic jams represents a direct positive impact in terms of urban mobility and air quality in cities. Moreover, heavy-duty vehicles (HDVs) and freight traffic release into the atmosphere large quantities of carbon emissions [57]. The automotive industry has put significant effort into developing more energy efficient powertrains (for example, hybrid electric vehicles). However, most HDVs are still fueled by diesel and providing optimal solutions to reduce the carbon emissions produced by these types of vehicles becomes a fundamental task [58].

Due to the COVID-19 pandemic, urban mobility underwent significant changes, such as a noticeable decrease in collective and individual transport, and with that a reduction in air pollution and carbon emissions [59]. This phenomenon has made governments and citizens consider future changes in post-lockdown mobility, to maintain cleaner environments in their cities. Applications in smart mobility include vehicle–vehicle (V2V), vehicle–infrastructure (V21), vehicle–pedestrian (V2P) [60], and vehicle–everything (V2X) connections (see Figure 5) [61].

Vehicles include several sensors needed for their proper operation, measuring several operational parameters of the vehicle, such as speed, energy consumption, atmospheric pressure, and ambient temperature [62,63]. Such parameters are used to optimize speed profiles to minimize vehicle energy consumption considering traffic condition and geographical information. To achieve this aim, a cloud architecture is implemented that retrieves information from vehicle sensors and external services. In [64], an eco-route planner is proposed to determine and communicate to the drivers of heavy-duty vehicles (HDVs) the eco-route that guarantees the minimum fuel consumption by respecting the travel time established by the freight companies. Additionally, in this case, a cloud computing system is proposed that determines the optimal eco-route and speed and gear profiles by integrating predictive traffic data, road topology, and weather conditions. Vehicle weight and speed regulation are also important to ensure road and passengers safety, helping in the avoidance of serious accidents [65]. Efforts in increasing pedestrian's safety are valuable contributions in improving urban mobility [66]. Regarding traffic, conventional traffic light systems are defined in a non-flexible structure, such that light transitions have defined delays and onsets [67]. Dynamic changes in traffic volume, congestions, accidents, and pedestrian confluence, should have been considered to provide an optimized traffic control [67].

Pedestrian´s movement and behavior in urban settings have been monitored mainly using cell phones, by monitoring call detail records (CDRs) [68], social media check-ins [37,69], MAC address reading [70], and smart cards detection in public transport [71]. Vehicle detection have been achieved by cement-based piezoelectric, induction loop sensors, measuring vehicle's weight-in-motion (WIM), and performing vehicle type classification [65], and ferromagnetic sensors buried in the asphalt for smart-parking solutions [72]. Vision-based sensors, such as infrared (IR) [67] and light detection and ranging (LiDAR) [73] have been used to detect the position of vehicles, pedestrians and buildings within a given proximity. Pedestrian–vehicle (P2V) oriented sensors also exist, such as the "smart car seat", a contact-free heart rate monitoring sensor oriented to ensure driver's well-being and safety [74]. Mobile phone apps have allowed P2V and V2P applications for collision prediction [66,75].

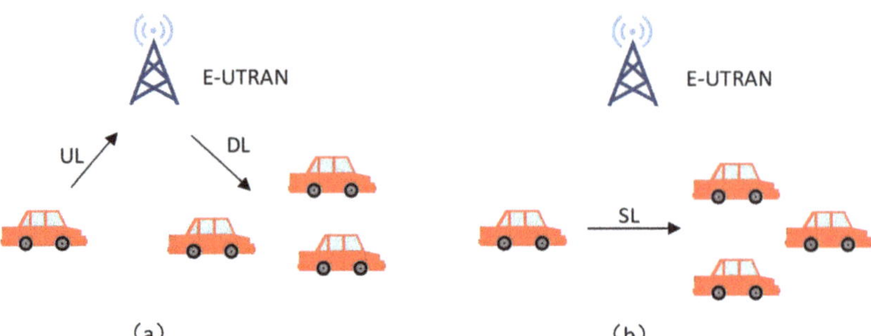

Figure 5. Two types of V2V connections: (**a**) from vehicles to vehicles with an intermediate transfer connection and; (**b**) directly from vehicles to other vehicles [61].

Recently, virtual sensors have been used to enhance innovative solutions especially in the electro-mobility sector. VSs have been introduced for operating in the sensor-cloud platform as an abstraction of the physical devices. In particular, a VS can logically reproduce one or more physical sensors, facilitating and increasing their functionalities, performing complex tasks that cannot be accomplished by physical sensors [76]. Differently from a real sensor, the VS is equipped with an intelligent component based on data processing algorithm to derive the required information elaborating the available input data from heterogeneous sources. Indeed, VS is typically used in services in which it is necessary to derive data and information that are not available or directly measurable from physical sensing instrumentation [77,78]. Although the use of such sensors has been explored in different domains or verticals of the smart city, the mobility sector is the one where they find large application. In the electric mobility domain, for instance, they are used to predict the personal mobility needs of the driver to estimate the duration and cost of the battery charging, to predict the energy demand of medium or long-range trips, etc. All these predictions are performed through ad-hoc algorithms able to process available input data from the electric vehicles, the users, and the charging stations [76–78].

3.3.3. Sensors for Security

The human and environmental security approaches are a very crucial ingredient to achieve sustainable development in smart cities. Security is referred to as a state of being free from danger or threat and for maintaining the stability of a system. Safety is a dynamic equilibrium, which consists in maintaining the parameters important for the existence of the system within the permissible limits of the norm. According to the United Nation's Human Security Handbook and Agenda 2030 Sustainable Development Goals (SDGs) [79] the types of insecurities endangering the sustainable development of humans and, hence, future cities are: food, cybernetic, health, environmental, personal, community, economic, and political, as the main core of a smart city.

Food security: The importance and technological challenges of the integration of urban food systems in smart city planning are discussed in [80]. High quality and sustainable production include smart hydroponics and gardening systems that gather information by sensors that measure pH, humidity, water and soil temperature, light intensity, and moisture [81]. Several methods are proposed to monitor the quality and safety of the food during production and distribution, including gas sensor array [82] for the analysis of chemical reaction occurred in spoiled food. Hybrid nanocomposites and biosensors have also been reported in food security context [83].

Cyber security: The main security challenges, including privacy preservation, securing a network, trustworthy data sharing practices, properly utilizing AI, and mitigating failures, as well as the new ways of digital investigation, are discussed in [11,40]. Design plane solutions are usually software-based and use diverse types of encryption techniques, including advanced encryption standard (AES) and elliptic curve cryptography (ECC) for crypto or level security and encryption, authentication, key management, and pattern analysis for the system-level security [84,85] (see Figure 6).

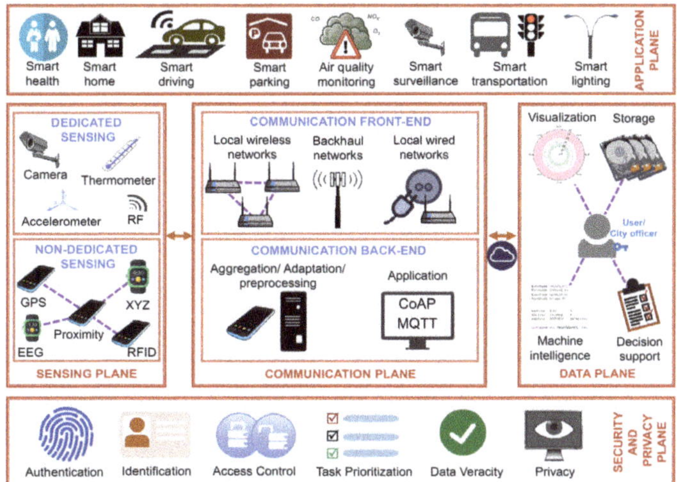

Figure 6. Smart city architecture defined in five planes: application (connects city and citizens), sensing (sensors measurements), communication (cloud services), data (processing and analysis) and security and privacy planes (assurance of security and privacy) [85].

Health security: In-body inserted devices are designed to communicate with health-centers and hospitals [55]. The privacy, security, and integrity of these sensors and the information on the health record concerning legal and moral issues are of great interest which is widely discussed in works such as [86,87].

Environmental security: In recent decades, the use of satellite remote sensing and in-orbit weather observation, disaster prediction systems have risen drastically. These tools are an integral sensing part of the future smart cities [88,89]. There is a wide range of sensors, including earthquake early detection systems which use vibration detection and monitoring soil moisture and density of the earth [90], radiation level detectors [91], tsunami inundation forecast methods assimilating ocean bottom pressure data [92]. These sensors are connected in a wireless network, offering a global prognosis of environmental threats. Continuous emission monitoring systems (CEMS) have helped develop market-based environmental policies to address air pollution [93]. CEMS are allowing better tracking of powerplant emissions in real time to inform decarbonization strategies for the grid [94]. Efforts to deploy cost-effective sensing capabilities so far have produced fragmented data, but new optimization and AI tools are being proposed to resolve this issue [95]. New smart sensing and visualization (e.g., satellite, LiDAR, etc.) capabilities are focusing on greenhouse gas emissions (GHG), for instance around the carbon capture potential of agriculture, forestry, and other land uses (e.g., natural climate solutions). Advanced monitoring, reporting, and verification (MRV) features will continue to play a major role in enhancing the transparency, environmental integrity, and credibility of subnational, national, and regional emissions trading systems (ETS) for the future integration of a global carbon market [96,97]. Regarding infrastructure and buildings, continuous monitoring to detect corrosion and minor damages to prevent a possible failure takes advantage of the integration of surveillance cameras, humidity, atmospheric, and stress sensors, among others [98]. A simple combination of vibration and tilt sensing devices provides one of the low-cost and high-efficiency techniques proposed for a wide range of structures [99].

Personal and community security: To detect anomalies, violence, and unauthorized actions, biometrics and surveillance cameras are widely used. Smart lighting systems are a useful and cost-effective tool that uses common sensors like light and motion detectors and can improve the security tasks [100]. Surveillance cameras, face-recognition systems, and global positioning systems (GPS), in combination with data handling systems, are increasingly common tools in the hand of law-enforcement agencies as smart public security strategy reported in [39]. Ref. [101] reviews the possible combination of different devices categorized in sensors, actuators, and network systems. The challenges presented by the growing use of such technologies and concerns for individual privacy is the topic of an emerging research area [102].

3.3.4. Sensors for Water Quality Monitoring

Water is an invaluable commodity and is necessary for any living being. Smart water management focuses mainly on making water distribution systems more efficient by applying sensors and telemetry for metering and communication [103,104]. It applies in three broad areas: fresh water, wastewater, and agriculture. Moreover, more holistic perspectives around shared resource systems, such as the water–energy–food nexus are also benefiting from new sensing capacities and smart management systems enabled by digital technologies to provide more sustainable, resource efficient use solutions [97]. The principal usefulness of smart water systems lies in controlling valves and pumps remotely [104] measuring quality [103], pressure, flow, and consumption [105].

Consumption monitoring includes metering and model applications to describe consumption patterns. Water loss management encompasses leakage detection and localization [105]. For water quality the focus is on measuring, analyzing, and maintaining a set of pre-established parameters. It is an integral real-time management involving stakeholders [106–108]. In the agriculture, the use of IoT devices is a common way to make irrigation more efficient and effortless [109–111]. Noise sensors and accelerometers are popular methods to detect leaks in water distribution infrastructure [105,106].

The use of electromagnetic and ultrasonic flow meters and sensors for measuring pressure are IoT technologies for water consumption rate analysis [103,108].

Sensors used to analyze the quality of the water are mainly applied for physical–chemical parameters such pH, temperature, electrical conductivity and dissolved oxygen [108,109], also oxidation-reduction potential and turbidity [112–114], and presence of toxic substances [115]. In some cases, novel probes, such as for residual chlorine [103] or nitrate and nitrite, were implemented (see Figure 7) [112]. Humidity sensors are applied to measure soil moisture and assist in managing the schedule programs of irrigation in agricultural lands [110,116,117].

3.3.5. Sensors for Waste Monitoring

Smart waste management consists of resolving the inherent problems of collection and transportation, storage, segregation, and recycling of the waste produced. Use of smartbins, solutions for the Vehicle Routing Problem (VRP) and waste management practices have been reported [41,118]. The use of smartbins refers to the implementation of different kind of sensors in the bins used to collect waste, which provide quantitative and qualitative information about the bin content [119–121].

For the VRP, the proposals are algorithms for optimization of the routes, considering social, environmental, economic factors, peak hours, infrastructure, type and capacity of the collection vehicles and others, in an effort to save resources like money, time, fuel, and labor [121–123].

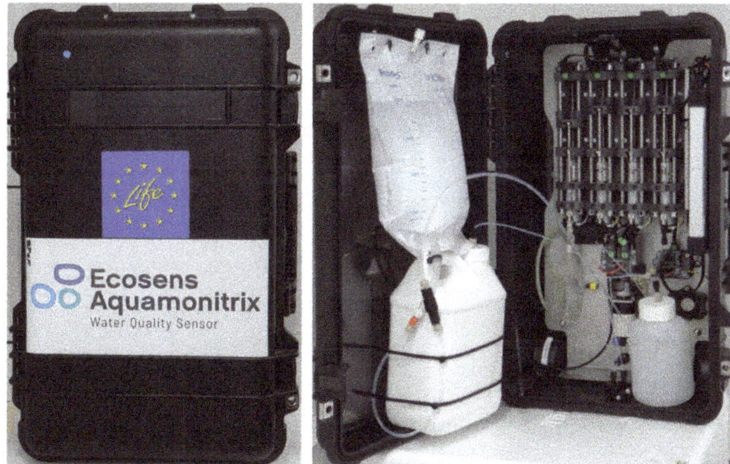

Figure 7. Outer and inner view of an integrated IoT sensor for water quality monitoring applications. The sensor consists of a nitrite and nitrate analyzer based on a novel ion chromatography method, used for detection of toxic substances [112].

With this, researchers aim for an integrated, real-time management that involves the communities and all the stakeholders [124,125]. The principal use of sensors in smartbins is monitoring the volume, weight, and content of the bins. For monitoring the filling level of the container, the main approaches that have been used are ultra-sound (US) [43,119], and in some cases IR sensors [121,126]. A load cell is also used to detect the weight of the bin [124,127]. In the literature, various sensors are used to detect harmful gases [126], movements near the container [120], and metal sensors to separate metallic waste [128], and to measure humidity [126,128], as well as temperature [43,123]. The My Waste Bin IoT container presented in [43] is shown in Figure 8.

Figure 8. Front and back view of the My Waste Bin, an IoT smart waste container, enabling real-time GPS tracking and weight monitoring [43].

3.3.6. Sensors for Energy Efficiency

Energy is an essential resource for the operation of the many activities occurring in cities [14]; therefore, the efficient use of this resource is paramount to reduce costs and promote environmental and economic sustainability [129].

The main sinks of energy consumption in urban communities are those associated with industrial and transport activities, buildings operations, and public lighting. In this section, we focus on the sensors used to monitor the usage of energy ground transport in buildings and public lighting. Since sensors for industrial activities were covered in previous sections.

Ground transportation represents the main sink of energy consumption ($\sim$45%) and the major source of air pollutants in urban centers [130]. Car manufacturers report the specific fuel consumption (SFC measured as L/km) of their vehicles using laboratory test protocols. However, they do not report these data for heavy-duty vehicles. Furthermore, the real vehicles' energy consumption is affected by human (driving), external (traffic, road, and weather conditions), and technological factors. For gasoline and diesel-fueled vehicles, the common strategies to measure real-world fuel consumption on a representative sample of vehicles are: (i) measuring the fuel's weight before and after a specific distance being driven (gravimetric method), and (ii) measuring instantaneous fuel consumption through the on-board diagnostic system (OBD method). This second alternative uses optical sensors to measure the engine RPM, pressure sensors to measure the inlet air flow. The engine computer unit (ECU) uses these measured data to determine the engine fuel injection time. In addition, a global position system (GPS) determines the vehicle's speed. Using all this information, the ECU reports via OBD the vehicle's instant fuel consumption. Currently, there are commercially available readers that read the OBD data and send the collected information to the cloud. Using these technologies, telematics companies monitor thousands of vehicles in operation [131–133]. Similar systems are available for electric vehicles. We recommend this OBD-based alternative to measure the real energy consumption in ground vehicles.

Buildings represent 40% of total energy consumption [134] and 30% of greenhouse gas (GHG) emissions [129,134]. The main physical and non-physical factors involved in indoor environment quality are shown in Figure 9 [134]. These factors are measured using wireless sensors [135,136], virtual sensors [137], and artificial neural networks [16]. Energy consumption in buildings is associated mainly with (i) thermal comfort (operation of heating, ventilation, and air conditioning-HVAC systems); (ii) indoor lighting; (iii) various electrical loads (operation of electric equipment); (iv) thermal loads (use of fuels for heating and cooking), and (v) indoor air quality (pollutant concentration, odor, and noise) [138,139]. Table 2 shows the variables used to grade these five aspects and the sensors frequently used to measure these variables. However, additional variables influence energy consumption, such as the occupation level [140], and the building's structural design [141], and outdoor conditions (temperature, humidity, pressure, and solar radiation). Therefore, additional sensors are used to measure them. Some research works have focused on designing intelligent building management systems (BMS) that use, in real-time, data from the sensors listed above and take actions oriented toward the reduction in energy consumption, such as turn lights off, closing doors and windows, etc. [142].

Public lighting systems represent almost 20% of world electricity consumption, and it is responsible for 6% of GHG [143]. Therefore, it is essential to centralize street lighting control and smart management to reduce energy consumption, maintain maximum visual comfort and occupant requirements [139]. The variable most used in lighting systems is the lighting power density (LPD) [138]. Neural networks, wireless sensors, algorithms, and statistical methods are used to estimate the energy consumption and the corresponding costs [129,135,136]. Environmental factors, pedestrians' flow, weather conditions, and brightness levels influence light intensity [129,144]. The urban space adopts the most advanced Information and ICTs to support value-added services to manage public affairs, connecting the city and its citizens while respecting their privacy [20,145].

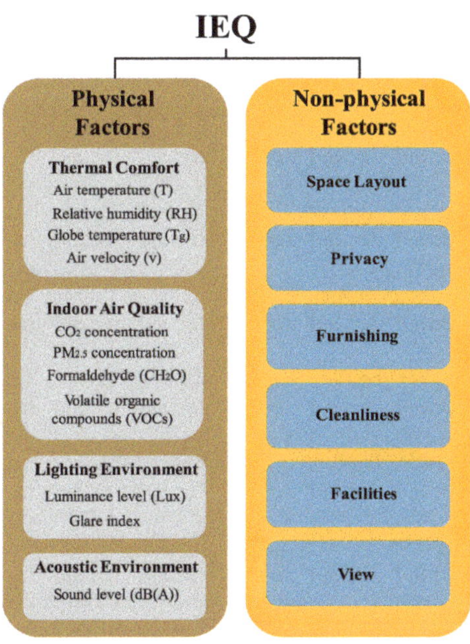

Figure 9. Physical and non-physical factors in IEQ studies [134].

Most of the time, the monitoring of the variables influencing energy consumption in buildings and public lighting is carried out wirelessly [18,129]. Various intelligent sensors [135], such as artificial vision, are used to measure temperature, relative humidity [17], electric current, gas flow, air quality [139], lighting, luminosity [129], solar radiation [139], and acoustic emission [146]. Measurement devices include light-dependent resistor (LDR) [139], IR radiation [140,147], semiconductors, magnets, and optic fiber. Intelligent sensors based on IoT are key to real-time monitoring of the many variables involved in energy management; these sensors can be adapted to microcontrollers and virtual sensors [129,137]. The main problem with wireless sensors is their battery life; therefore, alternative energy sources (thermal, solar, wind, mechanical, etc.) is vital, although these energies are usually available in minimal quantities [18].

Table 2. Sources of energy consumption in cities.

Sources of	Energy Consumption	Variable	Sensor	Application
Air Conditioning	Heat/Cooling	Temperature/Humidity	Thermohydrometer	Temperature and relative humidity
Indoor Air Quality	Air pollution/Concentration	CO_2/CO	NDIR (Non-Dispersive Infrared)	CO_2/CO Concentrations
		NO_2/NO		NO_2/NO Concentrations
		Air renovations	VQT airflow	HVAC system/equipment, building automation, vents
Lighting	Indoor/Outdoor	Presence	Passive Infrared (PIR)	Count/Occupation/Movement of people and vehicles/Temperature/Security
		Light	Light Depending Resistor (LDR)	Color/Resistance/Security alarms, Lighting On/Off
		Brightness	Phototransistor	Home networks, Wi-Fi LED luminaires/Indoor lighting apps/Proximity Light level
		Illumination	Photodiode	

3.4. Communication Technologies

Some of the most common communication technologies involve 3G, 4G LTE, and 5G Wi-Fi network connections [109,114]. In general, communication infrastructures can be classified as local area networks (LAN), wide area networks (WAN), and global area networks (GAN) [116]. LAN sensors can communicate with gateways through different protocols, such as: long range (LoRa) [107], Bluetooth [126], and low power radio [41]. Furthermore, new technologies have arisen in recent years, such as ZigBee [110] and narrowband internet of things (NB-IoT) [72]. On the other hand, gateways in WAN connections send the information to the cloud through SigFox, long range wide area network (LoRaWAN) [112], while GAN encompasses more complex technologies and all the cellular networks GSM-GPRS [121,148]. Other protocols include IP and API, TCP, Dash7 and MQTT [17], along with standards such as IEEE802.15.4, IEEE802.11.x, and IEEE1451 [135].

However, as the number of network nodes increase, interference problems can take place, particularly in very dynamic environments, such as health and mobility [149]. For instance, inter-body network problems depend on the number of sensors per network, as well as the body networks in a location, traffic, physical mobility of each body, and the location of nodes on the body [150]. In order to attenuate this problem, active and passive schemes have been proposed, which allow a high throughput and packet delivery ratio, as along with both a low average end-to-end delay and low average energy consumption for a single and multi-WBSNs [151,152]. Similarly, sharing data through connected vehicles can occur through incorporated or third-party systems [62], but accuracy and availability widely depend on the interface used [153]. As the number of users increases, the safety, access, and efficiency can be affected. Here, the vehicle itself is considered an integrated sensor platform [154], since it can work as a data sender, receiver, and router within the V2V or V2I communications used in ITS [19].

3.5. Applications

In the health section, sensors can be used in different multidisciplinary areas, for example: smart toilets, applications that monitor and direct physical exercise with the final objective of rehabilitation or prevention of injuries, and applications in health institutions [155,156]. For cycling sports, sensors are used most frequently in long-distance running and swimming [157]. There are different methods used to monitor the condition of patients, for example: monitoring the electrical activity of the heart using an array of electrodes placed in a sterile way on the body [158], monitoring of vital signs via wireless [159] for the tracking of the patient's location by issuing alerts if necessary, implementation of user-friendly interfaces to share information, based on wireless ECG sensors and pulse oximeters [159], implementation of a network-based multi-channel frequency EM for rehabilitation patients with exercises [160], design of a system with three sensors to monitor the ECG, body weight, and pulse of the patient [161], portable home health monitoring system using ECG, fall detection, and GPS to monitor people outside [162], e-health monitoring system based on the fusion of multisensory data to predict activities and promote the decision-making process about the health of the person [163]; all of these home health systems allow to monitor patient's activity in their homes.

In the mobility section, the most common ITS applications in reducing road traffic are related to accident detection [164,165] and prevention [40], identifying road traffic events [166] and studying the driving behavior and applying real-time feedback [167,168]. Interesting applications in urban mobility have been proposed as well, such as smart traffic lights [67], smart parking [72], collision prediction and avoidance [66], vehicle WIM detection [65], and mobility visualization through heatmap representations [169]. Heatmaps can also be obtained from analyzing mobility data from vehicles and pedestrians and reflect the behavior of different phenomena, such as average speed in the city [170], traffic jams location, frequency, and duration [171], and combined spatio-temporal traffic clustering and analysis [169].

For the water section, by analyzing parameters like soil humidity and nutrients, the design of efficient irrigation programs, including remote control or automatic irrigation systems, can be done, reducing water and energy consumption [42,127,172]. For example, a system that allows real-time monitoring of surface water quality in different aquatic environments [43], systems for monitoring the quality of the river that crosses a city [173], and the implementation of a system that could be used in pipelines networks to monitor the quality of water [122]. Distribution systems need to be regulated to ensure the required quantity and quality. The use of sensors to collect data in real-time can detect and locate leaks, which can affect the correct supply of water to a city or deteriorate the infrastructure around the leak [14]. The water distribution systems need to be monitored to ensure that correct water quality is distributed and for detecting pollutants [174].

In the waste section, the IoT system permits an onboard surveillance system, which raises the process of problem reporting and evidence good waste collection practices in real-time [135]. The use of mobile apps and software permit that truck drivers receive alerts from the smart bins that need attention, and also to get the optimal route to collect the garbage, reducing the effort and cost of waste collection [136–138]. With the collection of information about the filling level of the containers, it is possible to determine the best types and sizes of containers, areas that require greater or special collection capacity, and the timing of the collection [134–136,138]. The sensors can improve the automation in identifying and separating waste, allowing an increase in the processing speed for reuse and recycling to convert a smart city to a city with zero-waste [20,137]. All data collected from the bins and the analytics, in conjunction with the use of a GPS to know the coordinate position of bins, dumps, and fleet, can be used to manage in novel ways the collection and segregation of waste [17,129,136,138,141].

Finally, countries are looking to implement innovative technologies focused on minimizing energy consumption and improving their citizens' environment and welfare. For this reason, there are various applications in the areas of buildings, public lighting, and urban space such as counting, movement, and location of people and vehicles, security actions for citizens, fire detection in building enclosures [140], smart homes [135,147,175] and home networks [135,136], Light Emitted Diode-wireless (LED) indoor and outdoor lighting fixtures [18], geothermal technology [137], hygrothermal comfort [17], cybernetic cities, ubiquitous connectivity [176], HPCense (seismic activity), smart thermostat [18], indoor lighting apps [129], microgrids [139], structural health monitoring of buildings [18], ecological buildings [134], among others.

3.6. Study Cases

This section describes experimental results of study cases implemented by the research groups of the co-authors of this review, developed in Tecnologico de Monterrey, under the Campus City initiative. Figure 10 shows a visual representation of the study cases discussed in this section.

Health. In this study, by using EEG electrodes and Bluetooth, brain signals from students were recorded to assess learning outcomes under different modalities [177]. The aim of the work was to propose EEG sensing as a support in education by inferring the state of the brain. The results showed that machine learning models based on the EEG recordings were able to predict with 85% accuracy, the cognitive performance of the students, and it could also be used to identify unwanted conditions, such as mental fatigue, anxiety, and stress under different contexts in the healthcare sector.

Security. PiBOT is a multifunctional robot developed to monitor spaces and implement regulations in the context of the COVID-19 pandemic [178]. Such robot integrates video and thermal cameras, LiDAR, ultrasonic, and IR sensors to allow object and people detection, facemask recognition, temperature maps, and distance between persons estimation. It also integrates automated navigation algorithms, teleoperation control modes and cloud connection (IoT) protocols for data transmission. The robot is also able to generate and send real-time data to a web server about people count, facemask misuse, and safe distance violations.

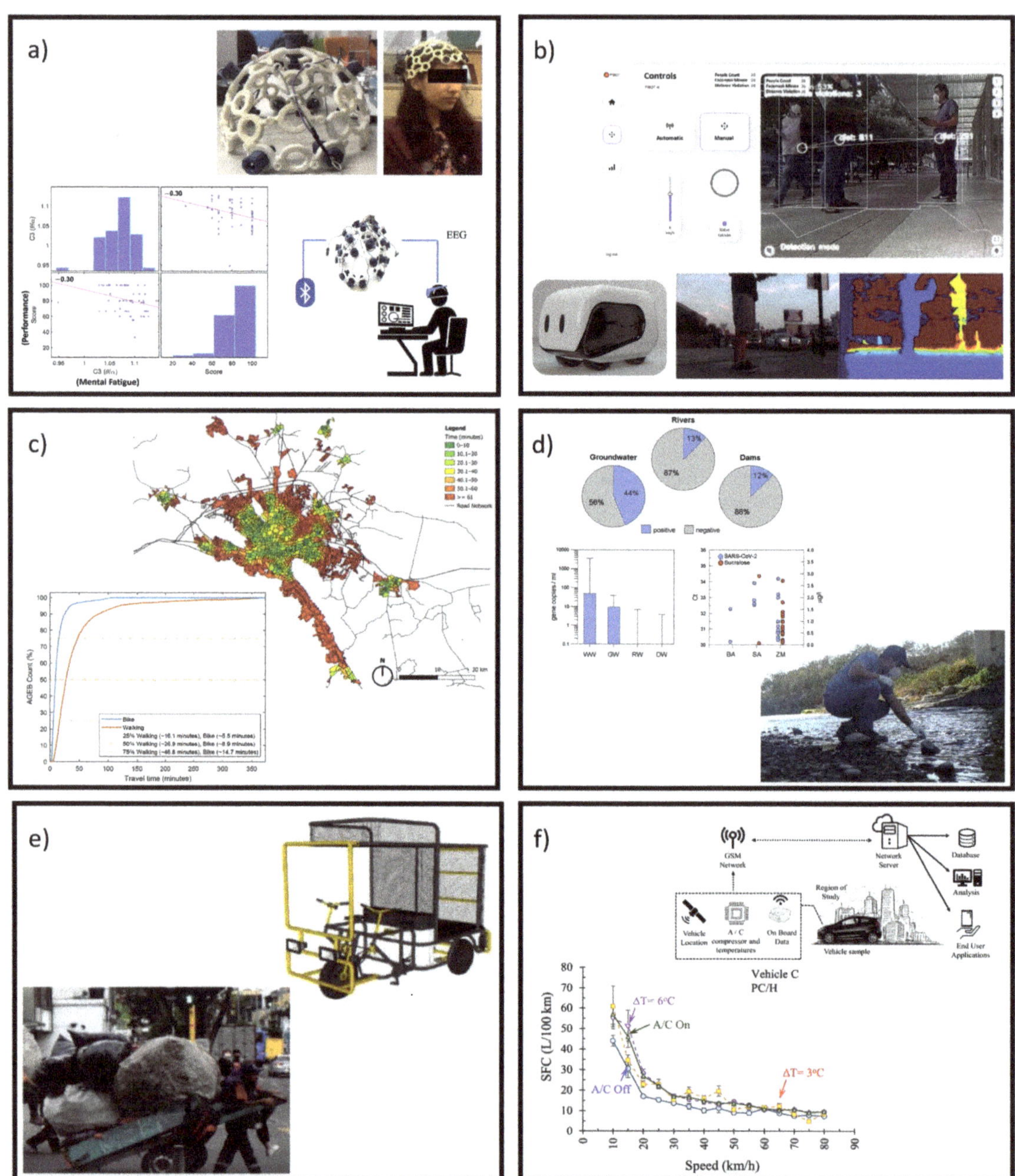

Figure 10. Study cases of smart cities implementations in Tecnologico de Monterrey, (**a**) Health: EEG monitoring for educational services; (**b**) Security: Space monitoring in the context of COVID-19 pandemic; (**c**) Mobility: Urban accessibility analysis in Monterrey City; (**d**) Water: Detection of SARS-CoV-2 RNA in freshwater environments from Monterrey City; (**e**) Waste: GPS monitoring and app for recycling vehicles; (**f**) Energy: Impact of A/C usage on light-duty vehicle's fuel consumption.

Mobility. The following study presents the results from the analysis of accessibility to different services (health and education) in the urban environment of the city of Monterrey, Mexico [179]. The software tools used in this study enabled to obtain quantitative representations of the accessibility of the city when using different transport modes (walking and cycling). The results from this work showed low accessibility to medical services, but acceptable accessibility to educational services in the city. The study found that the use of bicycles and other micro mobility vehicles can enhance the accessibility to services in the city.

Water. A recent study from one of the co-authors studies the presence of SARS-CoV-2 RNA in different freshwater environments (groundwater and surface water from rivers and dams) in urban settings [180] from the city of Monterrey, Mexico. The detection and quantification of the viral loads in such environments were determined by RT-qPCR in samples acquired from October 2020 to January 2021. The results of the study demonstrated the feasibility of the presence of SARS-CoV-2 in freshwater environments. It also found that viral loads variations in groundwater and surface water over time at the submetropolitan level reflected the reported trends in COVID-19 cases in the city of Monterrey.

Waste. A recent collaboration between Tecnologico de Monterrey and industrial partners from Smart City Colombia [181] and SmarTech [182] has started for the development of smart solutions for cities. This collaboration proposes the use of a smart recycling implementations using an urban mechanism for intelligent disposal (MUDI). The MUDI is a GPS monitored vehicle that establishes optimal routes for collection of recycled material, while informing both the recycler and the citizen through an app about the final disposal of said materials.

Energy. Using engine sensors, information about the fuel consumption associated to the operation of the A/C in light-duty vehicles was monitored via OBD for a five-month period [131]. Results showed that specific fuel consumption due to A/C usage is higher at lower speeds of the vehicle and it is lower at higher driving speeds; and shows the potential of proposing solutions towards vehicle energy efficiency by analyzing information coming from engine sensors through the OBD port.

4. Challenges and Opportunities

Advances in the use and implementation of sensors and their application for the development of smart cities will allow residents to access a better quality of life. Even though the use of wireless sensor networks provides valuable data that are used for a better management of resources, there are still areas of opportunity for improvement. Although different areas within the smart city face specific challenges, a common opportunity is the development of new sensors and new approaches to the problem of detection, prevention, or anticipation of the dangers which future smart cities can face.

A major challenge in health applications is privacy and the secure transmission of data, a concern for which different studies have been conducted. One example is a decentralized mobile-health system that leverages patients' verifiable attributes in order to run an authentication process and preserve the attribute and identity [183]. Another study designs two schemes that focus on the privacy of medical records. These schemes ensure that highly similar plaintexts can be transformed into distinctly different ciphertexts and resist ciphertext-only attacks. Nevertheless, important performance metrics, such as computation overhead, network connectivity, delay and power consumption, are ignored [184]. Low-cost wireless sensor networks also help to achieve a direct communication between a user's mobile terminals and wearable medical devices while enforcing privacy-preserving strategies [149]. In addition, a study presented a big health application system based on big data and the health internet of things (IoT). This study also proposed the cloud to end fusion big health application system architecture [185]. One last study modifies the design of sensor networks in a way that each one can manipulate four symbols (quaternary) instead of two (binary), resulting in more efficient systems [186].

With respect to mobility, ITS are used in smart cities, having a positive impact on saving resources, such as time and workforce, while reducing the use of fuels and emissions into the atmosphere. ITS image-based mobility applications are simple and inexpensive, but face decreased efficiency during lightning and weather changes [18]. Another challenge is faced by flexible traffic control, as well as collision avoidance systems as high speed detections and data exchange are needed for successful V2V, V2P, V2I, and V2X protocols [18,38,66]. This information exchange process is susceptible to security threats, such as malicious attacks or data leaks [61]. To address these challenges, more reliable sensors and faster data transfer protocols need to be implemented.

In smart security, there is a big effort on detection, prevention, or anticipation of the dangers that citizens and infrastructure of the smart cities can face. Sensors for security present a tendency to improve the sensibility, resolution, and precision of the current sensors. Almost all services in a smart city use digital data and are completely dependent on the security and integrity of that data. Due to this reason, sensors must be hardened with effective security solutions such as cryptography and advanced self-protection techniques.

Regarding smart water monitoring, sensors are used to measure water quantity and quality data continuously and consistently. The obtained data can be processed and visualized in real-time to the end-users, or forecasts can be developed for the water agencies. These technologies allow minimization of the risks associated to poor water quality and deficiencies in water supply. Future sensors need to be improved in cost and energy consumption to withstand long periods of measurements without intervention, in addition to an enhanced robustness, to resist adverse environmental conditions.

Solid waste management is crucial in any town or city, but take a new role in the smart city scheme, and it is focused on a more clean, tidy, and healthy environment for living, using sensors and IoT technologies to improve waste management [42,43]. Currently there are only sensors that can identify wet, dry, or metal garbage; however, it would be optimal to develop sensors that allow identifying waste in greater detail. For this reason, new sensors oriented to waste segregation need to be developed and implemented. Segregation is a key component in the waste management system, as it allows much of what is discarded to be recycled or reused, resulting in a reduction in the amount of garbage that reaches landfills [128,172].

Innovations in energy consumption monitoring in buildings, public lighting systems, and urban spaces using ICTs are an excellent option due to their adaptability. The literature proposes the implementation of virtual sensors by building information modeling (BIM), integrated with IoT devices including qualification tools to develop ecological buildings [134,187]. Intelligent lighting systems with sensors adaptable to weather conditions, hours of use and presence of people or vehicles [20] where the street lamps serve as Wi-Fi connection points, allowing interconnected networks over the entire urban area monitoring the quality of the environment, noise levels, and surveillance, among others. Battery or energy consumption, high volume of data storage and security, life span and replacement, size, cost, installation, and maintenance are the main deficiencies when designing a WSN. Studies have proposed the implementations of energy efficiency surveillance using multimodal sensors [188] and low power hardware systems [189]. The implementation of low-cost sensors and using energy harvesting are the most attractive technologies for buildings' sensors in the future. Artificial intelligence, big data, and machine learning [190,191] become essential due to the vast amount of data gathered and analyzed in the presented applications.

5. Conclusions

Following the reported trends of population growth and mobility to urban environments, it is clear that in the years to come, cities will face a constantly increasing need to satisfy the demands of their citizens [1]. Diverse strategies have been implemented in cities from all continents to move towards *smartness* as a means to enhance management of their resources, offer more efficient and trustable services, improve the liveability of the city, and promote government, academia, and citizens' engagement.

Geographically speaking, continents such as Europe and Asia have the highest amount of reported smart city implementations, followed by America, Oceania, and Africa. High-income countries such as United States and China presented a high number of smart city deployments at different cites. Although other continents and cities present fewer smart cities, it is only a matter of time for more to arise, as they follow the steps of their predecessors [28]. Several applications of sensors for smart cities were identified and described within this work. To summarize, most of the applications involved the sensing and sharing of data to offer on-demand services (medical records and check-ups, urban transportation, water, and energy consumption), while others are oriented to improve the liveability of the city (citizens' security, green spaces management, water quality, waste collection, public lighting). Among the main challenges of smart city implementations, each sector has its own; however, common factors such as the improvement of sensors, massive data analytics implementations (big data), and citizens' distrust in data sharing can be identified.

Although in future smart cities, the security of the society and community is to be ensured by digitalization and inter-institutional cooperation, human security, generally speaking, is guaranteed by the individual's behavior. It should not be forgotten that safety cannot be forced; it can only be educated, and there is a need to form an internal motive for safe and ethical behavior by creating and fostering a culture of safety in such a new digital environment. Finally, it is also important to consider that in order to implement the proposed smart city solutions, collaboration, and partnership with government agencies is imperative [3]. Deep understanding of each context of implementation and key interconnections between sectors (e.g., transport–energy, energy–water–food, resource efficiency and recovery, etc.), along with meaningful community engagement and involvement in the planning and use of new technologies in the urban infrastructure is essential to enhance political feasibility, transparency, equity, and financial sustainability.

As societies around the world begin to better understand how technological progress can improve quality of life and foster clean economic development, smarter communities will be driving the future of cities towards a more liveable, inclusive, net-zero carbon and sustainable future by mid-century [192].

Author Contributions: Conceptualization, J.d.J.L.-S. and R.A.R.-M.; methodology, M.A.R.-M., R.E.P.-G.; validation, J.d.J.L.-S., R.A.R.-M. and S.C.M.; investigation, M.A.R.-M., S.K., D.A.P.-R., E.R.-L., M.G.-M., M.C.H.-L., A.E.M., J.M., J.I.H., R.E.P.-G., R.A.R.-M., A.M.M., M.R., B.L.P.-H., S.C.M., J.d.J.L.-S.; writing—original draft preparation, M.A.R.-M., S.K., D.A.P.-R., E.R.-L., M.G.-M., M.C.H.-L., A.E.M., R.E.P.-G., R.A.R.-M., A.M.M., M.R., B.L.P.-H.; writing—review and editing, J.M., J.I.H., S.CM. and J.d.J.L.-S.; supervision, R.A.R.-M., S.C.M. and J.d.J.L.-S.; project administration, J.M., J.I.H. and J.d.J.L.-S.; funding acquisition, R.A.R.-M. All authors have read and agreed to the published version of the manuscript.

Funding: This project is funded by the Campus City initiative from Tecnologico de Monterrey. The APC was funded by Tecnologico de Monterrey.

Institutional Review Board Statement: Not applicable.

Informed Consent Statement: Not applicable.

Data Availability Statement: Not applicable.

Acknowledgments: Authors would like to thank the support of Tecnologico de Monterrey through the Campus City initiative and the California-Global Energy, Water & Infrastructure Innovation Initiative at Stanford University.

Conflicts of Interest: The authors declare no conflict of interest. The funders had no role in the writing of the manuscript.

References

1. Calzada, I. Metropolitan and city-regional politics in the urban age: Why does "(smart) devolution" matter? *Palgrave Commun.* **2017**, *3*, 1–17. [CrossRef]
2. Batista e Silva, F.; Freire, S.; Schiavina, M.; Rosina, K.; Marín-Herrera, M.A.; Ziemba, L.; Craglia, M.; Koomen, E.; Lavalle, C. Uncovering temporal changes in Europe's population density patterns using a data fusion approach. *Nat. Commun.* **2020**, *11*, 1–11. [CrossRef]
3. Joss, S. Future cities: Asserting public governance. *Palgrave Commun.* **2018**, *4*, 1–4. [CrossRef]
4. Pellicer, S.; Santa, G.; Bleda, A.L.; Maestre, R.; Jara, A.J.; Skarmeta, A.G. A global perspective of smart cities: A survey. In Proceedings of the 2013 Seventh International Conference on Innovative Mobile and Internet Services in Ubiquitous Computing, Taichung, Taiwan, 3–5 July 2013; pp. 439–444. [CrossRef]
5. Bai, X.; Nagendra, H.; Shi, P.; Liu, H. Cities: Build networks and share plans to emerge stronger from COVID-19. *Nature* **2020**, *584*, 517–520. [CrossRef]
6. Staletić, N.; Labus, A.; Bogdanović, Z.; Despotović-Zrakić, M.; Radenković, B. Citizens' readiness to crowdsource smart city services: A developing country perspective. *Cities* **2020**, *107*, 102883. [CrossRef]
7. Pranggono, B.; Arabo, A. COVID-19 pandemic cybersecurity issues. *Internet Technol. Lett.* **2021**, *4*, 4–9. [CrossRef]
8. He, Y.; Aliyu, A.; Evans, M.; Luo, C. Health care cybersecurity challenges and solutions under the climate of COVID-19: Scoping review. *J. Med. Internet Res.* **2021**, *23*, 1–18. [CrossRef]
9. Ahad, M.A.; Paiva, S.; Tripathi, G.; Feroz, N. Enabling technologies and sustainable smart cities. *Sustain. Cities Soc.* **2020**, *61*, 102301. [CrossRef]
10. Moustaka, V.; Theodosiou, Z.; Vakali, A.; Kounoudes, A.; Anthopoulos, L.G. Enhancing social networking in smart cities: Privacy and security borderlines. *Technol. Forecast. Soc. Chang.* **2019**, *142*, 285–300. [CrossRef]
11. Braun, T.; Fung, B.C.; Iqbal, F.; Shah, B. Security and privacy challenges in smart cities. *Sustain. Cities Soc.* **2018**, *39*, 499–507. [CrossRef]
12. Kadry, S. Safe drive in smart city. *Smart Solut. Future Cities* **2016**, 1–7. [CrossRef]
13. Moher, D.; Liberati, A.; Tetzlaff, J.; Altman, D.G.; Group, T.P. Preferred Reporting Items for Systematic Reviews and Meta-Analyses: The PRISMA Statement. *PLoS Med.* **2009**, *6*, 1–6. [CrossRef]
14. Yang, L.; Yan, H.; Lam, J.C. Thermal comfort and building energy consumption implications—A review. *Appl. Energy* **2014**, *115*, 164–173. [CrossRef]
15. Gunduz, M.Z.; Das, R. Cyber-security on smart grid: Threats and potential solutions. *Comput. Netw.* **2020**, *169*, 107094. [CrossRef]
16. Zhao, H.X.; Magoulès, F. A review on the prediction of building energy consumption. *Renew. Sustain. Energy Rev.* **2012**, *16*, 3586–3592. [CrossRef]
17. Carli, R.; Cavone, G.; Othman, S.B.; Dotoli, M. IoT based architecture for model predictive control of HVAC systems in smart buildings. *Sensors* **2020**, *20*, 781. [CrossRef]
18. Hancke, G.P.; de Silva, B.d.C.; Hancke, G.P. The role of advanced sensing in smart cities. *Sensors* **2013**, *13*, 393–425. [CrossRef]
19. Benevolo, C.; Dameri, R.P.; Auria, B.D. Smart Mobility in Smart City. In *Empowering Organizations: Enabling Platforms and Artefacts*; Springer: Cham, Switzerland, 2016; Volume 11, pp. 13–28. [CrossRef]
20. Zanella, A.; Bui, N.; Castellani, A.; Vangelista, L.; Zorzi, M. Internet of things for smart cities. *IEEE Internet Things J.* **2014**, *1*, 22–32. [CrossRef]
21. Capdevila, I.; Zarlenga, M.I. Smart City or Smart Citizens? The Barcelona Case *J. Strategy Manag.* **2015**, *8*, 266–282. [CrossRef]
22. Sovacool, B.K.; Griffiths, S. Culture and low-carbon energy transitions. *Nat. Sustain.* **2020**, *3*, 685–693. [CrossRef]
23. Yu, D.; Yin, J.; Wilby, R.; Lane, S. Disruption of emergency response to vulnerable populations during floods. *Nat. Sustain.* **2020**, *3*, 728–736. [CrossRef]
24. Ejaz, W.; Anpalagan, A. *Internet of Things for Smart Cities: Technologies, Big Data and Security*; Springer: Cham, Switzerland, 2019; pp. 1–15.
25. Okai, E.; Feng, X.; Sant, P. Smart Cities Survey. In Proceedings of the 2018 IEEE 20th International Conference on High Performance Computing and Communications; IEEE 16th International Conference on Smart City; IEEE 4th International Conference on Data Science and Systems (HPCC/SmartCity/DSS), Exeter, UK, 28–30 June 2018; pp. 1726–1730. [CrossRef]
26. Sanchez, L.; Muñoz, L.; Galache, J.A.; Sotres, P.; Santana, J.R.; Gutierrez, V.; Ramdhany, R.; Gluhak, A.; Krco, S.; Theodoridis, E.; et al. SmartSantander: IoT experimentation over a smart city testbed. *Comput. Netw.* **2014**, *61*, 217–238. [CrossRef]
27. Anthopoulos, L. Smart utopia VS smart reality: Learning by experience from 10 smart city cases. *Cities* **2017**, *63*, 128–148. [CrossRef]
28. Lu, D.; Tian, Y.; Liu, V.Y.; Zhang, Y. The performance of the smart cities in China-A comparative study by means of self-organizing maps and social networks analysis. *Sustainability* **2015**, *7*, 7604–7621. [CrossRef]
29. Joo, Y.M.; Tan, T.B. Smart cities in Asia: An introduction. In *Smart Cities in Asia: Governing Development in the Era of Hyper-Connectivity*; Edward Elgar Publishing: Cheltenham, UK, 2020; pp. 1–17. [CrossRef]
30. Useche, M.P.; Carlos, N.S.J.; Vilafañe, C. Medellin (Colombia): A Case of Smart City. In Proceedings of the 7th International Conference on Theory and Practice of Electronic Governance ICEGOV '13, Seoul, Korea, 22–25 October 2013; pp. 231–233. [CrossRef]

31. Gaffney, C.; Robertson, C. Smarter than Smart: Rio de Janeiro's Flawed Emergence as a Smart City. *J. Urban Technol.* **2018**, *25*, 47–64. [CrossRef]
32. Yigitcanlar, T.; Kamruzzaman, M. Smart Cities and Mobility: Does the Smartness of Australian Cities Lead to Sustainable Commuting Patterns? *J. Urban Technol.* **2019**, *26*, 21–46. [CrossRef]
33. Niculescu, A.I.; Wadhwa, B. Smart cities in south east Asia: Singapore concepts-An HCI4D perspective. In Proceedings of the ASEAN CHI Symposium'15, Seoul, Korea, 18–23 April 2015; pp. 20–23. [CrossRef]
34. Endesa, F. Smart Cities. Available online: https://www.fundacionendesa.org/es/recursos/a201908-smart-city (accessed on 13 November 2020).
35. Cities, B. Bright Cities. Available online: https://www.brightcities.city/ (accessed on 13 November 2020).
36. IMD Real Learning Real Impact; SCO Observatory. Smart City Index 2020. Available online: https://www.imd.org/smart-city-observatory/smart-city-index/ (accessed on 13 November 2020).
37. Wu, L.; Zhi, Y.; Sui, Z.; Liu, Y. Intra-urban human mobility and activity transition: Evidence from social media check-in data. *PLoS ONE* **2014**, *9*, e97010. [CrossRef]
38. Chen, S.; Hu, J.; Shi, Y.; Peng, Y.; Fang, J.; Zhao, R.; Zhao, L. Vehicle-to-Everything (v2x) Services Supported by LTE-Based Systems and 5G. *IEEE Commun. Stand. Mag.* **2017**, *1*, 70–76. [CrossRef]
39. Chang, C.Y.; Chien, L.C.; Chang, Y.H.; Kuo, E.C.; Hwang, Y.S. A smart public security strategy: The New Taipei City Technology defense plan. *Procedia Comput. Sci.* **2019**, *159*, 1715–1719. [CrossRef]
40. Baig, Z.A.; Szewczyk, P.; Valli, C.; Rabadia, P.; Hannay, P.; Chernyshev, M.; Johnstone, M.; Kerai, P.; Ibrahim, A.; Sansurooah, K.; et al. Future challenges for smart cities: Cyber-security and digital forensics. *Digit. Investig.* **2017**, *22*, 3–13. [CrossRef]
41. Anagnostopoulos, T.; Zaslavsky, A.; Kolomvatsos, K.; Medvedev, A.; Amirian, P.; Morley, J.; Hadjieftymiades, S. Challenges and Opportunities of Waste Management in IoT-Enabled Smart Cities: A Survey. *IEEE Trans. Sustain. Comput.* **2017**, *2*, 275–289. [CrossRef]
42. Aktemur, I.; Erensoy, K.; Kocyigit, E. Optimization of Waste Collection in Smart Cities with the use of Evolutionary Algorithms. In Proceedings of the 2nd International Congress on Human-Computer Interaction, Optimization and Robotic Applications HORA 2020, Ankara, Turkey, 26–27 June 2020; pp. 1–8. [CrossRef]
43. Pardini, K.; Rodrigues, J.J.; Diallo, O.; Das, A.K.; de Albuquerque, V.H.C.; Kozlov, S.A. A smart waste management solution geared towards citizens. *Sensors* **2020**, *20*, 2380. [CrossRef] [PubMed]
44. de Mattos, W.D.; Gondim, P.R. M-Health Solutions Using 5G Networks and M2M Communications. *IT Prof.* **2016**, *18*, 24–29. [CrossRef]
45. Mustapa, M.A.; Abu Bakar, M.H.; Mustapha Kamil, Y.; Syahir, A.; Mahdi, M.A. Bio-Functionalized Tapered Multimode Fiber Coated with Dengue Virus NS1 Glycoprotein for Label Free Detection of Anti-Dengue Virus NS1 IgG Antibody. *IEEE Sens. J.* **2018**, *18*, 4066–4072. [CrossRef]
46. Zheng, Y.; Rundell, A. Biosensor immunosurface engineering inspired by B-cell membrane-bound antibodies: Modeling and analysis of multivalent antigen capture by immobilized antibodies. *IEEE Trans. NanoBiosci.* **2003**, *2*, 14–25. [CrossRef]
47. Sekitani, T.; Someya, T. Stretchable organic integrated circuits for large-area electronic skin surfaces. *MRS Bull.* **2012**, *37*, 236–245. [CrossRef]
48. Kim, Y.S.; Lee, J.; Ameen, A.; Shi, L.; Li, M.; Wang, S.; Ma, R.; Jin, S.H.; Kang, Z.; Huang, Y.; et al. Multifunctional Epidermal Electronics Printed Directly Onto the Skin. *Adv. Mater.* **2013**, *25*, 2773–2778. [CrossRef]
49. Ko, H.; Stoykovich, M.; Song, J.; Malyarchuk, V.; Choi, W.; Yu, C.J.; Geddes, J.; Xiao, J.; Wang, S.; Huang, Y.; et al. A hemispherical electronic eye camera based on compressible silicon optoelectronics. *Nature* **2008**, *454*, 748–753. [CrossRef]
50. Wang, Y.; Yang, R.; Shi, Z.; Zhang, L.; Shi, D.X.; Wang, E.; Zhang, G. Super-Elastic Graphene Ripples for Flexible Strain Sensors. *ACS Nano* **2011**, *5*, 3645–3650. [CrossRef]
51. Pan, L.J.; Chortos, A.; Yu, G.; Wang, Y.; Isaacson, S.; Allen, R.; Shi, Y.; Dauskardt, R.; Bao, Z. An ultra-sensitive resistive pressure sensor based on hollow-sphere microstructure induced elasticity in conducting polymer film. *Nat. Commun.* **2014**, *5*, 3002. [CrossRef]
52. Chen, Y.; Lu, B.; Chen, Y.; Feng, X. Biocompatible and ultra-flexible inorganic strain sensors attached to skin for long-term vital signs monitoring. *IEEE Electron. Device Lett.* **2016**, *37*, 496–499. [CrossRef]
53. Zhang, Y.; Black, A.; Wu, N.; Cui, Y. Comparison of "Dry Sensing" and "Wet Sensing" of a Protein With a Graphene Sensor. *IEEE Sens. Lett.* **2018**, *2*, 1–4. [CrossRef]
54. Yazdi, E.; Willig, A.; Pawlikowski, K. On channel adaptation in IEEE 802.15.4 mobile body sensor networks: What can be Gained? In Proceedings of the IEEE International Conference on Networks, ICON, Singapore, Singapore, 12–14 December 2012; pp. 262–267. [CrossRef]
55. Liao, Y.; Leeson, M.S.; Higgins, M.D. Flexible quality of service model for wireless body area sensor networks. *Healthc. Technol. Lett.* **2016**, *3*, 12–15. [CrossRef]
56. Al-Turjman, F.; Lemayian, J.P. Intelligence, security, and vehicular sensor networks in internet of things (IoT)-enabled smart-cities: An overview. *Comput. Electr. Eng.* **2020**, *87*, 106776. [CrossRef]

57. Fanti, M.P.; Mangini, A.M.; Rotunno, G.; Fiume, G.; Favenza, A.; Gaetani, M. A Cloud Computing Architecture for Eco Route Planning of Heavy Duty Vehicles. In Proceedings of the 2018 IEEE 14th International Conference on Automation Science and Engineering (CASE), Munich, Bavaria, Germany, 20–24 August 2018; pp. 730–735. [CrossRef]

58. Difilippo, G.; Fanti, M.P.; Fiume, G.; Mangini, A.M.; Monsel, N. A Cloud Optimizer for Eco Route Planning of Heavy Duty Vehicles. In Proceedings of the 2018 IEEE Conference on Decision and Control (CDC), Miami, FL, USA, 17–19 December 2018; pp. 7142–7147. [CrossRef]

59. Koehl, A. Urban transport and COVID-19: Challenges and prospects in low- and middle-income countries. *Cities Health* **2020**, 1–6. [CrossRef]

60. Hanbyul, S.; Ki-Dong, L.; Shinpei, Y.; Ying, P.; Philippe, S. LTE Evolution for Vehicle-to-Everything Services. *IEEE Commun. Mag.* **2016**. *54*, 22–28.

61. Wang, J.; Shao, Y.; Ge, Y.; Yu, R. A survey of vehicle to everything (V2X) testing. *Sensors* **2019**, *19*, 334. [CrossRef]

62. Toglaw, S.; Aloqaily, M.; Alkheir, A.A. Connected, Autonomous and Electric Vehicles: The Optimum Value for a Successful Business Model. In Proceedings of the 2018 Fifth International Conference on Internet of Things: Systems, Management and Security, Valencia, Spain, 15–18 October 2018; pp. 303–308. [CrossRef]

63. Ozatay, E.; Onori, S.; Wollaeger, J.; Ozguner, U.; Rizzoni, G.; Filev, D.; Michelini, J.; Di Cairano, S. Cloud-Based Velocity Profile Optimization for Everyday Driving: A Dynamic-Programming-Based Solution. *IEEE Trans. Intell. Transp. Syst.* **2014**, *15*, 2491–2505. [CrossRef]

64. Fanti, M.P.; Mangini, A.M.; Favenza, A.; Difilippo, G. An Eco-Route planner for heavy duty vehicles. *IEEE/CAA J. Autom. Sin.* **2021**, *8*, 37–51. [CrossRef]

65. Zhang, J.; Lu, Y.; Lu, Z.; Liu, C.; Sun, G.; Li, Z. A new smart traffic monitoring method using embedded cement-based piezoelectric sensors. *Smart Mater. Struct.* **2015**, *24*, 1–8. [CrossRef]

66. Hussein, A.; García, F.; Armingol, J.M.; Olaverri-Monreal, C. P2V and V2P communication for pedestrian warning on the basis of autonomous vehicles. In Proceedings of the IEEE Conference on Intelligent Transportation Systems, Proceedings, ITSC, Rio de Janeiro, Brazil, 1–4 November 2016; pp. 2034–2039. [CrossRef]

67. Ghazal, B.; Elkhatib, K.; Chahine, K.; Kherfan, M. Smart traffic light control system. In Proceedings of the 3rd International Conference on Electrical, Electronics, Computer Engineering and Their Applications, EECEA 2016, Beirut, Lebanon, 21–23 April 2016; pp. 140–145. [CrossRef]

68. Ahas, R.; Aasa, A.; Yuan, Y.; Raubal, M.; Smoreda, Z.; Liu, Y.; Ziemlicki, C.; Tiru, M.; Zook, M. Everyday space–time geographies: Using mobile phone-based sensor data to monitor urban activity in Harbin, Paris, and Tallinn. *Int. J. Geogr. Inf. Sci.* **2015**, *29*, 2017–2039. [CrossRef]

69. Cuenca-Jara, J.; Terroso-Saenz, F.; Valdes-Vela, M.; Gonzalez-Vidal, A.; Skarmeta, A.F. Human mobility analysis based on social media and fuzzy clustering. In Proceedings of the Global Internet of Things Summit GIoTS 2017, Geneva, Switzerland, 1–6 June 2017; pp. 1–6. [CrossRef]

70. Fukuzaki, Y.; Murao, K.; Mochizuki, M.; Nishio, N. Statistical analysis of actual number of pedestrians for Wi-Fi packet-based pedestrian flow sensing. In Proceedings of the 2015 ACM International Joint Conference on Pervasive and Ubiquitous Computing and the Proceedings of the 2015 ACM International Symposium on Wearable Computers, UbiComp and ISWC 2015, Osaka, Japan, 7–11 September 2015; pp. 1519–1526. [CrossRef]

71. El Mahrsi, M.K.; Come, E.; Oukhellou, L.; Verleysen, M. Clustering Smart Card Data for Urban Mobility Analysis. *IEEE Trans. Intell. Transp. Syst.* **2017**, *18*, 712–728. [CrossRef]

72. Vlahogianni, E.I.; Kepaptsoglou, K.; Tsetsos, V.; Karlaftis, M.G. A Real-Time Parking Prediction System for Smart Cities. *J. Intell. Transp. Syst. Technol. Plan. Oper.* **2016**, *20*, 192–204. [CrossRef]

73. Zhao, J.; Xu, H.; Liu, H.; Wu, J.; Zheng, Y.; Wu, D. Detection and tracking of pedestrians and vehicles using roadside LiDAR sensors. *Transp. Res. Part C Emerg. Technol.* **2019**, *100*, 68–87. [CrossRef]

74. Walter, M.; Eilebrecht, B.; Wartzek, T.; Leonhardt, S. The smart car seat: Personalized monitoring of vital signs in automotive applications. *Pers. Ubiquitous Comput.* **2011**, *15*, 707–715. [CrossRef]

75. Rahimian, P.; O'Neal, E.E.; Zhou, S.; Plumert, J.M.; Kearney, J.K. Harnessing Vehicle-to-Pedestrian (V2P) Communication Technology: Sending Traffic Warnings to Texting Pedestrians. *Hum. Factors* **2018**, *60*, 833–843. [CrossRef]

76. Roccotelli, M.; Nolich, M.; Fanti, M.P.; Ukovich, W. Internet of things and virtual sensors for electromobility. *Internet Technol. Lett.* **2018**, *1*, e39. [CrossRef]

77. Fanti, M.P.; Mangini, A.M.; Roccotelli, M.; Nolich, M.; Ukovich, W. Modeling Virtual Sensors for Electric Vehicles Charge Services. In Proceedings of the 2018 IEEE International Conference on Systems, Man, and Cybernetics (SMC), Miyazaki, Japan, 7–10 October 2018; pp. 3853–3858. [CrossRef]

78. Fanti, M.P.; Mangini, A.M.; Roccotelli, M. An Innovative Service for Electric Vehicle Energy Demand Prediction. In Proceedings of the 2020 7th International Conference on Control, Decision and Information Technologies (CoDIT), Prague, Czech Republic, 19 June–2 July 2020; Volume 1, pp. 880–885. [CrossRef]

79. Human Security Unit of the United Nation. *Human Security Handbook an Integrated Approach for the Realization of the SDG's*; United Nations: New York, NY, USA, 2016; pp. 1–47.

80. Maye, D. 'smart food city': Conceptual relations between smart city planning, urban food systems and innovation theory. *City Cult. Soc.* **2019**, *16*, 18–24. [CrossRef]

81. Alipio, M.I.; Dela Cruz, A.E.M.; Doria, J.D.A.; Fruto, R.M.S. On the design of Nutrient Film Technique hydroponics farm for smart agriculture. *Eng. Agric. Environ. Food* **2019**, *12*, 315–324. [CrossRef]

82. Matindoust, S.; Baghaei-Nejad, M.; Abadi, M.H.S.; Zou, Z.; Zheng, L.R. Food quality and safety monitoring using gas sensor array in intelligent packaging. *Sens. Rev.* **2016**, *36*, 169–183. [CrossRef]

83. Lu, L.; Zhu, Z.; Hu, X. Hybrid nanocomposites modified on sensors and biosensors for the analysis of food functionality and safety. *Trends Food Sci. Technol.* **2019**, *90*, 100–110. [CrossRef]

84. Zhou, J.; Cao, Z.; Dong, X.; Vasilakos, A.V. Security and Privacy for Cloud-Based IoT: Challenges. *IEEE Commun. Mag.* **2017**, *55*, 26–33. [CrossRef]

85. Habibzadeh, H.; Soyata, T.; Kantarci, B.; Boukerche, A.; Kaptan, C. Sensing, communication and security planes: A new challenge for a smart city system design. *Comput. Netw.* **2018**. [CrossRef]

86. Ray, P.P.; Dash, D.; Kumar, N. Sensors for internet of medical things: State-of-the-art, security and privacy issues, challenges and future directions. *Comput. Commun.* **2020**, *160*, 111–131. [CrossRef]

87. Keshta, I.; Odeh, A. Security and privacy of electronic health records: Concerns and challenges. *Egypt. Inform. J.* **2020**. [CrossRef]

88. Kaku, K. Satellite remote sensing for disaster management support: A holistic and staged approach based on case studies in Sentinel Asia. *Int. J. Disaster Risk Reduct.* **2019**, *33*, 417–432. [CrossRef]

89. Clark, N.E. Towards a standard licensing scheme for the access and use of satellite earth observation data for disaster management. *Acta Astronaut.* **2017**, *139*, 325–331. [CrossRef]

90. Liu, L.; fan Li, C.; kun Sun, X.; Zhao, J. Event alert and detection in smart cities using anomaly information from remote sensing earthquake data. *Comput. Commun.* **2020**, *153*, 397–405. [CrossRef]

91. Miyata, E.; Miyata, H.; Fukasawa, E.; Kakizaki, K.; Abe, H.; Katsumata, M.; Sato, M. A Hybrid semiconductor radiation detectors using conductive polymers. *Nucl. Instrum. Methods Phys. Res. Sect. Accel. Spectrometers. Detect. Assoc. Equip.* **2020**. *955*, 1–7. [CrossRef]

92. Tanioka, Y.; Gusman, A.R. Near-field tsunami inundation forecast method assimilating ocean bottom pressure data: A synthetic test for the 2011 Tohoku-oki tsunami. *Phys. Earth Planet. Inter.* **2018**, *283*, 82–91. [CrossRef]

93. Henríquez, B.P. Information Technology: The Unsung Hero of Market-Based Environmental Policies. 2004. Available online: https://gspp.berkeley.edu/assets/uploads/research/pdf/RFF_Resources_152_infotech.pdf (accessed on 1 September 2021).

94. de Chalendar, J.A.; Taggart, J.; Benson, S.M. Tracking emissions in the US electricity system. *Proc. Natl. Acad. Sci. USA* **2019**, *116*, 25497–25502. [CrossRef]

95. Badii, C.; Bilotta, S.; Cenni, D.; Difino, A.; Nesi, P.; Paoli, I.; Paolucci, M. High Density Real-Time Air Quality Derived Services from IoT Networks. *Sensors* **2020**, *20*, 5435. [CrossRef]

96. Henríquez, B.P. *Environmental Commodities Markets and Emissions Trading: Towards a Low-Carbon Future*; Routledge: London, UK, 2012.

97. Pérez Henríquez, B. *Handbook on the Resource Nexus*; Chapter California Innovations @ the Water Energy Nexus (WEN); Routledge: London, UK, 2018; p. 18.

98. Yan, K.; Zhang, Y.; Yan, Y.; Xu, C.; Zhang, S. Fault diagnosis method of sensors in building structural health monitoring system based on communication load optimization. *Comput. Commun.* **2020**, *159*, 310–316. [CrossRef]

99. Ayyildiz, C.; Erdem, H.E.; Dirikgil, T.; Dugenci, O.; Kocak, T.; Altun, F.; Gungor, V.C. Structure Health Monitoring Using Wireless Sensor Networks on Structural Elements. *Ad Hoc Netw.* **2019**, *82*, 68–76; Erratum to **2020**, *105*, 68–76. [CrossRef]

100. Smart street lighting system: A platform for innovative smart city applications and a new frontier for cyber-security. *Electr. J.* **2016**, *29*, 28–35. [CrossRef]

101. Laufs, J.; Borrion, H.; Bradford, B. Security and the smart city: A systematic review. *Sustain. Cities Soc.* **2020**, *55*, 102023. [CrossRef]

102. Ross, A.; Banerjee, S.; Chowdhury, A. Security in smart cities: A brief review of digital forensic schemes for biometric data. *Pattern Recognit. Lett.* **2020**, *138*, 346–354. [CrossRef]

103. Mezzera, L.; Carminati, M.; Di Mauro, M.; Turolla, A.; Tizzoni, M.; Antonelli, M. A 7-Parameter Platform for Smart and Wireless Networks Monitoring On-Line Water Quality. In Proceedings of the 25th IEEE International Conference on Electronics Circuits and Systems, ICECS 2018, Bordeaux, Gironde, France, 9–12 December 2019; pp. 709–712. [CrossRef]

104. Gonçalves, R.; Soares, J.J.; Lima, R.M. An IoT-based framework for smartwater supply systems management. *Future Internet* **2020**, *12*, 114. [CrossRef]

105. Mohd Ismail, M.I.; Dziyauddin, R.A.; Salleh, N.A.A.; Muhammad-Sukki, F.; Bani, N.A.; Izhar, M.A.M.; Latiff, L.A. A review of vibration detection methods using accelerometer sensors for water pipeline leakage. *IEEE Access* **2019**, *7*, 51965–51981. [CrossRef]

106. Kodali, R.K.; Sarjerao, B.S. A low cost smart irrigation system using MQTT protocol. In Proceedings of the IEEE International Symposium on Technologies for Smart Cities TENSYMP 2017, Kochi, Kerala, India, 14–16 July 2017; pp. 1–5. [CrossRef]

107. Mamun, K.A.; Islam, F.R.; Haque, R.; Khan, M.G.; Prasad, A.N.; Haqva, H.; Mudliar, R.R.; Mani, F.S. Smart Water Quality Monitoring System Design and KPIs Analysis: Case Sites of Fiji Surface Water. *Sustainability* **2019**, *11*, 7110. [CrossRef]

108. Quadar, N.; Chehri, A.; Jeon, G.; Ahmad, A. Smart water distribution system based on IoT networks, a critical review. In *Smart Innovation, Systems and Technologies*; Springer: Singapore, 2021; Volume 189, pp. 293–303. [CrossRef]

109. Kulkarni, P.; Farnham, T. Smart City Wireless Connectivity Considerations and Cost Analysis: Lessons Learnt from Smart Water Case Studies. *IEEE Access* **2016**, *4*, 660–672. [CrossRef]

110. de Oliveira, K.V.; Esgalha Castelli, H.M.; José Montebeller, S.; Prado Avancini, T.G. Wireless Sensor Network for Smart Agriculture using ZigBee Protocol. In Proceedings of the 2017 IEEE First Summer School on Smart Cities (S3C), Natal, Brazil, 6–11 August 2017; pp. 61–66. [CrossRef]

111. Lopes, S.F.; Pereira, R.M.; Lopes, S.O.; Coutinho, M.; Malheiro, A.; Fonte, V. Yet a smarter irrigation system. In *Lecture Notes of the Institute for Computer Sciences, Social-Informatics and Telecommunications Engineering, LNICST*; Springer Cham: Cham, Switzerland, 2020; Volume 323, pp. 337–346. [CrossRef]

112. Martínez, R.; Vela, N.; el Aatik, A.; Murray, E.; Roche, P.; Navarro, J.M. On the use of an IoT integrated system for water quality monitoring and management in wastewater treatment plants. *Water* **2020**, *12*, 1096. [CrossRef]

113. Simitha, K.M.; Subodh Raj, M.S. IoT and WSN Based Water Quality Monitoring System. In Proceedings of the 3rd International Conference on Electronics and Communication and Aerospace Technology, ICECA 2019, Coimbatore, Tamil Nadu, India, 12–14 June 2019; pp. 205–210. [CrossRef]

114. Pasika, S.; Gandla, S.T. Smart water quality monitoring system with cost-effective using IoT. *Heliyon* **2020**, *6*, e04096. [CrossRef] [PubMed]

115. Yang, W.; Wei, X.; Choi, S. A Dual-Channel, Interference-Free, Bacteria-Based Biosensor for Highly Sensitive Water Quality Monitoring. *IEEE Sens. J.* **2016**, *16*, 8672–8677. [CrossRef]

116. Chen, Y.; Han, D. Water quality monitoring in smart city: A pilot project. *Autom. Constr.* **2018**, *89*, 307–316. [CrossRef]

117. Esmaeilian, B.; Wang, B.; Lewis, K.; Duarte, F.; Ratti, C.; Behdad, S. The future of waste management in smart and sustainable cities: A review and concept paper. *Waste Manag.* **2018**, *81*, 177–195. [CrossRef]

118. Folianto, F.; Low, Y.S.; Yeow, W.L. Smartbin: Smart waste management system. In Proceedings of the 2015 IEEE 10th International Conference on Intelligent Sensors, Sensor Networks and Information Processing, ISSNIP 2015, Singapore, 7–9 April 2015; pp. 1–2. [CrossRef]

119. Medvedev, A.; Fedchenkov, P.; Zaslavsky, A.; Anagnostopoulos, T.; Khoruzhnikov, S. Waste Management as an IoT-Enabled Service in Smart Cities. In Proceedings of the 15th International Conference, NEW2AN 2015, and 8th Conference, ruSMART 2015, St. Petesburg, Russia, 26–28 August 2015; pp. 104–115. [CrossRef]

120. Bharadwaj, a.S.; Rego, R.; Chowdhury, A. IoT Based Solid Waste Management System. In Proceedings of the 2016 IEEE Annual India Conference (INDICON), Bangalore, Karnataka, India, 16–18 December 2016; pp. 1–6. [CrossRef]

121. Abdullah, N.; Alwesabi, O.A.; Abdullah, R. IoT-based smart waste management system in a smart city. In *Advances in Intelligent Systems and Computing*; Springer International Publishing: Cham, Switzerland, 2019; Volume 843, pp. 364–371. [CrossRef]

122. Aiswatha, J.; Pankajakshan, A.; Nair, A.M.; Taha, A.B. Garbage monitoring robot. *Proc. AIP Conf.* **2020**, *2222*, 040019. [CrossRef]

123. Lozano, Á.; Caridad, J.; De Paz, J.F.; González, G.V.; Bajo, J. Smart waste collection system with low consumption LoRaWAN nodes and route optimization. *Sensors* **2018**, *18*, 1465. [CrossRef]

124. Jagtap, S.; Gandhi, A.; Bochare, R.; Patil, A.; Shitole, A. Waste Management Improvement in Cities using IoT. In Proceedings of the 2020 International Conference on Power Electronics and IoT Applications in Renewable Energy and Its Control, PARC 2020, Mathura, Uttar Pradesh, India, 28–29 February 2020; pp. 382–385. [CrossRef]

125. Kang, K.D.; Kang, H.; Ilankoon, I.M.; Chong, C.Y. Electronic waste collection systems using Internet of Things (IoT): Household electronic waste management in Malaysia. *J. Clean. Prod.* **2020**, *252*, 119801. [CrossRef]

126. Tripathi, D.K.; Dubey, S.; Agrawal, S.K. Survey on IOT based smart waste bin. In Proceedings of the 2020 IEEE 9th International Conference on Communication Systems and Network Technologies, CSNT 2020, Gwalior, India, 10–12 April 2020; pp. 140–144. [CrossRef]

127. Memon, S.K.; Shaikh, F.K.; Mahoto, N.A.; Memon, A.A. IoT based smart garbage monitoring & collection system using WeMos & Ultrasonic sensors. In Proceedings of the 2nd International Conference on Computing, Mathematics and Engineering Technologies (iCoMET), Sukkur, Pakistan, 30–31 January 2019; pp. 1–6.

128. Raj, J.R.; Rajula, B.I.P.; Tamilbharathi, R.; Srinivasulu, S. AN IoT Based Waste Segreggator for Recycling Biodegradable and Non-Biodegradable Waste. In Proceedings of the 6th International Conference on Advanced Computing and Communication Systems, ICACCS 2020, Coimbatore, Tamil Nadu, India, 6–7 March 2020; pp. 928–930. [CrossRef]

129. De Paz, J.F.; Bajo, J.; Rodríguez, S.; Villarrubia, G.; Corchado, J.M. Intelligent system for lighting control in smart cities. *Inf. Sci.* **2016**, *372*, 241–255. [CrossRef]

130. International Energy Agency. *CO_2 Emissions From Fuel Combustion—Overview 2017*; IEA: Washington, DC, USA, 2017; ISBN 978-92-64-27819-6.

131. Mogro, A.; Huertas, J. Assessment of the effect of using air conditioning on the vehicle's real fuel consumption. *Int. J. Interact. Des. Manuf.* **2021**, *15*, 271–285. [CrossRef]

132. Quirama, L.F.; Giraldo, M.; Huertas, J.I.; Jaller, M. Driving cycles that reproduce driving patterns, energy consumptions and tailpipe emissions. *Transp. Res. Part D Transp. Environ.* **2020**, *82*, 102294. [CrossRef]

133. Giraldo, M.; Huertas, J.I. Real emissions, driving patterns and fuel consumption of in-use diesel buses operating at high altitude. *Transp. Res. Part D Transp. Environ.* **2019**, *77*, 21–36. [CrossRef]

134. Geng, Y.; Ji, W.; Wang, Z.; Lin, B.; Zhu, Y. A review of operating performance in green buildings: Energy use, indoor environmental quality and occupant satisfaction. *Energy Build.* **2019**, *183*, 500–514. [CrossRef]

135. Suryadevara, N.K.; Mukhopadhyay, S.C.; Kelly, S.D.T.; Gill, S.P.S. WSN-based smart sensors and actuator for power management in intelligent buildings. *IEEE/ASME Trans. Mechatron.* **2015**, *20*, 564–571. [CrossRef]

136. Ejaz, W.; Muhammad, N.; Shahid, A.; Jo, M. Efficient Energy Management for Internet of Things in Smart Cities. *IEEE Commun. Mag.* **2016**, *55*, 84–91. [CrossRef]
137. Ploennigs, J.; Ahmed, A.; Hensel, B.; Stack, P.; Menzel, K. Virtual sensors for estimation of energy consumption and thermal comfort in buildings with underfloor heating. *Adv. Eng. Inform.* **2011**, *25*, 688–698. [CrossRef]
138. Li, H.; Hong, T.; Lee, S.H.; Sofos, M. System-level key performance indicators for building performance evaluation. *Energy Build.* **2020**, *209*, 109703. [CrossRef]
139. Kumar, A.; Singh, A.; Kumar, A.; Singh, M.K.; Mahanta, P.; Mukhopadhyay, S.C. Sensing Technologies for Monitoring Intelligent Buildings: A Review. *IEEE Sens. J.* **2018**, *18*, 4847–4860. [CrossRef]
140. Akhter, F.; Khadivizand, S.; Siddiquei, H.R.; Alahi, M.E.E.; Mukhopadhyay, S. Iot enabled intelligent sensor node for smart city: Pedestrian counting and ambient monitoring. *Sensors* **2019**, *19*, 3374. [CrossRef] [PubMed]
141. Jo, O.; Kim, Y.K.; Kim, J. Internet of Things for Smart Railway: Feasibility and Applications. *IEEE Internet Things J.* **2018**, *5*, 482–490. [CrossRef]
142. Xiong, J.; Li, F.; Zhao, N.; Jiang, N. Tracking and recognition of multiple human targets moving in a wireless pyroelectric infrared sensor network. *Sensors* **2014**, *14*, 7209–7228. [CrossRef]
143. International Energy Agency, I. International Energy Agency. Available online: https://www.iea.org/ (accessed on 13 November 2020).
144. Lau, S.P.; Merrett, G.V.; Weddell, A.S.; White, N.M. A traffic-aware street lighting scheme for Smart Cities using autonomous networked sensors. *Comput. Electr. Eng.* **2015**, *45*, 192–207. [CrossRef]
145. Schaffers, H.; Komninos, N.; Pallot, M.; Trousse, B.; Nilsson, M.; Oliveira, A. Smart Cities and the Future Internet: Towards Cooperation Frameworks for Open Innovation. *Proc. Future Internet Conf.* **2011**, *6656*, 431–446. [CrossRef]
146. Higuera, J.; Hertog, W.; Perálvarez, M.; Polo, J.; Carreras, J. Smart lighting system ISO/IEC/IEEE 21451 compatible. *IEEE Sens. J.* **2015**, *15*, 2595–2602. [CrossRef]
147. Kelly, S.D.T.; Suryadevara, N.K.; Mukhopadhyay, S.C. Towards the implementation of IoT for environmental condition monitoring in homes. *IEEE Sens. J.* **2013**, *13*, 3846–3853. [CrossRef]
148. Idwan, S.; Mahmood, I.; Zubairi, J.A.; Matar, I. Optimal Management of Solid Waste in Smart Cities using Internet of Things. *Wirel. Person. Commun.* **2020**, *110*, 485–501. [CrossRef]
149. Huang, H.; Gong, T.; Ye, N.; Wang, R.; Dou, Y. Private and Secured Medical Data Transmission and Analysis for Wireless Sensing Healthcare System. *IEEE Trans. Ind. Inform.* **2017**, *13*, 1227–1237. [CrossRef]
150. Cao, B.; Ge, Y.; Kim, C.W.; Feng, G.; Tan, H.; Li, Y. An Experimental Study for Inter-User Interference Mitigation in Wireless Body Sensor Networks. *Sens. J. IEEE* **2013**, *13*, 3585–3595. [CrossRef]
151. Kim, B.; Choi, J.h.; Cho, J. A Hybrid Channel Access Scheme for Coexistence Mitigation in IEEE 802.15.4-based WBAN. *IEEE Sens. J.* **2017**, *17*, 7189–7195. [CrossRef]
152. Velusamy, B.; Pushpan, S.C. An Enhanced Channel Access Method to Mitigate the Effect of Interference among Body Sensor Networks for Smart Healthcare. *IEEE Sens. J.* **2019**, *19*, 7082–7088. [CrossRef]
153. Baltusis, P. On-board vehicle diagnostics. In *Convergence Transportation Electronics Association*; SAE International: Warrendale, PA, USA, 2004; p. 15.
154. Gerla, M.; Lee, E.K.; Pau, G.; Lee, U. Internet of Vehicles: From Intelligent Grid to Autonomous Cars and Vehicular Clouds The Genesis of IOT. In Proceedings of the 2014 IEEE World Forum on Internet of Things (WF-IoT), Seoul, Korea, 6–8 March 2015; pp. 241–246.
155. Gruwsved, A.; Söderback, I.; Fernholm, C. Evaluation of a vocational training programme in primary health care rehabilitation: A case study. *Work* **1996**, *7*, 47–61. [CrossRef]
156. Silva, A.S.M. Wearable Sensors Systems for Human Motion Analysis. Ph.D. Thesis, Universidade do Porto, Porto, Portugal, 2014.
157. Camomilla, V.; Bergamini, E.; Fantozzi, S.; Vannozzi, G. In-Field Use of Wearable Magneto-Inertial Sensors for Sports Performance Evaluation. In Proceedings of the 33rd International Conference on Biomechanics in Sports, Poitiers, Vienne, France, 29 June–3 July 2015; pp. 1–4.
158. Dey, N.; Ashour, A.S.; Shi, F.; Fong, S.J.; Sherratt, R.S. Developing residential wireless sensor networks for ECG healthcare monitoring. *IEEE Trans. Consum. Electron.* **2017**, *63*, 442–449. [CrossRef]
159. Teaw, E.; Hou, G.; Gouzman, M.; Tang, K.W.; Kesluk, A.; Kane, M.; Farrell, J. A wireless health monitoring system. In Proceedings of the 2005 IEEE International Conference on Information Acquisition, Hefei, China, 21–25 June 2005; p. 6.
160. Park, J.; Cho, J.; Choi, J.; Nam, T. A zigbee network-based multi-channel heart rate monitoring system for exercising rehabilitation patients. In Proceedings of the TENCON 2007—2007 IEEE Region 10 Conference, Taipei, Taiwan, 30 October–2 November 2007; pp. 1–4. [CrossRef]
161. Becher, K.; Figueiredo, C.; Muhle, C.; Ruff, R.; Mendes, P.; Hoffmann, K.P. Design and Realization of a Wireless Sensor Gateway for Health Monitoring. In Proceedings of the Annual International Conference of the IEEE Engineering in Medicine and Biology Society, Buenos Aires, Argentina, 1–4 September 2010; pp. 374–377. [CrossRef]
162. Wang, L.H.; Hsiao, Y.M.; Xie, X.Q.; Lee, S.Y. An outdoor intelligent healthcare monitoring device for the elderly. *IEEE Trans. Consum. Electron.* **2016**, *62*, 128–135. [CrossRef]
163. Yan, H.; Huo, H.; Xu, Y.; Gidlund, M. Wireless sensor network based e-health system - implementation and experimental results. *IEEE Trans. Consum. Electron.* **2010**, *56*, 2288–2295. [CrossRef]

164. Zaldivar, J.; Calafate, C.T.; Cano, J.C.; Manzoni, P. Providing accident detection in vehicular networks through OBD-II devices and Android-based smartphones. In Proceedings of the 2011 IEEE 36th Conference on Local Computer Networks, Bonn, North Rhine-Westphalia, Germany, 4–7 October 2011; pp. 813–819. [CrossRef]

165. Fernandes, B.; Alam, M.; Gomes, V.; Ferreira, J.; Oliveira, A. Automatic accident detection with multi-modal alert system implementation for ITS. *Veh. Commun.* **2016**, *3*, 1–11. [CrossRef]

166. Sohail, A.M.; Khattak, K.S.; Iqbal, A.; Khan, Z.H.; Ahmad, A. Cloud-based detection of road bottlenecks using obd-ii telematics. In Proceedings of the 22nd International Multitopic Conference, INMIC 2019, Islamabad, Pakistan, 29–30 November 2019; pp. 1–7. [CrossRef]

167. Rolim, C.; Baptista, P.; Duarte, G.; Farias, T.; Shiftan, Y. Quantification of the Impacts of Eco-driving Training and Real-time Feedback on Urban Buses Driver's Behaviour. *Transp. Res. Procedia* **2014**, *3*, 70–79.

168. Chen, S.H.; Pan, J.S.; Lu, K. Driving behavior analysis based on vehicle OBD information and adaboost algorithms. In Proceedings of the International Multiconference of Engineers and Computer Scientists, Hong Kong, China, 18–20 March 2015; Volume 1, pp. 102–106.

169. von Landesberger, T.; Brodkorb, F.; Roskosch, P.; Andrienko, N.; Andrienko, G.; Kerren, A. MobilityGraphs: Visual Analysis of Mass Mobility Dynamics via Spatio-Temporal Graphs and Clustering. *IEEE Trans. Vis. Comput. Graph.* **2016**, *22*, 11–20. [CrossRef]

170. Andrienko, N.; Andrienko, G.; Rinzivillo, S. Leveraging spatial abstraction in traffic analysis and forecasting with visual analytics. *Inf. Syst.* **2016**, *57*, 172–194. [CrossRef]

171. Andrienko, G.; Andrienko, N.; Bak, P.; Keim, D.; Wrobel, S. *Visual Analytics of Movement*; Springer: Berlin, Germany, 2013; Volume 1, pp. 1–387. [CrossRef]

172. Jaikumar, K.; Brindha, T.; Deepalakshmi, T.K.; Gomathi, S. IOT Assisted MQTT for Segregation and Monitoring of Waste for Smart Cities. In Proceedings of the 2020 6th International Conference on Advanced Computing and Communication Systems (ICACCS), Coimbatore, Tamil Nadu, India, 6–7 March 2020; pp. 887–891. [CrossRef]

173. Rao, P.V.; Azeez, P.M.A.; Peri, S.S.; Kumar, V.; Devi, R.S.; Rengarajan, A.; Thenmozhi, K.; Praveenkumar, P. IoT based waste management for smart cities. In Proceedings of the 2020 International Conference on Computer Communication and Informatics, ICCCI 2020, Coimbatore, Tamil Nadu, India, 22–24 June 2020; pp. 20–24. [CrossRef]

174. Cerchecci, M.; Luti, F.; Mecocci, A.; Parrino, S.; Peruzzi, G.; Pozzebon, A. A low power IoT sensor node architecture for waste management within smart cities context. *Sensors* **2018**, *18*, 1282. [CrossRef] [PubMed]

175. Cook, D.J. Learning Setting-Generalized Activity Models for Smart Spaces. *IEEE Intell. Syst.* **2010**, *2010*, 1. [CrossRef]

176. Albino, V.; Berardi, U.; Dangelico, R.M. Smart cities: Definitions, dimensions, performance, and initiatives. *J. Urban Technol.* **2015**, *22*, 3–21. [CrossRef]

177. Ramírez-Moreno, M.A.; Díaz-Padilla, M.; Valenzuela-Gómez, K.D.; Vargas-Martínez, A.; Tudón-Martínez, J.C.; Morales-Menendez, R.; Ramírez-Mendoza, R.A.; Pérez-Henríquez, B.L.; Lozoya-Santos, J.d.J. EEG-Based Tool for Prediction of University Students' Cognitive Performance in the Classroom. *Brain Sci.* **2021**, *11*, 698. [CrossRef] [PubMed]

178. Ramirez, M.; TEC. PiBot, the Multifunctional Robot Developed at Tec. Available online: https://transferencia.tec.mx/en/2021/0 6/24/pibot-el-robot-multifuncional-desarrollado-en-el-tec/ (accessed on 21 August 2021).

179. Gaxiola-Beltrán, A.L.; Narezo-Balzaretti, J.; Ramírez-Moreno, M.A.; Pérez-Henríquez, B.L.; Ramírez-Mendoza, R.A.; Krajzewicz, D.; Lozoya-Santos, J.d.J. Assessing Urban Accessibility in Monterrey, Mexico: A Transferable Approach to Evaluate Access to Main Destinations at the Metropolitan and Local Levels. *Appl. Sci.* **2021**, *11*, 7519. [CrossRef]

180. Mahlknecht, J.; Alonso-Padilla, D.; Ramos, E.; Reyes, L.; Álvarez, M. The presence of SARS-CoV-2 RNA in different freshwater environments in urban settings determined by RT-qPCR: Implications for water safety. *Sci. Total Environ.* **2021**, *784*, 147183. [CrossRef]

181. Colombia, S.C. Smart City Colombia. Available online: https://www.smartcitycolombia.org/ (accessed on 21 August 2021).

182. SmarTech. SmarTech, Thinking of the Future. Available online: https://smartech.biz/indexE.html (accessed on 21 August 2021).

183. Lu, R.; Lin, X.; Shen, X. SPOC: A Secure and Privacy-Preserving Opportunistic Computing Framework for Mobile-Healthcare Emergency. *Parallel Distrib. Syst. IEEE Trans.* **2013**, *24*, 614–624. [CrossRef]

184. Chan, A. Symmetric-Key Homomorphic Encryption for Encrypted Data Processing. In Proceedings of the 2009 IEEE International Conference on Communications, Dresden, Saxony, Germany, 14–18 June 2009; pp. 1–5. [CrossRef]

185. Ma, Y.; Wang, Y.; Yang, J.; Miao, Y.; Li, W. Big Health Application System based on Health Internet of Things and Big Data. *IEEE Access* **2016**, *5*, 7885–7897. [CrossRef]

186. Saleh, N.; Kassem, A.; Haidar, A.M. Energy-efficient architecture for wireless sensor networks in healthcare applications. *IEEE Access* **2018**, *6*, 6478–6496. [CrossRef]

187. Tang, S.; Shelden, D.; Eastman, C.; Pishdad-Bozorgi, P.; Gao, X. A review of building information modeling (BIM) and the internet of things (IoT) devices integration: Present status and future trends. *Autom. Constr.* **2019**, *101*, 127–139. [CrossRef]

188. Jelicic, V.; Magno, M.; Brunelli, D.; Bilas, V.; Benini, L. Benefits of Wake-Up Radio in Energy-Efficient Multimodal Surveillance Wireless Sensor Network. *Sens. J. IEEE* **2014**, *14*, 3210–3220. [CrossRef]

189. Mekonnen, T.; Porambage, P.; Harjula, E.; Ylianttila, M. Energy Consumption Analysis of High Quality Multi-Tier Wireless Multimedia Sensor Network. *IEEE Access* **2017**, *5*, 15848–15858. [CrossRef]

190. Allam, Z.; Dhunny, Z.A. On big data, artificial intelligence and smart cities. *Cities* **2019**, *89*, 80–91. [CrossRef]

191. Davenport, T.H.; Ronanki, R. Artificial intelligence for the real world. *Harv. Bus. Rev.* **2018**. Available online: https://hbr.org/webinar/2018/02/artificial-intelligence-for-the-real-world (accessed on 1 September 2021).
192. Pérez Henríquez, B. *Transportation, Land Use, and Environmental Planning*; Chapter Energy Sources for Sustainable Transportation and Urban Development; Elsevier: Amsterdam, The Netherlands, 2019; pp. 281–298. [CrossRef]

Review

A Systematic Review of Technologies, Control Methods, and Optimization for Extended-Range Electric Vehicles

David Sebastian Puma-Benavides [1,†,‡,§], Javier Izquierdo-Reyes [1,2], Juan de Dios Calderon-Najera [2,*] and Ricardo A. Ramirez-Mendoza [1,†,‡,§]

1   School of Engineering and Science, Tecnologico de Monterrey, Monterrey 64849, Mexico; sebastian.puma@tec.mx (D.S.P.-B.); jizquierdo.reyes@tec.mx (J.I.-R.); ricardo.ramirez@tec.mx (R.A.R.-M.)
2   Microsystems Technology Laboratories, Massachusetts Institute of Technology, Cambridge, MA 02139, USA
*   Correspondence: jcalderon@tec.mx
†   Current address: Toluca Campus, Av. Eduardo Monroy Cárdenas 2000 San Antonio Buenavista, Toluca de Lerdo 50110, Mexico.
‡   Current address: Mexico City Campus, Calle Puente 222, AMSA, Tlalpan, Mexico City 14380, Mexico.
§   Current address: Mex and Monterrey City Campus, Av. Eugenio Garza Sada 2501 Sur, Tecnológico, Monterrey 64849, Mexico.

Citation: Puma-Benavides, D.S.; Izquierdo-Reyes, J.; Calderon-Najera, J.d.D.; Ramirez-Mendoza, R.A. A Systematic Review of Technologies, Control Methods, and Optimization for Extended-Range Electric Vehicles. *Appl. Sci.* **2021**, *11*, 7095. https://doi.org/10.3390/app11157095

Academic Editors: Michele Roccotelli and Agostino Marcello Mangini

Received: 30 June 2021
Accepted: 21 July 2021
Published: 31 July 2021

Abstract: For smart cities using clean energy, optimal energy management has made the development of electric vehicles more popular. However, the fear of range anxiety—that a vehicle has insufficient range to reach its destination—is slowing down the adoption of EVs. The integration of an auxiliary power unit (APU) can extend the range of a vehicle, making them more attractive to consumers. The increased interest in optimizing electric vehicles is generating research around range extenders. These days, many systems and configurations of extended-range electric vehicles (EREVs) have been proposed to recover energy. However, it is necessary to summarize all those efforts made by researchers and industry to find the optimal solution regarding range extenders. This paper analyzes the most relevant technologies that recover energy, the current topologies and configurations of EREVs, and the state-of-the-art in control methods used to manage energy. The analysis presented mainly focuses on finding maximum fuel economy, reducing emissions, minimizing the system's costs, and providing optimal driving performance. Our summary and evaluation of range extenders for electric vehicles seeks to guide researchers and automakers to generate new topologies and configurations for EVs with optimized range, improved functionality, and low emissions.

Keywords: extended range electric vehicle; technologies; optimization methods; EREV key components; level optimization

1. Introduction

Extended-range electric vehicles (EREVs), commonly known as series hybrid electric vehicles (Series-HEV), have better autonomy than electric vehicles (EV) without range extenders (REs). EREVs can go from one city to another or make long journeys in general. In recent years, EREVs have attracted considerable attention because of the necessity to improve autonomy using new and different technologies to generate extra energy for EVs. Today, fossil fuels meet the needs of the transportation sector to a significant extent, but bring on various adverse effects, such as air pollution, noise, and global warming. Compared to internal combustion engine vehicles (ICEVs), EREVs reduce emissions and are considered a favorable alternative [1,2]. EREVs, compared with EV, not only have the advantage of "zero fuel consumption and zero emissions"; they also effectively solve the problem of having an inadequate driving range due to power storage limitations in batteries [3]. This paper presents a systematic review on the subject of EREVs. First, an explanation of all the technologies used to extend the ranges of electric vehicles is presented and compared, considering the characteristics of each technology. The next stage reviews all the possible topologies for an EREV, analyzing the components and their interactions. The different control methods and their applications are also analyzed. The last part the analysis of how an EREV can be optimized. All the information is organized and presented in graphs and tables. As a contribution of our own, we propose a method for selecting the components of an EREV and designing its architecture based on the final application and use.

2. Extended Range Electric Vehicle Technology

A range extender (RE) is a small electricity generator (APU) which operates when needed as a solution to increase autonomy in EVs. The main components of the RE are the generator and internal or external combustion engine; the internal or external combustion engine is coupled to the generator in a series configuration. The primary function of the RE for an EV is to extend the vehicle's mileage. Operation of the range extender is initiated if the SOC (state of charge) of the EVs battery drops below a specified level. In this situation, the engine provides electricity by recharging the battery or directly driving the EV during travel and continues the vehicle's operation [4]. The difference in a plug-in hybrid electric vehicle (PHEV) is that the electric motor always propels the wheels. The engine acts as a generator to recharge the vehicle's battery when it depletes or as it propels the vehicle [5]. A series configuration is used as the main system, which is considered an APU. The system is connected to several subsystems, such as the generator, battery, electronic management system, and electric motor. The electric motor converts electrical energy from the battery to mechanical power. It propels the wheels while the APU generates electric energy to recharge the battery. Finally, the electronic management system controls all the systems for optimal functioning. The EREV has two operation modes: pure electric vehicle and extended-range mode. If the distance is short, the vehicle operates in pure electric vehicle mode without the RE. If the distance is long, the vehicle operates in extended-range electric vehicle mode.

The RE is off as long as there is sufficient energy in the battery for purely electric driving, and activated whenever the SOC drops below a certain level. The RE works until the desired SOC is achieved. The battery power manager gives this function. Figure 1 shows an EREV and energy flow configuration: (a) charge sustaining period and (b) depletion period. There are many technical and social challenges ahead for EVs coming up against the conventional ICEVs. Range anxiety is the most challenging problem facing EV drivers due to their shorter driving ranges compared to ICEVs. Range anxiety stems from the limited energy density in the current batteries (0.565 MJ/kg for Li-ion battery), which is very low as compared to fossil fuel (43.48 MJ/kg) [6].

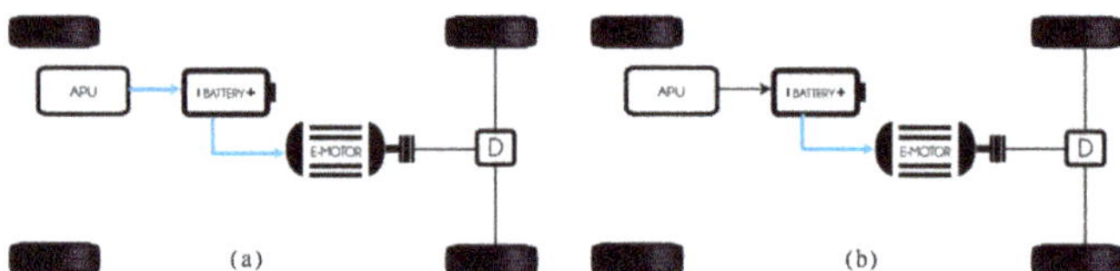

Figure 1. Configuration of an EREV and energy flow: (**a**) charge sustaining period; (**b**) depleting period.

2.1. Technological Classification of EREV

The electric propulsion system is the heart of an EREV. It consists of the motor drive, a transmission (optional) device, and wheels. There are three kinds of electric motors: direct or alternating current and in-wheel motors (also called wheel motors). The primary requirements of the EREV motor are summarized as follows:

1. High instant power and high power density.
2. High torque at low speeds for starting and climbing, and high power at high speeds for cruising.

3.	An extensive speed range including constant-torque and constant-power regions. In this case, the APU, when it is on, needs to operate in the same regions.
4.	Fast torque response.
5.	High efficiency over a large speed and torque ranges.
6.	High reliability and robustness for various vehicle operating conditions.
7.	Reasonable cost.

2.1.1. Internal Combustion Engine Extended Range (ICE-ER)

The range extender comprises a fuel tank, an internal combustion engine, and a permanent magnet synchronous generator [3,6–34], as shown in Figure 2. The structure is mechanically decoupled between the RE and the wheels of the EV. This configuration leads to a strong point whereby the output characteristics of the RE are not related to the vehicle's traction performance, and the output power only needs to meet the driving requirements. Therefore, one of the main objectives is to keep the RE operating in the high-efficiency region. The engine and the generator should be matched to achieve this common operating region [8]. As another solution, several studies have focused on energy harvesting using other types of fuels, such as natural gas [35] or diesel [36,37], to reduce pollution levels.

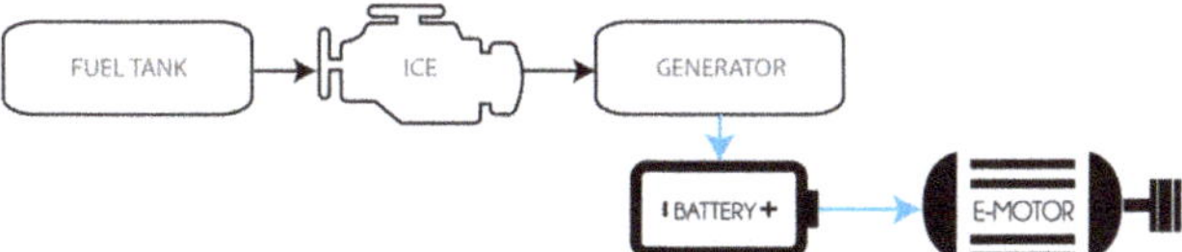

Figure 2. A diagram of the configuration of ICE-ER.

2.1.2. Regenerative Shock Absorber Extended Range (RSA-ER)

The shock absorber is a crucial component of the vehicle suspension and is combined with the suspension spring to filter vibrations when driving on rough roads. Typically, energy from vibrational sources is dissipated through hydraulic friction and heat via shock absorbers [38]. Currently, there are three categories for RSAs. The first type directly uses an electromagnetic method to generate electric power. The schemes of operation can be linear or rotary [39]. A linear electromagnetic RSA converts the kinetic energy of vertical oscillations into electricity by electromagnetic induction.

The second is the hydraulic RSA. This RSA can harvest energy by employing oscillatory motion to drive the power generator. Some studies reformed the existing hydraulic shock absorber and utilized the oil in the shock absorber to flow into a parallel oil circuit. They usually used the flowing fluid to drive a hydraulic motor connected in parallel to a DC/AC generator [40].

The third category is the mechanical RSA, which was developed quickly because of its greater efficiency and average power [38]. The general architecture of said RSA using supercapacitors, which are applied to extend the battery endurance, has four main parts: (1) the suspension vibration input module, (2) the transmission module, (3) the generator module, and (4) the power storage module, as shown in Figure 3.

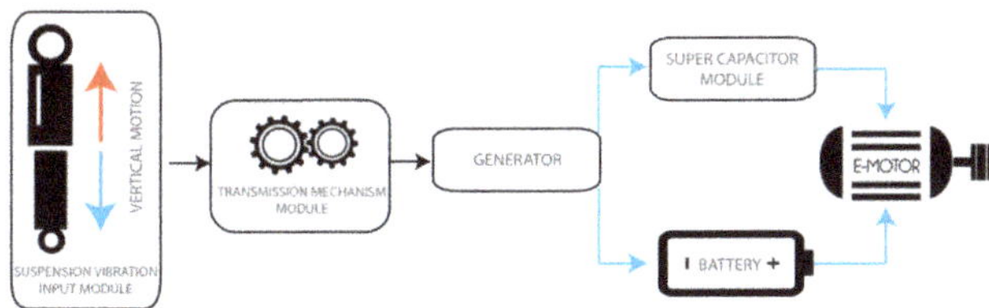

Figure 3. A diagram of the configuration of an RSA-ER.

2.1.3. Regenerative Braking Extended Range RB-ER

The EV's motor can work as a generator under deceleration procedures, charging the battery and exerting regenerative braking torque on the axle simultaneously, as shown in Figure 4. A regenerative braking system (RBS) can capture the kinetic energy of an electric vehicle during the deceleration process, thereby improving the EV's energy efficiency. For an EV, friction and regenerative brakes generate brake torque, either separately or together [41]. In addition, the use of an adequate control strategy helps to reduce the loss rate; for example, the revised regenerative braking control strategy (RRBCS) can reduce inefficiency at the expense of slight braking energy recovery loss. Thus it positively affects prolonging the battery's life while ensuring braking safety and maximal recovery energy [42].

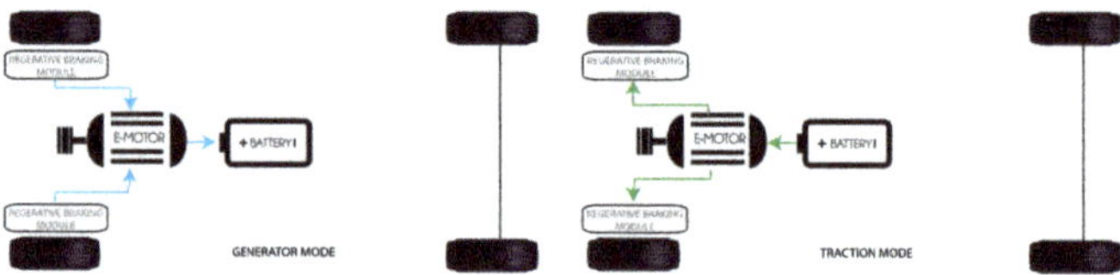

Figure 4. The configuration of an RB-ER. Energy flow: left generator mode and right traction mode.

An EV with automatic mechanical transmission (AMT) has higher transmission efficiency because it uses a composite braking process. The braking force of the motor varies according to the transmission gears at the same speed. Therefore, when the vehicle is braking, the transmission gear can be shifted reasonably according to the vehicle's condition. This mechanism improves the EV economy, since it allows the motor to work efficiently while recovering the braking energy to the maximum extent. An appropriate strategy can effectively improve the energy recovery rate and ensure braking safety and stability [43].

2.1.4. Fuel Cell Extended Range (FC-ER)

Fuel cells (FC) are electrochemical energy conversion devices that convert chemical energy directly into electrical energy and heat [44]. The electrochemical transformation is a chemical reaction of oxidant and reductant to produce electricity and water in stack output [45].

An anode, a cathode, and an electrolyte are the main components of an FC; water and electrical energy are the products. In the process, the anode supplies hydrogen, and the cathode terminal supplies oxygen [46]. During the reaction, hydrogen decomposes into positive protons and negative ions on the anode side. The resulting positive particles reach the cathode tip through the electrolyte, allowing only the positively charged particles to pass. The electrons, the negative ions at the end of the anode, tend to reunite with the positively charged particles and pass to the cathode side through an external circuit. This electron flow in the external circuit generates electricity. The electrons passing to the cathode side combine with positively charged particles and oxygen to produce pure water and heat, as shown in Figure 5 [1,47–53].

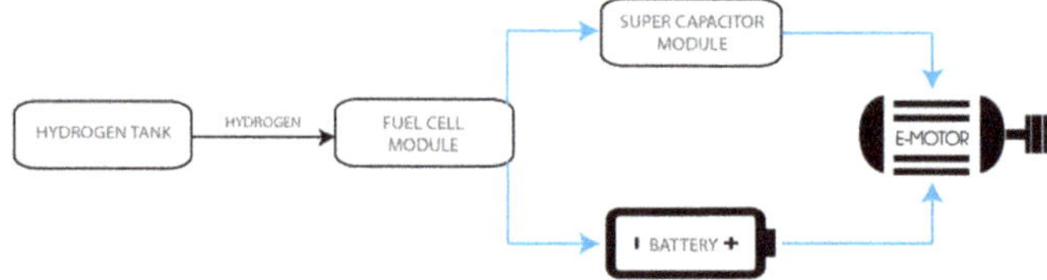

Figure 5. A diagram of the configuration of FC-ER.

2.1.5. Micro Gas Turbine Extended Range (MGT-ER)

Microturbines (MGT) are small gas turbines with output power levels of 30 to 500 kW [54]. A MGT mainly consists of a single-stage radial compressor, a radial turbine section, and a recuperator. They usually use foil bearings (air bearings). The typical cycle of an MGT consists of four processes:

1. A radial compressor compresses the inlet air.
2. Air is pre-heated in the recuperator using heat from the turbine exhaust.
3. Heated air from the recuperator is mixed with fuel in the combustion chamber and burned.
4. Hot gas expands in turbine stages, and the gas's energy is converted into mechanical energy to drive the air-compressor and the drive equipment (usually generator).

Automakers claim that the gas turbine is the most efficient solution that is on its way. In particular, MGT can be an alternative to the ICE as a RE for EVs. The MGT produces less raw exhaust gaseous emissions, such as hydrocarbons and carbon monoxide, and has more static applications compared to the ICE. In addition, any MGT is lighter than the equivalent ICE, and it provides a potential reduction in the level of carbon dioxide produced [55–57]. Figure 6 shows the configuration of an MGT-ER connected to a generator and in series with a battery.

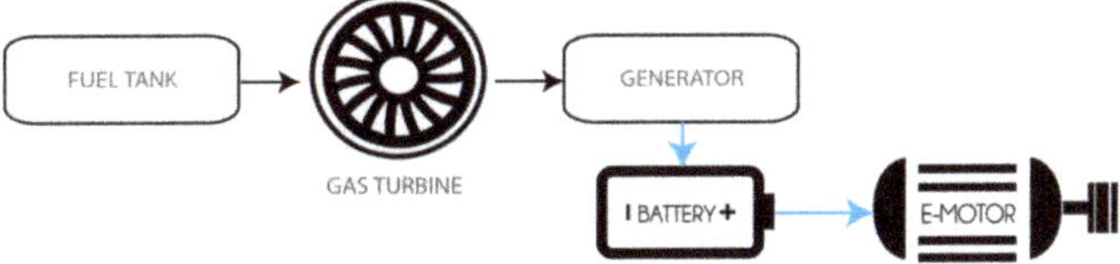

Figure 6. A diagram of the configuration of MGT-ER.

2.1.6. Thermoacoustic Engine Extended Range (TAE-ER)

Most vehicles waste nearly a third of their fuel's energy through the exhaust. Therefore, an efficient waste heat recovery process would undoubtedly improve fuel efficiency and reduce greenhouse gas emissions. Multiple waste heat recovery proposals exist. One of these is the thermoacoustic converter (TAC). The engine exhaust (hot side) and the coolant (cold side) produce a temperature differential that the TAC uses to produce electricity. Essentially, the TAC converts exhaust waste heat into electricity in two steps:

1. The exhaust heat is converted to acoustic energy (mechanical);
2. The acoustic energy is converted to electrical energy [58].

Three main stages illustrate how a TAE functions. The first is fuel burning in the combustion compartment, containing the combustion chamber blower and the combustion chamber. The second is the hot exhaust gases heading into the hot heat exchanger, transferring heat to the working fluids through a heat pipe. The third is the stack, which is the TAE's thermal module area and is surrounded by a hot and cold reservoir on each side, exchanging heat. The cold reservoir exchanges heat through the cold HEX with the ambient air [59,60]. Figure 7 shows a diagram with the main components in a TAE-ER.

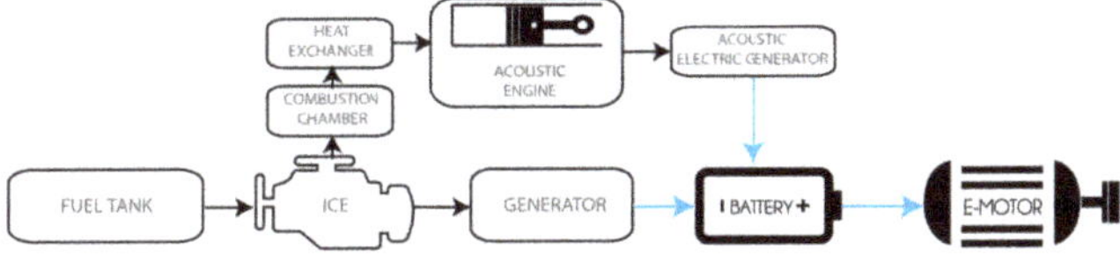

Figure 7. A diagram of the configuration of TAE-ER.

2.1.7. Flywheel Energy Storage Extended Range (FES-ER)

A flywheel energy storage (FES) system has fast charge/discharge, is infinitely clean, and is highly efficient. The system consists of three energy storage components: a flywheel, a battery, and an ultra-capacitor. A flywheel is a rotating disk used as a mechanical energy storage device [61]. Two classes of materials are commonly used to fabricate the flyWheel, steel and composite materials. The difference is their rotational stress limitations. A composite-based flywheel can support higher speeds and rotational stress thresholds than a steel-based one. Therefore, composite materials can be used at high speeds (up to 100.00 rpm), whereas lower speeds (up to 10.000 rpm) apply to steel-based flywheels, which are heavier than the composite ones. The main limitation for the use of the composite material is its cost [62]. As a kind of short-term energy storage system, the FES system cannot be the primary power source of the vehicle. Therefore, a FES system with high power density is often used as an APU for a vehicle. The FES works while the vehicle brakes; it absorbs the RB energy. When the vehicle needs to accelerate, the FES system and the battery provide energy to the vehicle [31], as shown in Figure 8.

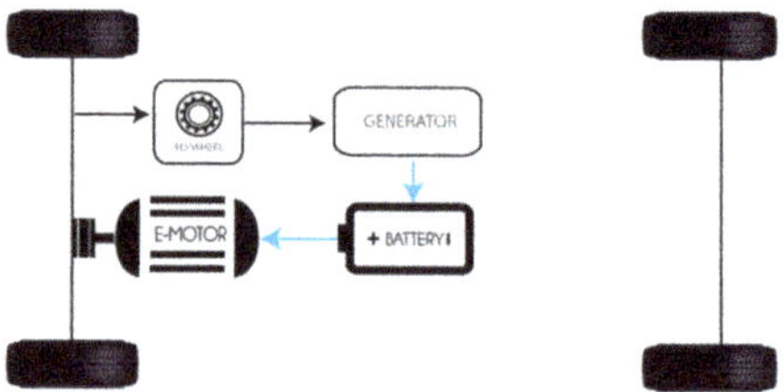

Figure 8. A diagram of the configuration of FES-ER.

2.1.8. Solar Energy Storage Extended Range (SES-ER)

Solar photovoltaic cells (PVCs) generate electricity by absorbing sunlight and converting it to electric current. Vehicle companies favor solar energy storage (SES) systems for their cleanliness, safety, and economic performance. Studying efficient and stable SES systems has become critical for many automobile enterprises [63]. A car using a PVC improves autonomy by about 10% when used in a city [64]. Ezzat et al. [65] proposed a novel comprehensive energy storage system for EREV. The energy storage system proposed consists of PVCs, an FC, and batteries. The results showed that the addition of solar cells to the energy storage system of EREV could improve energy efficiency, perfectly complementing a range extender. Figure 9 shows the configuration of an FES-ER.

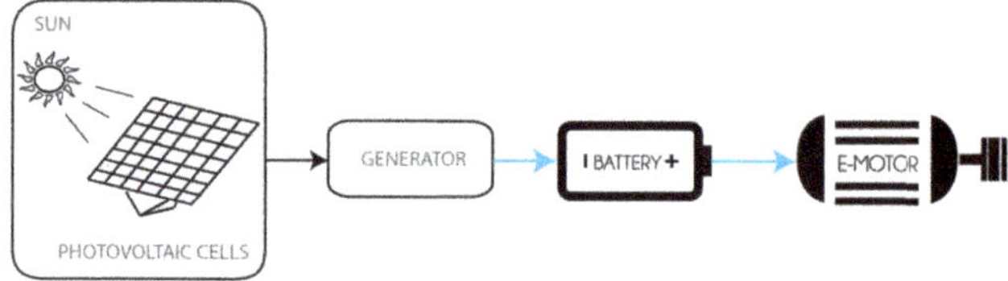

Figure 9. A diagram of the configuration of SES-ER.

2.1.9. Rotatory Engine Extended Range (RE-ER)

Rotary engines (REs) are small in size, and their high power output makes multiple electrification technology solutions possible via a shared packaging layout. The rotary-powered range extender takes advantage of their compatibility with gaseous fuels: it works by burning liquefied petroleum gas to provide a source of electricity. A rotary engine has only two moving parts: the rotor and the shaft are inherently balanced with no oscillating components and produce minimal vibrations [66]. Each crankshaft revolution produces one rotor revolution, a complete engine cycle in each of the four chambers, and four power strokes [67]. Figure 10 shows the configuration of RE-ER.

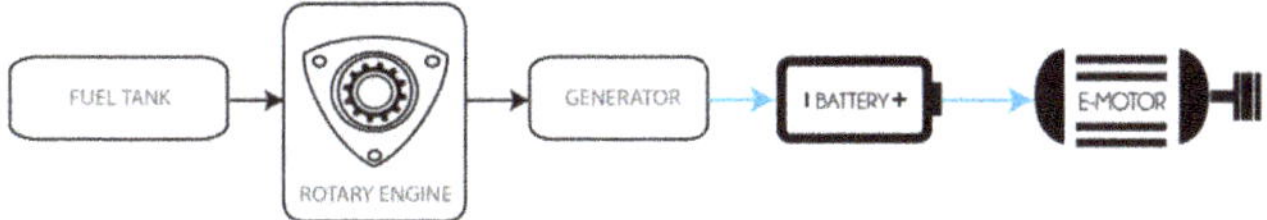

Figure 10. A diagram of the configuration of RE-ER.

Hydrogen combustion in a RE was carried out by Zambolov et al. [68], which complements the research about REs.

2.1.10. Wind Turbine Extended Range (WT-ER)

When a vehicle moves, it experiences wind resistance in two different forms—frictional drag and form drag. Frictional drag arises due to the viscosity of air, and form drag arises due to air pressure variation in the front and rear sides of the vehicle [69]. Suppose this wind energy is used to extract some power, not to create any component of force or thrust opposite the direction of the vehicle's propulsion. In that case, the energy can produce electricity to charge up the EV battery itself. [66,70,71] presented a conceptual design of harnessing wind's power to generate extra energy. Those designs can provide energy which could be stored or directly used to power electronic devices and vehicle instrumentation in a car or truck in movement. The use of wind energy in an EV can have different configurations, depending highly on the vehicle. One must not increase the drag coefficient to an extent that risks recovering the energy inefficiently. Figure 11 shows the configuration of WT-ER.

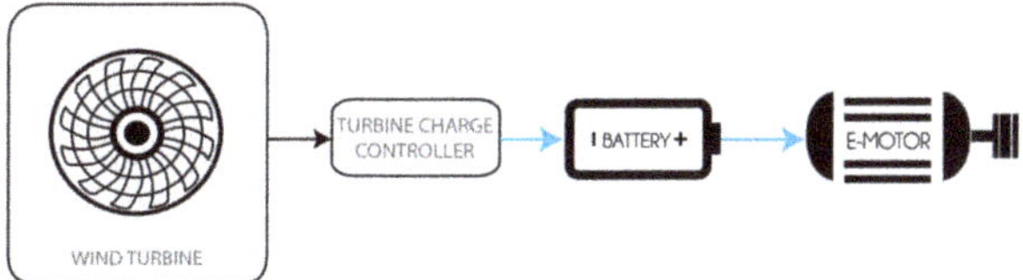

Figure 11. A diagram of the configuration of wind energy extended range.

3. A Comparison of the Technology Used in EREVs

When we design an extended range electric vehicle, we start by comparing the different technologies, analyzing their possible configurations, and analyzing the relationships among all their components.

The selection of the type of range extender at the time of vehicle design will depend on certain system characteristics. Thus, range extender systems are compared here. The criteria by which the systems are evaluated are as follows.

- System power;
- Amount of extra range;
- Global system efficiency;
- Emissions.

Vehicle concepts can have diverse sets of specifications. For example, the battery size and user profile determine the amount of time a range extender is used. This defines the importance of the efficiency. Furthermore, depending on the type of vehicle, the packing weight can be a more or less important criterion for choice of technology.

A brief overview of the collected information is presented in Table 1. It is clearly possible to identify that there are major differences between some of the concepts.

Table 1. A comparison of the extended range systems.

Extended Range System	System Power	Extra Range	Efficiency	Emissions
ICE-ER	30 kW [12] 35 kW [14] 5.5 kW [35] 111 kW [72]	232.79% [7] 430 km [12] 51–139 km [35] 380 km [73] 330 km [74] 676 [72]	20–40% [75,76] 31% [12]	Low
Fuel cell-ER	20 kW [77] 85–83 kW [78] 1200 W [47] 25 kW [79] 128 kW [80]	500 km [77] 650 km [80] 665 km [81] 594 km [81] 1500 km [82]	70% [77] 63.6–72.4% [83] 43% [47] 55.21% [79]	No
Rotary engine-ER	3.8 kW [84] 20 kW [85]	80 km [85] 321 km [86]	73% [87] 78% [84] 77% [85]	Low
RB-ER	14.8 kW [88] 55.75–82.66 kJ [89] 298.75 kJ [90]	32.1–47.7% of the total recoverable energy [89] 1.18% SOC improve [90]	79–94% [88] 30–60% [26] 47% [89]	No
MGT-ER	32 kW [91] 100 kW [92] 63.3 kW [93]	370 km [94]	47.2% [95] 28% [91] 30% [92] 35% [94] 38% [93]	Low
PVC-ER	68.2–300 W [96]	19.6 km [96]	91.2% [96] 20.2–23% [96]	No
WT-ER	2.64 kW [97] 0.1–1.1 kW [70]	add up to 10% [98] 7.27 km [97]	75% [97] 75–90% [70]	Low
FES-ER	40 kW to 1.6 MW [99] 60–101 kW [100] 1–20 kW [101]	50% milage over [100] 1.17% milage over [102]	60% [100] 90–95% [103] 70–90% [104]	No
TAE-ER	710 W [105] 1029 W [106] 58 W [107] 1.5 kW [108]	80% fuel consumtion savings [59]	33.8–38.7% [60] 30% [105] 5.4% [106] 18% [78] 16% [108]	Low
RSA-ER	8–40 W [100] 0.74–0.78 kW [109] 19.2–67.5 W [110] 4.3 W [38]	Can power an 8 W lidar for 323 days or a 2 W camera for 1292 days [100]	70–80% [100] 71–84% [111] 33–63% [110] 87% [38] 16% [38]	No

The increasing demand for extended driving range in electric vehicles is the most prominent factor driving the EREV market. As a result, the global EREV market is projected to reach more than 500,000 units and more than $1500 million by 2026, at a compound annual growth rate (CAGR) of between 9.0% and 11.89% [112]. The Asia–Pacific market is estimated to reach more than $750 million by 2026, with a CAGR of 8.1%. North American market is estimated to reach more than $340 million by 2026, at a significant CAGR of 10.6% [113,114].

Major industry participants are further investing a substantial amount into research and development processes to create advanced range extender products using various technologies. The top 10 automotive component manufacturers include the following: Magna offers a hydrogen fuel-cell platform as a range extender for battery electric vehicles. MAHLE offers an electric range extender engine for EVs to original equipment manufacturers (OEMs). The company's range extender engine offers an electric range extension of 65 km. Rheinmetall offers heater/cooler modules for extending the driving ranges of electric buses. Plug Power is developing a fuel cell range extender electric vehicle that can extend the driving range by approximately 136 km. AVL offers the entire range of powertrain systems for extended-range electric vehicles. The five other leading players are Ballard Power Systems, FEV, Delta Motorsport, Ceres Power, Nissan, General Motors, and BMW [112].

Many OEMs have integrated some of the RE technologies into mass production vehicles. Companies such as Chevrolet with its Volt model [72], BMW with its I3 and I3Rex models [73,74], and Honda with its Clarity model [115], have incorporated ICEs as range extenders. Brands such as Toyota with its Mirai model [80], and Hyundai with its Nexo model [81], have incorporated fuel cells as range extenders in light vehicles; NIKOLA MOTORS has included fuel cells in trucks, giving them autonomy for around 1500 km [82]. Mazda incorporated a rotary engine in its model Mazda 2 [86] and plans to incorporate the same technology in SUVs in the future. The heavy duty truck OEM MACK is using an MGT to extend the ranges of their vehicles. In its Karma model, Fisker placed photovoltaic cells over the entire roof of the vehicle to recover energy. It is worth mentioning that regenerative braking is used in almost all mass-produced light EVs and even in trucks such as the MACk LR model with a power output of 780 kW [116].

The market share of hybrid vehicles with alternative energy represents 0.7% of the total sales of all cars marketed until the middle of 2021. As of the end of 2020, sales were 0.6% in North America. Referring to EVs with ICEs to extend their ranges, Honda's Clarity and Chevrolet's Volt have market shares of 4% and 0.1% respectively. Furthermore, concerning EVs with fuel cells to extend the range, there is Toyota's Mirai model and Hyundai's Nexo model with 3.1% and 0.3% [117,118].

3.1. EREV Configuration

The system for power transmission in an EREV includes an APU, an electric motor, and a battery. Other components complement the system, such as the DC-DC converter and an electronic controller, but these components do not define the vehicle's configuration. The electric motor is the main component that defines the configuration of an electric vehicle or extended-range electric vehicle. The configuration of the vehicle is determined by the position of the electric motor and by the technology. The electric motor can be in a longitudinal or lateral position, and it can be at the front or the rear of the vehicle. The electric motor transforms electrical energy into mechanical energy and transmits it to the wheels, but it can also have a gearbox and a mechanical differential. If the car has an electric motor for each wheel, or if the motors are inside the wheels, the differential system must be electronic.

An EV and its APU contribute three factors of major importance:

1. The traction force, which can be forward, backward, or four-wheel drive.
2. The position of the engine.
3. If it has a gearbox and the type of differential, mechanical or electronic.

The type of configuration is studied and analyzed in order to optimize the topology better. The batteries are left aside and only are under consideration for the type of traction, the direction that the EV will have, the gearbox, and the final transmission ratio. The APU can be positioned regardless of the vehicle's configuration. The electric motor is the only one that provides power to the wheels.

EREV Topological Configurations

The EREV configuration combines the ICEV and EV configurations. It attempts to integrate the best parts of each one, and it is relatively flexible. This flexibility is due to several factors unique to the EV. An EREV works like an EV. First, the energy flow is mainly via flexible electrical wires rather than rigid and mechanical links, achieving distributed subsystems. Second, different EREV propulsion arrangements produce significant differences in system configuration. Third, different energy sources (such as auxiliary power units) have different characteristics and refueling systems. In general, the EREV consists of three major subsystems. Electric propulsion, which comprises an electronic controller, a power converter, an electric motor, mechanical transmission, a final drive, and driving wheels. An energy source, which involves the energy source, energy management unit, and charger. An auxiliary power unit, which consists of the generator, and depending on the technology, an ICE, a fuel cell, regenerative braking, regenerative shock absorbers, a flywheel, a thermoacoustic engine, photovoltaic cells, a gas turbine, a rotary engine, and a wind turbine/refueling unit. The energy management unit cooperates with the electronic controller to control regenerative energy and its energy recovery. It also works with the energy charger and monitors the usability of the energy source.

At present, there are many possible EREV configurations due to the variations in electric propulsion and energy sources. Thirteen alternatives focus on electric propulsion variations; some are in typical vehicles, and others are in high-performance vehicles.

(1) All-wheel drive (AWD; Figure 12a) is the first alternative, a direct extension of the existing ICEV adopting a longitudinal front engine. It consists of an electric motor, a gearbox, differential, an APU, battery, and a BMS connected to the electric motor; this configuration has two differentials to transmit the power to both axles.

Figure 12b shows a typical configuration for pick-ups and high-performance sedan vehicles. The electric motor changes its orientation from longitudinal to cross-wise, maintaining a gearbox, two differentials, an APU, a battery, and a BMS. Figure 12c shows one electric motor in each axle; this configuration keeps a gearbox coupled with the motor. The gearbox can be single-gear or two-gear to multiply and divide torque and revolutions. Figure 12d shows an electric motor for each wheel. The mechanical differential is replaced by an electronic differential that controls the speed differentiation of one wheel concerning the other. This type of configuration uses small, high-efficiency electric motors. Figure 12e shows a configuration with in-wheel motors. Again, one variant can have no gearbox. In this configuration, the electric motor has high efficiency and high velocity. The electric motor is connected directly to the wheel, and the other components are like those in the previous configurations. Figure 12f similarly uses in-wheel motors but simultaneously incorporates an electric differential, maintaining an APU, a battery, and a BMS; each wheel efficiently transmits power and reduces weight compared to other configurations.

(2) Front-wheel drive (FWD; Figure 12g). In this configuration, the electric motor changes from longitudinal to cross while maintaining the same components as the all-wheel-drive configuration. The difference is that the mechanical power transmission is only to the front wheels, and it can have an electric motor for each wheel with or without a gearbox. Figure 12h shows a configuration in which an electronic differential replaces the mechanical one. Figure 12i shows a similar configuration using electric motor in-wheel technology while keeping the battery, the BMS, and the APU.

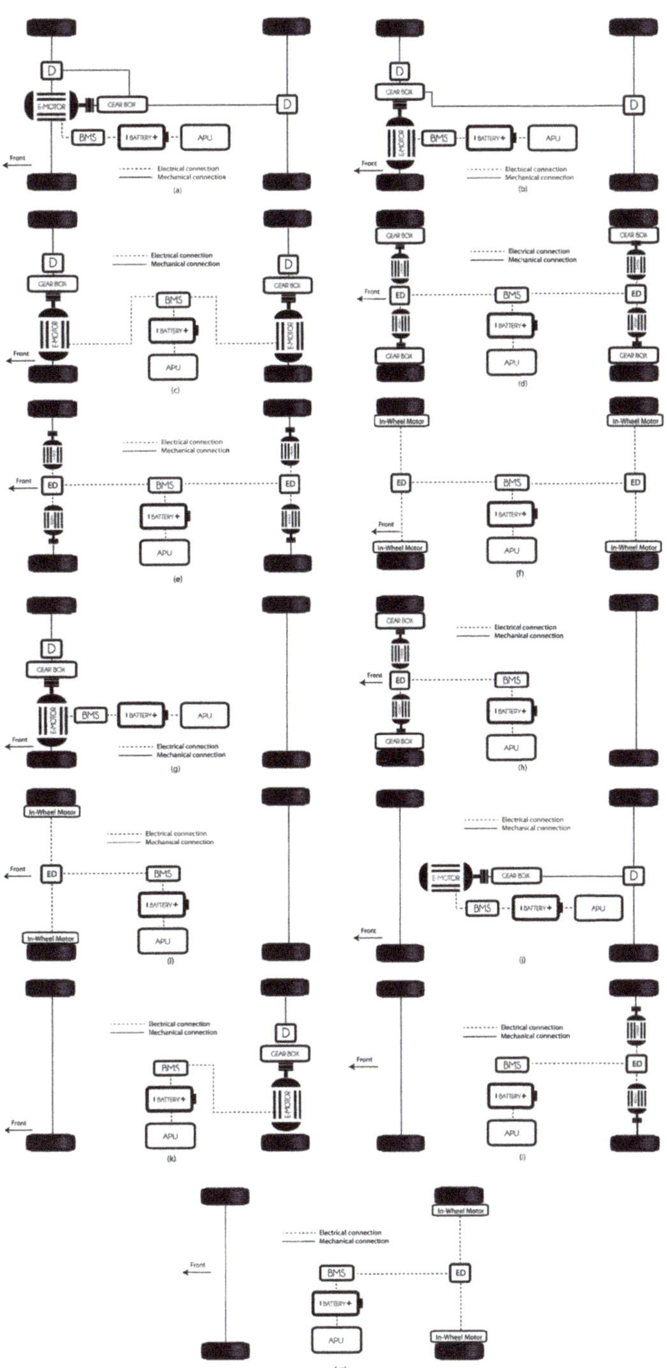

Figure 12. All possible EREV configurations.

(3) Rear-wheel drive (RWD) maintains a longitudinal electric motor in Figure 12j, but the rear wheels receive the mechanical power. In Figure 12k, the significant from

with the previous one is the position of the electric motor; it changes from longitudinal to cross; the rest of the components are the same. A configuration that improves the performance is having an electric motor in each wheel, with or without a gearbox, as shown in Figure 12l. Figure 12m shows the last configuration, which uses the electric motor in-wheel with electric differential. The selection of the configuration will depend on the size and application of the EREV. The primary criteria for selection are compactness, performance, weight, and cost.

Using driving cycles lets us know the emissions released and the energy consumption. However, due to the complexity and diversity of the vehicles, it is not easy to simulate a whole vehicle fleet using a physical approach [119], so we can use some software to estimate the emissions and energy consumption. Software such as MATLAB-Simulink, AVL Cruise, ADAMS, ANSYS, ADVISOR, ANSOFT, and MAPLESim can simulate driver/vehicle systems.

3.2. Key Components of an EREV

First, vehicle weight directly affects performance, especially the range and gradeability. Lightweight materials such as aluminum and composite materials for the body and chassis help with weight reduction. Second, achieving a low drag coefficient with the body design effectively reduces aerodynamic resistance, significantly extending the range of the EREV on highways or when cruising. The aerodynamic resistance can be reduced by tapering front and rear ends, and adopting a flat, covered, low-floor design. One can also optimize the airflow around the front and rear windows while using this flow to cool the batteries to minimize battery losses efficiently. Third, low rolling resistance tires effectively reduce running resistance at low and medium driving speeds and play an essential role in extending the range of EREVs in city driving. The design of an EREV requires considering the interactions of all the components that it may have. Figure 13 shows all the main components and the interactions that they may have with each other.

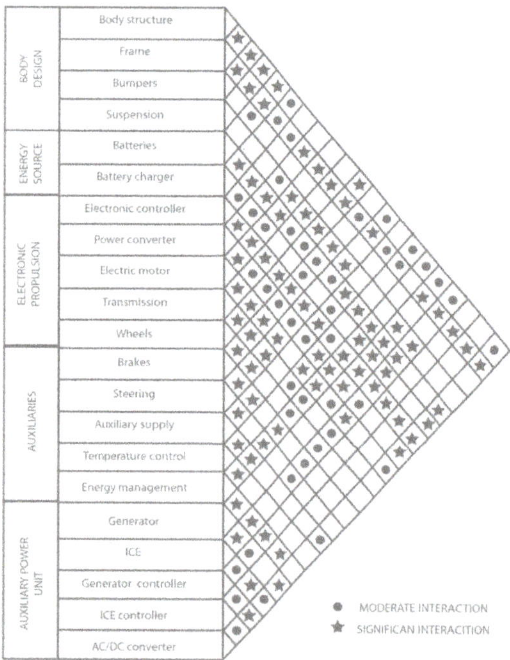

Figure 13. Key components and interactions.

4. Control and Management

The control issue of power electronic interface converters plays a vital role in the efficient and safe operation of EREVs. There are many studies on the control of power flow and energy management. Table 2 shows studies about energy management methods. The control and management strategies are focused on (1) minimization of fuel consumption and loss of energy, (2) simplifying the structure, (3) increasing the maximum efficiency, and (4) ensuring robustness and satisfactory driving performance. Energy management: Adopting an intelligent energy management system (EMS) helps maximize onboard stored energy. Using sensor inputs, such as air temperature and currents and voltages for the motors and batteries, among other data, the EMS can perform the following functions:

1. Optimize the system's energy flow;
2. Predict the remaining energy and hence the residual driving cycle;
3. Turn on the APU to charge and improve the autonomy with a suitable control method;
4. Suggest more efficient driving behavior;
5. Direct energy regenerated from braking to receptive energy sources such as the batteries;
6. Modulate temperature control as a response to external climate;
7. Propose battery charging;
8. Analyze the operation history of the energy source, especially the battery;
9. Diagnose any incorrect behavior or defective components of the energy source, or malfunctions of any component.

Table 2. A summary of control methods and strategies for EREVs.

Control Methods/Strategy	Author	Controlled System	Technology	Purpose	Application
Constant power control strategy	[3]	BMS	ER-ICE	The lowest permissible level of SOC after the drive charge the vehicle	Charging Management Arterial Roads
A power follower control strategy	[3]	BMS	ER-ICE	The lowest permissible level of SOC after the drive charge the vehicle	Charging Management Express Way
Proportional resonant control strategy	[8]	Generator	ER-ICE	To maintain the efficient region of the generator	Generate more energy
Partial power following control strategy	[8]	ICE	ER-ICE	To maintain the efficient region to operate the ICE	Reduce fuel consumption
A control strategy based on Pontryagins Minimum Principle (PMP)	[9]	ICE	ER-ICE	Monitors the current SOC of the battery	Minimizes the energy consumed during driving
Predictive control-based energy management	[90]	Fuel Cells	ER-FC	Forecasted speed	Minimize hydrogen consumption
Regenerative Braking control strategy (RRBCS)	[28]	Regenerative braking	ER-RB	The better capacity of the regenerative braking energy consider slip ratio of the tire	Coordinate regenerative braking torque and mechanical friction to maximize energy recovery and to ensure the braking efficiency

Table 2. *Cont.*

Control Methods/Strategy	Author	Controlled System	Technology	Purpose	Application
A normal control strategy based on a state of charge (SOC)	[12]	ICE	ER-ICE	Monitors the battery state of charge (SOC)	Reduce CO_2 emissions
Automatic Mechanical Transmission (AMT) Shift control strategy	[43]	Regenerative braking	ER-RB	Identify the braking intention and transmission shifts correctly	Improve the braking energy recovery rate, and ensure the braking safety a stability
Start-stop control strategy	[33]	ICE-Generator	ER-ICE	Reduce the start-stop times and running time	Fuel economy
Adaptive power management strategy PMS	[18]	ICE	ER-ICE	Asses the battery SOC and vehicle speed	Improve energy savings, the fuel, and electrical consumption
Method of quantitative estimation	[120]	Generator	ER-ICE	Optimize the design parameters aiming at the maximum efficiency in the continuos rated	Find maximum torque per ampere
Charge-deplete-charge-sustain (CDCS) strategy	[15]	ICE-Generator	ER-ICE	Asses the battery SOC and vehicle speed	Energy efficiency, reduce energy consumption and reduce costs of operation
Thermal management system to battery cooling strategy	[22]	Battery	ER-ICE	Quantify the heat generation sources and accurately predicting cell temperatures	Improve longevity, safety, and overall performance
A data driving behavior predictive control strategy	[23]	Driving behavior	ER-ICE	Predict the EV power requests and optimize their control inputs	Improving the driving range and battery life while maintaining thermal comfort for the passengers
Mixed-integer convex program	[37]	Powertrain	ER-ICE	Formulate an economic optimization	All the quantities to minimize are expressed as a monetary variable
The convex optimal control problem	[36]	Powertrain	ER-ICE	Optimization over the entire driving cycle is computed offline	Achieve the best possible energy consumption.
The optimal operation curve control strategy	[29]	ICE	ER-ICE	Research and control the vehicle required torque	Control the power allocation of APU and batteries to reduce fuel consumption and obtain good fuel economy
Multi-objective hierarchical prediction energy management strategy	[52]	Fuel cells	ER-FC	Propose a Global state of charge rapid planning method based only on the expected driving distance	Achieve optimal fuel cell life economy and energy consumption economy
A novel energy-aware velocity planning	[32]	ICE	ER-ICE	Propose energy-aware velocity planning	Improve electric vehicle fuel efficiency
Pseudospectral optimal control	[34]	APU	ER-ICE	Maintain engine speed constant is better for the dynamic characteristics of APU	Different limits of the APU power changing rate significantly influence the fuel consumption
Model predictive control	[121,122]	Powertrain	ER-ICE	Propose a computationally tractable model prediction control (MPC)	Prediction horizon so that energy consumption is minimized

5. System Optimization

As mentioned before, EREVs have complex architectures that contain multidisciplinary technologies. Since the EREV performance can be affected by many multidisciplinary, interrelated factors, computer simulations are the most critical technology with which to optimize performance and reduction costs. Additionally, EREV simulations can help manufactures to minimize prototyping costs and rapidly evaluate their concepts. Since an EREV's system consists of various subsystems clustered together by mechanical, electrical, control, and thermal links, the simulation should be a parameterized mixed-signal one. Hence, optimization is at the system level, at which there are many tradeoffs among various subsystem criteria. The preferred system criteria generally involve numerous iterative processes. In summary, the system-level simulation and optimization of EREVs should consider the following key issues.

1. As the interactions among various subsystems significantly affect the performances of EREVs, the significance of those interactions should be analyzed and taken into account.
2. The model's accuracy is usually correlated with the model's complexity, but the latter may run counter to usability, tradeoffs among the accuracy, complexity, usability, and simulation time should be considered.
3. The system voltage generally causes contradictory issues for EREV design. For example, the battery weight (higher voltage requires more battery modules in series, and hence more weight for the battery case). Similarly, motor drive voltage and current ratings, auxiliary power unit range, energy generated, acceleration performance, driving range, and safety should be optimized at the system level.
4. The adoption of multiple energy sources helps to increase the driving range. For the EREV case, the APU should be optimized based on the vehicle's performance and cost requirements.

5.1. Controller Optimization for Plant

In an EREV, a plant could be an ICE, an electric motor, or a battery. Different strategies can optimize the plant and its controller: sequential, iterative, bi-level, and simultaneous strategies [123]. Sequential optimization often leads to non-optimal system designs due to plant/controller optimization coupling. Iterative plant/controller optimization strategies attempt to improve the initial design by first improving the plant design without compromising control performance and optimizing the controller design without compromising plant performance. In a bi-level plant/controller optimization strategy, two nested optimization loops are used. The outer loop optimizes the scalar-substituted objective function by changing only the plant's design. The role of the inner loop is to generate the optimal controller for each plant selected by the outer loop. The simultaneous strategy can be mathematically and computationally challenging for several reasons. The simultaneous plant/controller optimization problem is a hybrid static/variational problem. Even when the plant and controller optimization subproblems are convex, the collaborative problem is not guaranteed to be convex. Figure 14 shows the strategies for plant/controller optimization.

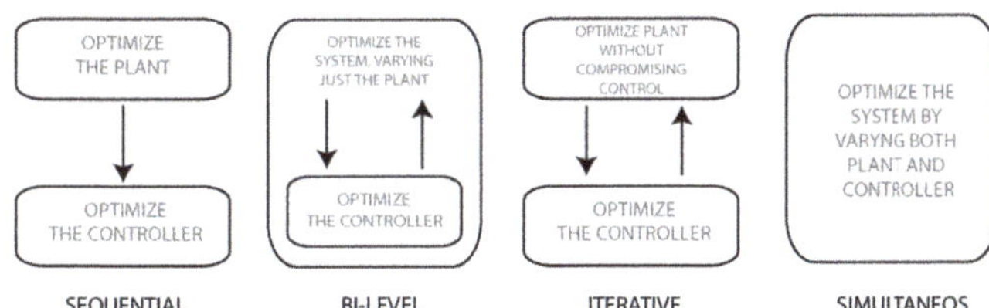

Figure 14. Strategies for plant/controller optimization.

5.2. Multidisciplinary Optimization

It is necessary to capture the effects of each energy domain on the dynamics of the other domains to optimize a system. For example, when analyzing extended-range electric vehicles, the generator and internal combustion engine coupling should be optimized to provide the energy needed to charge the batteries and increase the vehicle's autonomy.

EREV power trains reduce design space for the remainder of the physical system and increase the complexity of the control. The coupling (dependency) among the parameters of the physical system (e.g., topology) and the control parameters transforms the problem into a multi-level problem, as depicted in Figure 14. If solved sequentially, it is by definition sub-optimal [124]. Therefore, the physical system and the control should be designed in an integrated manner to obtain an optimal system. Due to the oversized dimensions of the design space, computer simulations of dynamical systems have become more important as a preliminary step to building prototypes, e.g., for different architectures and component sizes. Computer simulations significantly speed up the control synthesis of a given design and topology. However, even with computer systems, finding the optimal vehicle design that provides the best control performance is typically intractable. It is not feasible (cost or time-wise), given design space, to build all possible vehicles and evaluate which configuration and parameters provide the best control performance.

5.3. Optimal Control

There is tight integration between the physical design and an element's control from a dynamics perspective. A sequential design process is often used to design the physical system, which is followed by control-system design [125]. In EREVs, we need to know the functions of all components and their interactions to generate the best possible control systems. The simulations are important because they predict system behavior given the specifications provided. However, a simulation can also be used for the inverse task: identifying system specifications that produce the desired behavior.

5.4. Size Optimization

It is essential to optimize the sizing of propulsion system components without reducing performance to reduce the manufacturing costs and emissions associated with transport platforms. In a conventional vehicle, the size of the ICE is directly associated with the maximum power required for the vehicle. Similarly, in the case of pure electric vehicles, the range depends on the size of the battery. In general, it is the energy rather than the power that conditions the sizing of the batteries. There is not total freedom of design in either platform, because a single energy source powers the vehicle in both cases.

Extended-range electric vehicles have at least one degree of design freedom, that is, the sizes of their energy sources, because more than one energy source is integrated. The sizing of an EREV platform consists specifically of defining the size requirements in terms of power and/or energy. Then, the sizes of the energy sources and the vehicle's power requirements implicitly define the sizes of the power converters that make up the propulsion system. The choice of component size significantly affects vehicle's performance in terms of power availability, energy efficiency, manufacturing costs, and component life. Figure 15 shows the strategies for plant/controller optimization.

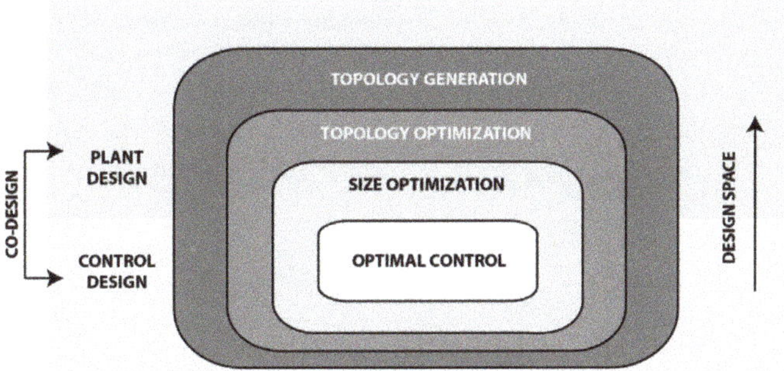

Figure 15. Optimization levels of an EREV.

6. The Process of Designing an EREV

Here we present a guide for designing an extended-range electric vehicle in general terms.

Once the kinds of technology have been analyzed and compared, after knowing the possible configurations and the interactions of the components, a conceptual design of an extended range electric vehicle can be generated. However, first, we must define the type of vehicle we are going to design. There are three types that cover the vast majority of vehicles, and they are compacts, pick ups, and trucks.

As a second step, depending on the type of vehicle we want to design, we select the type of traction that this may have, which varies depending on the type of vehicle; the three main types are FWD, RWD, and AWD.

As a third step, we define the technology of the electric motor that the EV will have and its position for its configuration. In-wheel motors are commonly used in compact vehicles.

As a fourth step, we have the selection of the range extender system. We can make our selection under certain criteria, and we are helped by Figure 16, where we compare these technologies. For practice, we define two selection criteria: the amount of power and the emissions. Large vehicles need high levels of autonomy; small vehicles can have medium or low levels of autonomy. Our selection criteria are the extra range needed, a high, medium, or low amount; and the space and weight that can be sacrificed.

As a fifth step, the controller selection (independent or general controllers) will depend on the control strategy to use; we can consult Table 1.

We can also optimize the controller selection (see Section 5) using the strategies shown, and our components' sizes.

Then, we will finally have our extended range electric vehicle, which should have ideal performance, optimal control, and optimized components. All steps are shown in Figure 16.

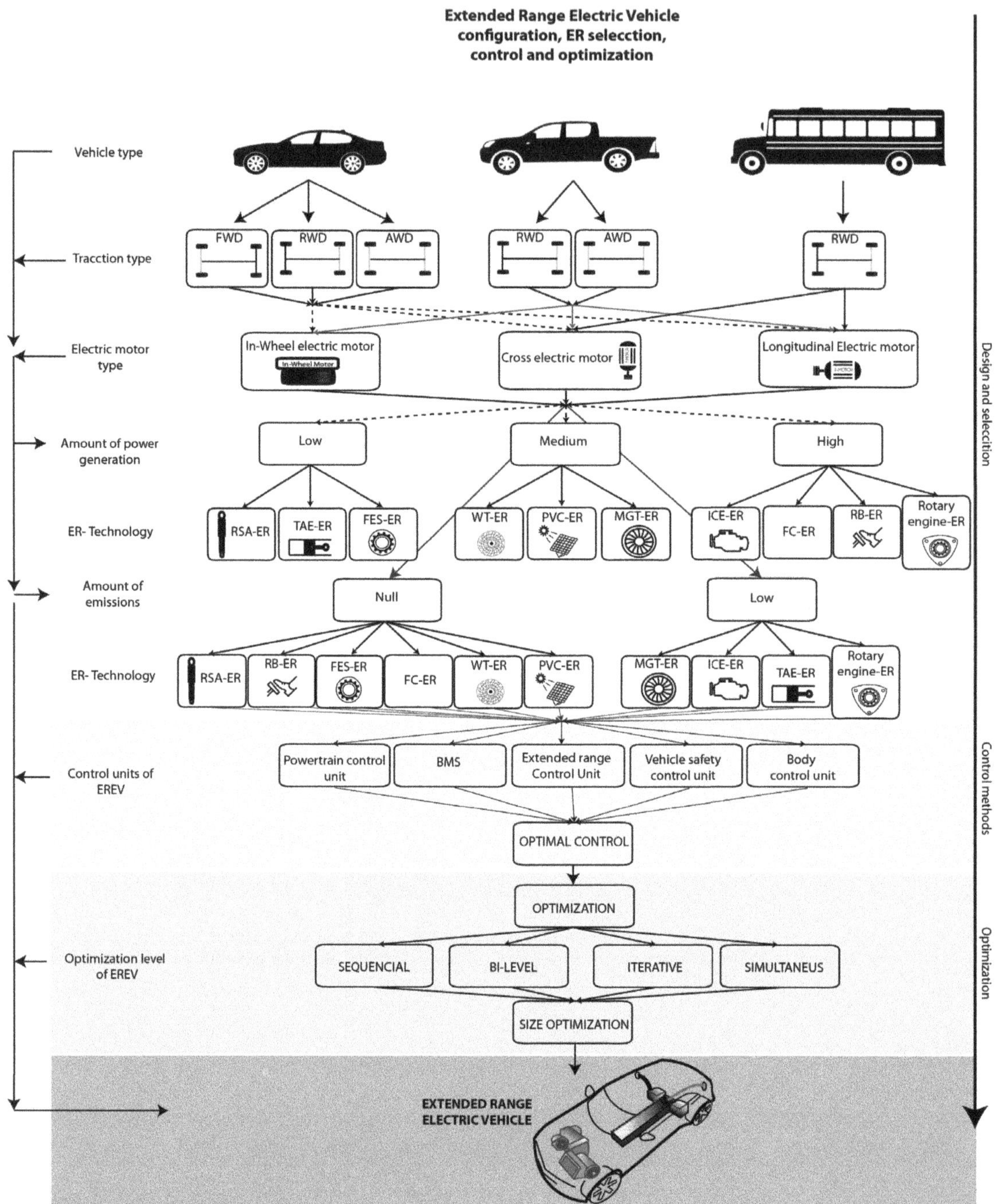

Figure 16. The process of designing an EREV.

7. Discussion

This paper reviewed the current technologies, control methods, optimization methods, and design methods for EREV vehicles, including the architecture, key components and their interactions with each other, the sizing of components, and methods to find the optimal system-level design. Although at first glance, there seem many different configurations, the most commonly used have an electric motor in the central position. However, the use of an in-wheel motor is typical if the vehicle has high performance. The central aspect to consider with the technologies used to recover energy and increase autonomy is the cost of implementing them. Some technologies are more economical and easier to control. Additionally, the fuel or resource cost for the range extender to work is a limiting factor for integration and profitability; it helps the electric vehicle concerning autonomy. All the technologies presented in this work aid electric vehicles. Depending on the budget that each researcher or research center has, it can implement and carry out tests to verify and optimize the implementation of the selected technologies. Currently, the combustion engine is the most common technology used to increase autonomy. That is why most researchers seek to reduce ICEs' fuel consumption to increase energy efficiency and recover energy. By analyzing the literature, we can conclude that the use of optimization methods will depend on the scope of the research, but usually involves finding the most efficient configuration of all components, thereby solving different optimization layers for design. These could be further used in more extended coordination methods to include the selection of topologies and technologies. For instance, these extended coordination methods might include: (i) simultaneous topology and sizing design, alternating with controller design; (ii) controller design nested for simultaneous topology and sizing, (iii) topology alternating with sizing or control; or (iv) simultaneous topology, sizing, and control design. We have presented a guide for determining critical components and the interactions between them in order to design a new topology and optimize all levels depending on the technologies used.

To substantially reduce the computational burden, the introduction of approximations of the original problem should shorten the driving cycle used for design, or one should use parallel computing. Driving cycles used as input for the control (energy management strategy) or any simulation should be short, realistic, and representative of realistic driving types. In some cases, it is necessary to create a personalized driving cycle to analyze the behavior of the extended-range electric vehicle concerning energy consumption, range, and emissions. A problem that remains open is how to address multiple topologies with a large variety in terms of component types and quantities in a more intuitive way. In addition, optimization problems and automatic construction of topologies spur on the development of control algorithms that automatically handle various topologies. Optimization objectives can be defined to include, in addition to fuel, also cost, emissions, and performance aspects to solve the design problem at the system level, to find a competitive EREV configuration for the market. User-friendly methodologies are needed to help developers so that the industry at large achieves the best designs early in HEV development.

Author Contributions: Conceptualization, D.S.P.-B. and J.d.D.C.-N.; writing—original draft preparation, D.S.P.-B.; writing—review and editing, J.I.-R. and R.A.R.-M.; supervision, J.d.D.C.-N. and R.A.R.-M. All authors have read and agreed to the published version of the manuscript.

Funding: This research was funded by Tecnologico de Monterrey, grant number a01366354, and by the National Council for Science and Technology (CONACYT), grant number 862836.

Institutional Review Board Statement: Not applicable.

Informed Consent Statement: Not applicable.

Data Availability Statement: Not applicable.

Acknowledgments: We thank the CIMA lab, Toluca Campus, for the valuable collaboration.

Conflicts of Interest: The authors declare no conflict of interest.

Abbreviations

The following abbreviations are used in this manuscript:

EV	Electric vehicle
EREV	Extended range electric vehicle
APU	Auxiliary power unit
Series HEV	Series hybrid electric vehicles
RE	Range extender
ICEV	Internal combustion engine vehicle
SOC	State of charge
PHEV	Plug-in hybrid electric vehicle
FWD	Forward wheel drive
RWD	Rear-wheel drive
AWD	All wheel drive
ICE	Internal combustion engine
ICE-ER	Internal combustion engine extended range
RSA-ER	Regenerative shock absorber extended range
RB-ER	Regenerative braking extended range
RBS	Regenerative braking system
RRBCS	Revised regenerative braking control strategy
AMT	Automatic mechanical transmission
FC	Fuel cell
FC-ER	Fuel cell extended range
MGT	Micro gas turbine
MGT-ER	Micro gas turbine extended range
TAE	Thermoacoustic engine
TAE-ER	Thermoacoustic engine extended range
TAC	Thermoacoustic converter
FES	Flywheel energy storage
FES-ER	Flywheel energy storage extended range
PVC	Photovoltaic cell
SES	Solar energy storage
RE	Rotary engine
RE-ER	Rotary engine extended range
WT	Wind turbine
WT-ER	Wind turbine extended range
GHG	Greenhouse gases
EMS	Energy management system
BMS	Battery management system
OEM	Original equipment manufacturer
CAGR	Compound annual growth rate

References

1. İnci, M.; Büyük, M.; Demir, M.H.; İlbey, G. A review and research on fuel cell electric vehicles: Topologies, power electronic converters, energy management methods, technical challenges, marketing and future aspects. *Renew. Sustain. Energy Rev.* **2021**, *137*, 110648. [CrossRef]
2. Lou, G.; Ma, H.; Fan, T.; Chan, H.K. Impact of the dual-credit policy on improvements in fuel economy and the production of internal combustion engine vehicles. *Resour. Conserv. Recycl.* **2020**, *156*, 104712. [CrossRef]
3. Song, W.; Zhang, X.; Tian, Y.; Xi, L.h. A charging management-based intelligent control strategy for extended-range electric vehicles. *J. Zhejiang Univ. Sci. A* **2016**, *17*, 903–910. [CrossRef]
4. Tate, E.D.; Harpster, M.O.; Savagian, P.J. The Electrification of the Automobile: From Conventional Hybrid, to Plug-in Hybrids, to Extended-Range Electric Vehicles. *SAE Int. J. Passeng. Cars-Electron. Electr. Syst.* **2008**, *1*, 2008-01-0458. [CrossRef]
5. Song, K.; Zhang, J.; Zhang, T. Design and Development of a pluggable PEMFC extended Range Electric Vehicle. In Proceedings of the 2011 Second International Conference on Mechanic Automation and Control Engineering, Inner Mongolia, China, 15–17 July 2011; pp. 1144–1147. [CrossRef]
6. Xian, T.F.; Soon, C.M.; Rajoo, S.; Romagnoli, A. A parametric study: The impact of components sizing on range extended electric vehicle's driving range. In Proceedings of the 2016 Asian Conference on Energy, Power and Transportation Electrification (ACEPT), Singapore, 25–27 October 2016; pp. 1–6. [CrossRef]

7. Wahono, B.; Nur, A.; Santoso, W.B.; Praptijanto, A. A comparison study of range-extended engines for electric vehicle based on vehicle simulator. *J. Mech. Eng. Sci.* **2016**, *10*, 1803–1816. [CrossRef]
8. Wang, X.; Lv, H.; Sun, Q.; Mi, Y.; Gao, P. A Proportional Resonant Control Strategy for Efficiency Improvement in Extended Range Electric Vehicles. *Energies* **2017**, *10*, 204. [CrossRef]
9. Lee, W.; Jeoung, H.; Park, D.; Kim, N. An Adaptive Concept of PMP-Based Control for Saving Operating Costs of Extended-Range Electric Vehicles. *IEEE Trans. Veh. Technol.* **2019**, *68*, 11505–11512. [CrossRef]
10. Tang, T.Q.; Chen, L.; Yang, S.C.; Shang, H.Y. An extended car-following model with consideration of the electric vehicle's driving range. *Phys. A Stat. Mech. Appl.* **2015**, *430*, 148–155. [CrossRef]
11. Xu, X.; Lv, M.; Chen, Z.; Ji, W.; Gao, R. Analysis of electric vehicle extended range misalignment based on rigid-flexible dynamics. *IOP Conf. Ser. Earth Environ. Sci.* **2017**, *61*, 012018. [CrossRef]
12. Hall, J.; Marlok, H.; Bassett, M.; Warth, M. Analysis of Real World Data from a Range Extended Electric Vehicle Demonstrator. *SAE Int. J. Altern. Powertrains* **2014**, *4*, 2887. [CrossRef]
13. Dost, P.; Spichartz, P.; Sourkounis, C. Charging behaviour of users utilising battery electric vehicles and extended range electric vehicles within the scope of a field test. In Proceedings of the 2015 International Conference on Renewable Energy Research and Applications (ICRERA), Palermo, Italy, 22–25 November 2015; Volume 54, pp. 1162–1167. [CrossRef]
14. Figueroa-García, J.; López-Santana, E.; Villa-Ramírez, J.; Ferro-Escobar, R. Applied Computer Sciences in Engineering. *Commun. Comput. Inf. Sci.* **2017**, *742*, 65–71. [CrossRef]
15. Du, J.; Chen, J.; Song, Z.; Gao, M.; Ouyang, M. Design method of a power management strategy for variable battery capacities range-extended electric vehicles to improve energy efficiency and cost-effectiveness. *Energy* **2017**, *121*, 32–42. [CrossRef]
16. Guan, J.C.; Chen, B.C.; Wu, Y.Y. Design of an Adaptive Power Management Strategy for Range Extended Electric Vehicles. *Energies* **2019**, *12*, 1610. [CrossRef]
17. Shi, L.; Guo, Y.; Xiao, D.; Han, Z.; Zhou, X. Design, optimization, and study of a rare-earth permanent-magnet generator with new consequent-pole rotor for extended-range electric vehicle. *IEEJ Trans. Electr. Electron. Eng.* **2019**, *14*, 917–923. [CrossRef]
18. Du, J.; Chen, J.; Gao, M.; Wang, J. Energy Efficiency Oriented Design Method of Power Management Strategy for Range-Extended Electric Vehicles. *Math. Probl. Eng.* **2016**, *2016*, 1–9. [CrossRef]
19. Sneha Angeline, P.; Newlin Rajkumar, M. Evolution of electric vehicle and its future scope. *Mater. Today Proc.* **2020**, *33*, 3930–3936. [CrossRef]
20. Lijewski, P.; Kozak, M.; Fuć, P.; Rymaniak, Ł.; Ziółkowski, A. Exhaust emissions generated under actual operating conditions from a hybrid vehicle and an electric one fitted with a range extender. *Transp. Res. Part D Transp. Environ.* **2020**, *78*, 102183. [CrossRef]
21. Guo, R.; Cao, C.; Mi, Y.; Huang, Y. Experimental investigation of the noise, vibration and harshness performances of a range-extended electric vehicle. *Proc. Inst. Mech. Eng. Part D J. Automob. Eng.* **2016**, *230*, 650–663. [CrossRef]
22. Ramotar, L.; Rohrauer, G.L.; Filion, R.; MacDonald, K. Experimental verification of a thermal equivalent circuit dynamic model on an extended range electric vehicle battery pack. *J. Power Sources* **2017**, *343*, 383–394. [CrossRef]
23. Vatanparvar, K.; Faezi, S.; Burago, I.; Levorato, M.; Al Faruque, M.A. Extended Range Electric Vehicle with Driving Behavior Estimation in Energy Management. *IEEE Trans. Smart Grid* **2019**, *10*, 2959–2968. [CrossRef]
24. Yan, X.; Fleming, J.; Lot, R. Extending the range of Plug-in Hybrid Electric Vehicles by CVT transmission optimal management. *Energy Procedia* **2018**, *151*, 17–22. [CrossRef]
25. Kim, T.Y.; Lee, S.H. Combustion and emission characteristics of wood pyrolysis oil-butanol blended fuels in a DI diesel engine. *Int. J. Automot. Technol.* **2015**, *16*, 903–912. [CrossRef]
26. Wu, D.; Feng, L. On-Off Control of Range Extender in Extended-Range Electric Vehicle using Bird Swarm Intelligence. *Electronics* **2019**, *8*, 1223. [CrossRef]
27. Lee, G.S.; Kim, D.H.; Han, J.H.; Hwang, M.H.; Cha, H.R. Optimal Operating Point Determination Method Design for Range-Extended Electric Vehicles Based on Real Driving Tests. *Energies* **2019**, *12*, 845. [CrossRef]
28. Liu, H.; Lei, Y.; Fu, Y.; Li, X. Parameter matching and optimization for power system of range-extended electric vehicle based on requirements. *Proc. Inst. Mech. Eng. Part D J. Automob. Eng.* **2020**, *234*, 3316–3328. [CrossRef]
29. Jiang, Y. Parameters Optimization for Extended-range Electric Vehicle Based on Improved Chaotic Particle Swarm Optimization. *Int. J. Grid Distrib. Comput.* **2016**, *9*, 1–10. [CrossRef]
30. Kopczyński, A.; Krawczyk, P.; Lasocki, J. Parameters selection of extended-range electric vehicle supplied with alternative fuel. *E3S Web Conf.* **2018**, *44*, 00073. [CrossRef]
31. Ren, G.; Shan, S.; Ma, G.; Shang, X.; Zhu, S.; Zhang, Q.; Yang, T. Review of energy storage technologies for extended range electric vehicle. *J. Appl. Sci. Eng.* **2019**, *22*, 69–82. [CrossRef]
32. Baek, D.; Chang, N. Runtime Power Management of Battery Electric Vehicles for Extended Range With Consideration of Driving Time. *IEEE Trans. Very Large Scale Integr. (VLSI) Syst.* **2019**, *27*, 549–559. [CrossRef]
33. Zhao, J.b.; Han, B.y.; Bei, S.y. Start-Stop Moment Optimization of Range Extender and Control Strategy Design for Extended -Range Electric Vehicle. *IOP Conf. Ser. Mater. Sci. Eng.* **2017**, *241*, 012025. [CrossRef]
34. Li, J.; Wang, Y.; Chen, J.; Zhang, X. Study on energy management strategy and dynamic modeling for auxiliary power units in range-extended electric vehicles. *Appl. Energy* **2017**, *194*, 363–375. [CrossRef]
35. Chambon, P.; Curran, S.; Huff, S.; Love, L.; Post, B.; Wagner, R.; Jackson, R.; Green, J. Development of a range-extended electric vehicle powertrain for an integrated energy systems research printed utility vehicle. *Appl. Energy* **2017**, *191*, 99–110. [CrossRef]

36. Pozzato, G.; Formentin, S.; Panzani, G.; Savaresi, S. Least Costly Energy Management for Extended-Range Electric Vehicles with Noise Emissions Characterization. *IFAC-PapersOnLine* **2019**, *52*, 586–591. [CrossRef]
37. Pozzato, G.; Formentin, S.; Panzani, G.; Savaresi, S.M. Least costly energy management for extended-range electric vehicles: An economic optimization framework. *Eur. J. Control* **2020**, *56*, 218–230. [CrossRef]
38. Zhang, Z.; Zhang, X.; Chen, W.; Rasim, Y.; Salman, W.; Pan, H.; Yuan, Y.; Wang, C. A high-efficiency energy regenerative shock absorber using supercapacitors for renewable energy applications in range extended electric vehicle. *Appl. Energy* **2016**, *178*, 177–188; Corrigendum in **2019**, *254*, 113634. [CrossRef]
39. Satpute, N.V.; Singh, S.; Sawant, S.M. Energy Harvesting Shock Absorber with Electromagnetic and Fluid Damping. *Adv. Mech. Eng.* **2015**, *6*, 693592–693592. [CrossRef]
40. Zhang, Y.; Zhang, X.; Zhan, M.; Guo, K.; Zhao, F.; Liu, Z. Study on a novel hydraulic pumping regenerative suspension for vehicles. *J. Frankl. Inst.* **2015**, *352*, 485–499. [CrossRef]
41. Qiu, C.; Wang, G.; Meng, M.; Shen, Y. A novel control strategy of regenerative braking system for electric vehicles under safety critical driving situations. *Energy* **2018**, *149*, 329–340. [CrossRef]
42. Liu, H.; Lei, Y.; Fu, Y.; Li, X. An Optimal Slip Ratio-Based Revised Regenerative Braking Control Strategy of Range-Extended Electric Vehicle. *Energies* **2020**, *13*, 1526. [CrossRef]
43. Zhao, X.; Xu, S.; Ye, Y.; Yu, M.; Wang, G. Composite braking AMT shift strategy for extended-range heavy commercial electric vehicle based on LHMM/ANFIS braking intention identification. *Clust. Comput.* **2019**, *22*, 8513–8528. [CrossRef]
44. Aygen, M.S.; İnci, M. Zero-sequence current injection based power flow control strategy for grid inverter interfaced renewable energy systems. *Energy Sources Part Recover. Util. Environ. Eff.* **2020**, 1–22. [CrossRef]
45. Ozden, A.; Shahgaldi, S.; Li, X.; Hamdullahpur, F. A review of gas diffusion layers for proton exchange membrane fuel cells—With a focus on characteristics, characterization techniques, materials and designs. *Prog. Energy Combust. Sci.* **2019**, *74*, 50–102. [CrossRef]
46. İnci, M. Active/reactive energy control scheme for grid-connected fuel cell system with local inductive loads. *Energy* **2020**, *197*, 117191. [CrossRef]
47. Zhou, Y.; Li, H.; Ravey, A.; Péra, M.C. An integrated predictive energy management for light-duty range-extended plug-in fuel cell electric vehicle. *J. Power Sources* **2020**, *451*, 227780. [CrossRef]
48. Chu, Y.; Wu, Y.; Chen, J.; Zheng, S.; Wang, Z. Design of energy and materials for ammonia-based extended-range electric vehicles. *Energy Procedia* **2019**, *158*, 3064–3069. [CrossRef]
49. Dimitrova, Z.; Maréchal, F. Environomic design for electric vehicles with an integrated solid oxide fuel cell (SOFC) unit as a range extender. *Renew. Energy* **2017**, *112*, 124–142. [CrossRef]
50. Wu, X.L.; Xu, Y.W.; Zhao, D.q.; Zhong, X.B.; Li, D.; Jiang, J.; Deng, Z.; Fu, X.; Li, X. Extended-range electric vehicle-oriented thermoelectric surge control of a solid oxide fuel cell system. *Appl. Energy* **2020**, *263*, 114628. [CrossRef]
51. Karaoğlan, M.U.; Kuralay, N.S.; Colpan, C.O. Investigation of the effects of battery types and power management algorithms on drive cycle simulation for a range-extended electric vehicle powertrain. *Int. J. Green Energy* **2019**, *16*, 1–11. [CrossRef]
52. Liu, Y.; Li, J.; Chen, Z.; Qin, D.; Zhang, Y. Research on a multi-objective hierarchical prediction energy management strategy for range extended fuel cell vehicles. *J. Power Sources* **2019**, *429*, 55–66. [CrossRef]
53. Geng, C.; Jin, X.; Zhang, X. Simulation research on a novel control strategy for fuel cell extended-range vehicles. *Int. J. Hydrogen Energy* **2019**, *44*, 408–420. [CrossRef]
54. Almasi, A. Bright future of micro-turbines: Nuclear emergency cooling and miniature generator. *Aust. J. Mech. Eng.* **2013**, *11*, 169–175. [CrossRef]
55. Karvountzis-Kontakiotis, A.; Andwari, A.M.; Pesyridis, A.; Russo, S.; Tuccillo, R.; Esfahanian, V. Application of Micro Gas Turbine in Range-Extended Electric Vehicles. *Energy* **2018**, *147*, 351–361. [CrossRef]
56. Bou Nader, W.S.; Mansour, C.J.; Nemer, M.G. Optimization of a Brayton external combustion gas-turbine system for extended range electric vehicles. *Energy* **2018**, *150*, 745–758. [CrossRef]
57. Reine, A.; Bou Nader, W. Fuel consumption potential of different external combustion gas-turbine thermodynamic configurations for extended range electric vehicles. *Energy* **2019**, *175*, 900–913. [CrossRef]
58. Sahoo, D.; Kotrba, A.; Steiner, T.; Swift, G. Waste Heat Recovery for Light-Duty Truck Application Using ThermoAcoustic Converter Technology. *SAE Int. J. Engines* **2017**, *10*, 0153. [CrossRef]
59. Nader, W.B.; Chamoun, J.; Dumand, C. Optimization of the thermodynamic configurations of a thermoacoustic engine auxiliary power unit for range extended hybrid electric vehicles. *Energy* **2020**, *195*. [CrossRef]
60. Bou Nader, W.; Chamoun, J.; Dumand, C. Thermoacoustic engine as waste heat recovery system on extended range hybrid electric vehicles. *Energy Convers. Manag.* **2020**, *215*, 112912. [CrossRef]
61. Arani, A.K.; Karami, H.; Gharehpetian, G.; Hejazi, M. Review of Flywheel Energy Storage Systems structures and applications in power systems and microgrids. *Renew. Sustain. Energy Rev.* **2017**, *69*, 9–18. [CrossRef]
62. Itani, K.; De Bernardinis, A.; Khatir, Z.; Jammal, A. Comparative analysis of two hybrid energy storage systems used in a two front wheel driven electric vehicle during extreme start-up and regenerative braking operations. *Energy Convers. Manag.* **2017**, *144*, 69–87. [CrossRef]
63. ElNozahy, M.; Salama, M. Studying the feasibility of charging plug-in hybrid electric vehicles using photovoltaic electricity in residential distribution systems. *Electr. Power Syst. Res.* **2014**, *110*, 133–143. [CrossRef]

64. Keshri, R.; Bertoluzzo, M.; Buja, G. Integration of a Photovoltaic Panel with an Electric City Car. *Electr. Power Compon. Syst.* **2014**, *42*, 481–495. [CrossRef]
65. Ezzat, M.; Dincer, I. Development, analysis and assessment of a fuel cell and solar photovoltaic system powered vehicle. *Energy Convers. Manag.* **2016**, *129*, 284–292. [CrossRef]
66. Zahir Hussain, M.; Anbalagan, R.; Jayabalakrishnan, D.; Naga Muruga, D.; Prabhahar, M.; Bhaskar, K.; Sendilvelan, S. Charging of car battery in electric vehicle by using wind energy. *Mater. Today Proc.* **2020**. [CrossRef]
67. King, P. The Development of the Szorenyi Four-Chamber Rotary Engine. In Proceedings of the SAE International Powertrains, Fuels and Lubricants Meeting, Heidelberg, Germany, 17–19 September 2018. [CrossRef]
68. Zambalov, S.D.; Yakovlev, I.A.; Maznoy, A.S. Effect of multiple fuel injection strategies on mixture formation and combustion in a hydrogen-fueled rotary range extender for battery electric vehicles. *Energy Convers. Manag.* **2020**, *220*, 113097. [CrossRef]
69. Ferdous, S.; Bin Khaled, W.; Ahmed, B.; Salehin, S.; Ghani Ovy, E. Electric Vehicle with Charging Facility in Motion using Wind Energy. In Proceedings of the World Renewable Energy Congress, Linköping, Sweden, 8–13 May 2011; Volume 57, pp. 3629–3636. [CrossRef]
70. Goushcha, O.; Felicissimo, R.; Danesh-Yazdi, A.H.; Andreopoulos, Y. Exploring harnessing wind power in moving reference frames with application to vehicles. *Adv. Mech. Eng.* **2019**, *11*, 168781401986568. [CrossRef]
71. Hossain, M.F.; Fara, N. Integration of wind into running vehicles to meet its total energy demand. *Energy Ecol. Environ.* **2017**, *2*, 35–48. [CrossRef]
72. General Motors. 2019 Chevrolet Volt Brochure. Available online: https://www.chevrolet.ca/content/dam/chevrolet/na/canada/english/index/download-a-brochure/02-pdfs/my19-cdn-volt-brochure-en.pdf (accessed on 1 December 2020).
73. BMW. BMW i3s REX Sport 2020. Available online: https://www.guideautoweb.com/en/makes/bmw/i3/2020/specifications/s-rex/ (accessed on 1 December 2020).
74. BMW i3 (94 Ah) REX Dynamic Eléctrico 2017. Available online: https://www.bmw.com.mx/content/dam/bmw/marketMX/bmw_com_mx/Descargas/Fichas-Tecnicas-2018/i3/FichaTécnicaBMWi3(94Ah)REXDynamic2017.pdf (accessed on 1 December 2020).
75. Leach, F.; Kalghatgi, G.; Stone, R.; Miles, P. The scope for improving the efficiency and environmental impact of internal combustion engines. *Transp. Eng.* **2020**, *1*, 100005. [CrossRef]
76. Menzel, G.; Hennings Och, S.; Cocco Mariani, V.; Mauro Moura, L.; Domingues, E. Multi-objective optimization of the volumetric and thermal efficiencies applied to a multi-cylinder internal combustion engine. *Energy Convers. Manag.* **2020**, *216*, 112930. [CrossRef]
77. Ouyang, T.; Zhao, Z.; Wang, Z.; Zhang, M.; Liu, B. A high-efficiency scheme for waste heat harvesting of solid oxide fuel cell integrated homogeneous charge compression ignition engine. *Energy* **2021**, *229*, 120720. [CrossRef]
78. Hahn, S.; Braun, J.; Kemmer, H.; Reuss, H.C. Optimization of the efficiency and degradation rate of an automotive fuel cell system. *Int. J. Hydrogen Energy* **2021**. [CrossRef]
79. Hoeflinger, J.; Hofmann, P. Air mass flow and pressure optimisation of a PEM fuel cell range extender system. *Int. J. Hydrogen Energy* **2020**, *45*, 29246–29258. [CrossRef]
80. Toyota. Second Generation Toyota Mirai. Still Zero Emissions. Available online: https://www.toyota-europe.com/world-of-toyota/articles-news-events/2019/new-mirai-concept (accessed on 1 December 2020).
81. Hyundai NEXO de pila de combustible | Hyundai | Hyundai Motor España. Available online: https://www.hyundai.com/es/modelos/nexo.html (accessed on 1 December 2020).
82. Nikola Two. Available online: https://nikolamotor.com/two (accessed on 1 December 2020)
83. Taner, T. The novel and innovative design with using H_2 fuel of PEM fuel cell: Efficiency of thermodynamic analyze. *Fuel* **2021**, *302*, 121109. [CrossRef]
84. Ji, C.; Wang, H.; Shi, C.; Wang, S.; Yang, J. Multi-objective optimization of operating parameters for a gasoline Wankel rotary engine by hydrogen enrichment. *Energy Convers. Manag.* **2021**, *229*, 113732. [CrossRef]
85. Sadiq, G.A.; Al-Dadah, R.; Mahmoud, S. Development of rotary Wankel devices for hybrid automotive applications. *Energy Convers. Manag.* **2019**, *202*, 112159. [CrossRef]
86. Mazda 2 RE Range-Extender: Return of the Rotary-News-Car and Driver. Available online: https://www.caranddriver.com/news/a15366035/mazda-2-re-range-extender-return-of-the-rotary/ (accessed on 1 November 2020).
87. Martins, J.J.; Uzuneanu, K.; Ribeiro, B.S.; Jasasky, O. Thermodynamic Analysis of an Over-Expanded Engine. *SAE Tech. Pap.* **2004**, *113*, 476–490. [CrossRef]
88. Zhao, J.; Ma, Y.; Zhang, Z.; Wang, S.; Wang, S. Optimization and matching for range-extenders of electric vehicles with artificial neural network and genetic algorithm. *Energy Convers. Manag.* **2019**, *184*, 709–725. [CrossRef]
89. Zhang, J.; Li, Y.; Lv, C.; Yuan, Y. New regenerative braking control strategy for rear-driven electrified minivans. *Energy Convers. Manag.* **2014**, *82*, 135–145. [CrossRef]
90. Yang, Y.; He, Q.; Chen, Y.; Fu, C. Efficiency Optimization and Control Strategy of Regenerative Braking System with Dual Motor. *Energies* **2020**, *13*, 400–411. [CrossRef]
91. Duan, J.; Fan, S.; An, Q.; Sun, L.; Wang, G. A comparison of micro gas turbine operation modes for optimal efficiency based on a nonlinear model. *Energy* **2017**, *134*, 400–411. [CrossRef]

92. Nicolosi, F.F.; Renzi, M. Effect of the Regenerator Efficiency on the Performance of a Micro Gas Turbine Fed with Alternative Fuels. *Energy Procedia* **2018**, *148*, 687–694. [CrossRef]
93. Ezzat, M.F.; Dincer, I. Development and exergetic assessment of a new hybrid vehicle incorporating gas turbine as powering option. *Energy* **2019**, *170*, 112–119. [CrossRef]
94. Ji, F.; Zhang, X.; Du, F.; Ding, S.; Zhao, Y.; Xu, Z.; Wang, Y.; Zhou, Y. Experimental and numerical investigation on micro gas turbine as a range extender for electric vehicle. *Appl. Therm. Eng.* **2020**, *173*, 115236. [CrossRef]
95. Barakat, A.A.; Diab, J.H.; Badawi, N.S.; Bou Nader, W.S.; Mansour, C.J. Combined cycle gas turbine system optimization for extended range electric vehicles. *Energy Convers. Manag.* **2020**, *226*, 113538. [CrossRef]
96. Fathabadi, H. Utilizing solar and wind energy in plug-in hybrid electric vehicles. *Energy Convers. Manag.* **2018**, *156*, 317–328. [CrossRef]
97. Fathabadi, H. Recovering waste vibration energy of an automobile using shock absorbers included magnet moving-coil mechanism and adding to overall efficiency using wind turbine. *Energy* **2019**, *189*, 116274. [CrossRef]
98. de Dios Javier Roldán, I.C.M. Meet Eolo, A Wind-Powered EV From Colombia. Available online: https://www.iamrenew.com/green-transportation/meet-eolo-wind-powered-ev-colombia/#:~:text=The%20Eolo%2C%20Colombia's%20only%20homegrown,it%20is%20powered%20by%20wind (accessed on 1 January 2021).
99. Boyes, J.D.; Clark, N.H. Technologies for energy storage flywheels and super conducting magnetic energy storage. In Proceedings of the 2000 Power Engineering Society Summer Meeting (Cat. No. 00CH37134), Seattle, WA, USA, 16–20 July 2000; Volume 3, pp. 1548–1550. [CrossRef]
100. Li, J.; Zhao, J. Energy recovery for hybrid hydraulic excavators: Flywheel-based solutions. *Autom. Constr.* **2021**, *125*, 103648. [CrossRef]
101. Erdemir, D.; Dincer, I. Assessment of Renewable Energy-Driven and Flywheel Integrated Fast-Charging Station for Electric Buses: A Case Study. *J. Energy Storage* **2020**, *30*, 101576. [CrossRef]
102. Wang, W.; Li, Y.; Shi, M.; Song, Y. Optimization and control of battery-flywheel compound energy storage system during an electric vehicle braking. *Energy* **2021**, *226*, 120404. [CrossRef]
103. Vazquez, S.; Lukic, S.M.; Galvan, E.; Franquelo, L.G.; Carrasco, J.M. Energy storage systems for transport and grid applications. *IEEE Trans. Ind. Electron.* **2010**, *57*, 3881–3895. [CrossRef]
104. Puddu, P.; Paderi, M. Hydro-pneumatic accumulators for vehicles kinetic energy storage: Influence of gas compressibility and thermal losses on storage capability. *Energy* **2013**, *57*, 326–335. [CrossRef]
105. Chen, B.; Yousif, A.A.; Riley, P.H.; Hann, D.B. Development and Assessment of Thermoacoustic Generators Operating by Waste Heat from Cooking Stove. *Engineering* **2012**, *04*, 894–902. [CrossRef]
106. Backhaus, S.; Swift, G.W. A thermoacoustic-Stirling heat engine: Detailed study. *J. Acoust. Soc. Am.* **2000**, *107*, 3148–3166. [CrossRef]
107. Mumith, J.A.; Makatsoris, C.; Karayiannis, T.G. Design of a thermoacoustic heat engine for low temperature waste heat recovery in food manufacturing: A thermoacoustic device for heat recovery. *Appl. Therm. Eng.* **2014**, *65*, 588–596. [CrossRef]
108. Blanc-benon, P.; Jondeau, E.; Comte-Bellot, G. Analysis of the acoustic field in a thermoacoustic system using time-resolved particle image velocimetry and constant-voltage anemometry. In Proceedings of the 3rd International Workshop on Thermoacoustics, Enschede, The Netherlands, 26–27 October 2015.
109. Khoshnoud, F.; Zhang, Y.; Shimura, R.; Shahba, A.; Jin, G.; Pissanidis, G.; Chen, Y.K.; De Silva, C.W. Energy Regeneration from Suspension Dynamic Modes and Self-Powered Actuation. *IEEE/ASME Trans. Mechatron.* **2015**, *20*, 2513–2524. [CrossRef]
110. Li, Z.; Zuo, L.; Luhrs, G.; Lin, L.; Qin, Y.X. Electromagnetic energy-harvesting shock absorbers: Design, modeling, and road tests. *IEEE Trans. Veh. Technol.* **2013**, *62*, 1065–1074. [CrossRef]
111. Maravandi, A.; Moallem, M. Regenerative shock absorber using a two-leg motion conversion mechanism. *IEEE/ASME Trans. Mechatron.* **2015**, *20*, 2853–2861. [CrossRef]
112. Electric Vehicle Range Extender Market Forecast Report, 2018–2025. Available online: https://www.marketsandmarkets.com/Market-Reports/electric-vehicle-range-extender-market-208131566.html (accessed on 1 July 2021).
113. Electric Vehicle (EV) Range Extender Market Statistics, Trends, Demand, Size by 2026. Available online: https://www.alliedmarketresearch.com/electric-vehicle-range-extender-market-A06026 (accessed on 1 July 2021).
114. Electric Vehicle Range Extender Market to Witness 10.2% CAGR By 2025 Owing to Rising Usage of Commercial & Fully Electric Passenger Vehicles to Lower CO_2 Emissions | Million Insights. Available online: https://www.prnewswire.com/news-releases/electric-vehicle-range-extender-market-to-witness-10-2-cagr-by-2025-owing-to-rising-usage-of-commercial--fully-electric-passenger-vehicles-to-lower-co2-emissions--million-insights-301220939.html (accessed on 1 July 2021).
115. 2021 Honda Clarity Plug-In Hybrid—The Versatile Hybrid | Honda. Available online: https://automobiles.honda.com/clarity-plug-in-hybrid (accessed on 1 July 2021).
116. Mack Trucks to Evaluate Wrightspeed Route Powertrain in Mack LR Model | Mack Trucks. Available online: https://www.macktrucks.com/mack-news/2016/mack-trucks-to-evaluate-wrightspeed-route-powertrain-in-mack-lr-model/ (accessed on 1 July 2021).
117. Rover, L. Quarterly Light-Vehicle—Q1 2021. 2021. Available online: https://www.coxautoinc.com/wp-content/uploads/2021/04/Q1.Kelley-Blue-Book-Electrfication-Deck-2021.revised-4.26.pdf (accessed on 1 July 2021).
118. Sales, Q.; Ev, B. Electrified-Vehicle 2021. pp. 1–5. Available online: https://www.coxautoinc.com/wp-content/uploads/2021/01/Kelley-Blue-Book-Salesand-Data-Report-for-2020-Q4.pdf (accessed on 1 July 2021).

119. Masclans Abelló, P.; Medina Iglesias, V.; de los Santos López, M.A.; Álvarez-Flórez, J. Real drive cycles analysis by ordered power methodology applied to fuel consumption, CO_2, NO_x and PM emissions estimation. *Front. Environ. Sci. Eng.* **2021**, *15*, 4. [CrossRef]
120. Lee, D.; Park, G.; Son, B.; Jung, H. Efficiency improvement of IPMSG in the electric power generating system of a range-extended electric vehicle. *IET Electr. Power Appl.* **2019**, *13*, 943–950. [CrossRef]
121. Han, K.; Nguyen, T.W.; Nam, K. Battery energy management of autonomous electric vehicles using computationally inexpensive model predictive control. *Electronics* **2020**, *9*, 1277. [CrossRef]
122. Hu, X.; Zhang, X.; Tang, X.; Lin, X. Model predictive control of hybrid electric vehicles for fuel economy, emission reductions, and inter-vehicle safety in car-following scenarios. *Energy* **2020**, *196*, 117101. [CrossRef]
123. Fathy, H.; Reyer, J.; Papalambros, P.; Ulsov, A. On the coupling between the plant and controller optimization problems. In Proceedings of the 2001 American Control Conference. (Cat. No.01CH37148), Arlington, VA, USA, 25–27 June 2001; Volume 3, pp. 1864–1869. [CrossRef]
124. Silvas, E.; Hofman, T.; Murgovski, N.; Etman, P.; Steinbuch, M. Review of Optimization Strategies for System-Level Design in Hybrid Electric Vehicles. *IEEE Trans. Veh. Technol.* **2016**, *66*. [CrossRef]
125. Allison, J.T.; Herber, D.R. Special Section on Multidisciplinary Design Optimization: Multidisciplinary Design Optimization of Dynamic Engineering Systems. *AIAA J.* **2014**, *52*, 691–710. [CrossRef]

Article

A Novel Application Based on a Heuristic Approach for Planning Itineraries of One-Day Tourist

Agostino Marcello Mangini [†], Michele Roccotelli *,[†] and Alessandro Rinaldi [†]

Department of Electric and Information Engineering, Faculty of Engineering, Politecnico di Bari, Via Orabona 4, 70126 Bari, Italy; agostinomarcello.mangini@poliba.it (A.M.M.); alessandro.rinaldi@poliba.it (A.R.)
* Correspondence: michele.roccotelli@poliba.it
† These authors contributed equally to this work.

Abstract: Technological innovations have revolutionized the lifestyle of the society and led to the development of advanced and intelligent cities. Smart city has recently become synonymous of a city characterized by an intelligent and extensive use of Information and Communications Technologies (ICTs) in order to allow efficient use of information. In this context, this paper proposes a new approach to optimize the planning of itineraries for one-day tourist. More in detail, an optimization approach based on Graph theory and multi-algorithms is provided to determine the optimal tourist itinerary. The aim is to minimize the travel times taking into account the tourist preferences. An Integer Linear Programming (ILP) problem is introduced to find the optimal outward and return paths of the touristic itinerary and a multi-algorithms strategy is used to maximize the number of attractions (PoIs) to be visited in the paths. Finally, a case study focusing on cruise tourist in the city of Bari, demonstrates the efficiency of the approach and the user interaction in the determination of the itinerary.

Keywords: itinerary planning; smart tourism; graph theory; heuristic approach

Citation: Mangini, A.M.; Roccotelli, M.; Rinaldi, A. A Novel Application Based on a Heuristic Approach for Planning Itineraries of One-Day Tourist. *Appl. Sci.* **2021**, *11*, 8989. https://doi.org/10.3390/app11198989

Academic Editor: Paola Pellegrini

Received: 9 August 2021
Accepted: 17 September 2021
Published: 27 September 2021

Publisher's Note: MDPI stays neutral with regard to jurisdictional claims in published maps and institutional affiliations.

1. Introduction

The planning of touristic itineraries is a typical decision making process for tourists visiting a city in a limited time period. The selection of the most valuable Points of Interests (PoIs) is not simple.

In the last years mobile applications are offering a variety of services from vacation planning to mobile tourist guides and tourism recommender systems [1,2]. The design of flexible, efficient, and user-friendly applications for mobile devices has a great interest from both a commercial and a research point of view. The authors in [3] propose a mobile application based on a hybrid multiobjective genetic algorithm to smartly generate feasible itineraries. The algorithm incorporates an advanced heuristic to build a route, with a start and an arrival time passing from a set of locations each characterized by a score measuring its attractiveness, an opening and a closing time, and visit duration. Vansteenwegen et al. [4,5] present an advanced mobile tourist guide, capable of suggesting a near-optimal and feasible selection of attractions and a route passing among them. The related optimization problem is solved by using a combination of guided local search metaheuristics. Booth et al. [6] develop a data model for trip planning in multimodal transportation systems and Navabpour [7] plans a trip with multimodal transportation based on Service Oriented Architecture (SOA). In addition, Andre et al. [8] design a journey planning system based on safety, weather and specific travel time for individual user. Gonzalez et al. [9] propose a fastest-path computation system on a road network using a traffic mining approach. However, while the above papers mainly focus on the mobile application architecture, the following two subsections analyze existing contributions focusing on methodology and parameters.

1.1. Related Works: Methodology-Based Classification

This section groups and analyses research works that provide rigorous description of heuristic and metaheuristic approaches to solve the Tourist Trip Design Problem (TTDP). These approaches result the only viable methods to efficiently optimize the travel itinerary, by analyzing the problem from different perspectives, with different problem variables and constraints. The objective in TTDP modeling is to identify a set of near-optimal itineraries to maximize tourist satisfaction. The baseline combinatorial optimization problem for TTDP is the orienteering problem (OP). The OP can be used to model the TTDP where the PoIs are associated with a profit and the goal is to find a single tour that maximizes the profit collected within a given time budget. In the OP, given a starting node s, a terminal node t and a positive time limit (budget), the goal is to find a path from s to t such that the total profit of the visited nodes is maximized. In the related literature, Garcia et al. [10] propose an intelligent routing system that defines an optimization problem including multiple paths to move from one location to another. Such a system, by exploiting an iterated local search metaheuristic method, suggests a personalized tour combining information about the local attractions, weather forecasting and public transportation. Gavalas analyzes the models, algorithmic approaches and methodologies about tourist trip design problems [11]. Recent approaches are reported aiming at taking into account a multitude of realistic PoIs attributes and user constraints. In this context, Gunawan et al. focus on the most recent works about the Orienteering Problem (OP) and its variants [12]. The authors focus on a comprehensive and thorough survey of recent variants of the OP, including the proposed solution approaches. The work reports the new variants of the OP, such as the Stochastic OP, the Generalized OP, the Arc OP, the Multi-agent OP, the Clustered OP and others. The authors summarize several interesting applications which are related to the mobile crowdsourcing problem, the Tourist Trip Design Problem, the theme park navigation problem and others. The authors in [13] provide a detailed explanation about operation on tour routes only qualitatively. An optimizer is proposed in [14], where a multiobjective evolutionary algorithm is used to identify the near-optimal solutions to the planning of multiple-day routes in a reasonable computational time. In the contribution [15], the authors present a heuristic procedure for the generalization of a optimization problem to plan personalized recommendations for daily sightseeing itineraries for mobile tourist guides.

Extensions of the OP have been applied to model more complex versions of TTDP: the OP with time windows (OPTW) considers visits to locations within a predefined time window; this allows modeling opening days/hours of PoIs. The time-dependent OP (TDOP) considers time dependency in the estimation of time required to move from one location to another; therefore, it is suitable for modeling multimodal transports among PoIs. In particular, Cotfas considers a more complex variant of the tourist trip design problem i.e., the time-dependent in the estimation of the time required to move from one location to another for planning daily tours according to tourist's preferences [16].

The team orienteering problem (TOP) is the extension of the OP to multiple tours. The TOP with time windows (TOPTW) has been mostly commonly studied among the aforementioned OP variants since it is useful for modeling several real-life optimization problems. Vansteenwegen et al. propose a metaheuristic algorithm to tackle a more effective extension of the optimization problem [17]. The proposed algorithm performs a planning of a multipleday tour by considering a set of PoIs, a visiting duration, and a set of multiple opening and closing times per day combined with the trip constraints of the tourist.

Other studies propose approaches based on Graph Theory [18–21], applied in tourism. The authors in [19] deal with typical tourist attractions in urban destinations, as pedestrian zones, market areas or urban areas of architectural, cultural and scenic value rather than only visiting sites of restricted access or taking the fastest route to move among city landmarks. Herein, the authors introduce Scenic Athens, a context- aware mobile city guide for Athens (Greece) which provides personalized tour planning services to tourists.

Scenic Athens derives near-optimal sequencing of PoIs along recommended tours, taking into account a multitude of travel restrictions and PoI properties, so as to best utilize time available for sightseeing. The authors in [20] define tourist routes by means of graph theory. The authors also calculate some relative indexes (e.g., the circle number, circle ratio, line-point ratio etc.) to make quantitative evaluation of tourist routes. Chen et al. apply the Graph theory to optimize tour path and tour flows to provide practical solutions to tourist guides [21].

1.2. Related Works: Parameter-Based Classification

Another classification that can be proposed is based on the works that emphasizes the study of the effect of key parameters of the TTDP on the final solution, such as: (i) the selection of transport modes to reach the different PoIs; (ii) the choice of PoIs; (iii) the number of tours to be generated, on the basis of visiting duration; (iv) the visit duration of a PoI; (v) the travel times among PoIs; (vi) the daily time budget that a tourist wishes to spend on visiting a PoI; (vii) the weather conditions.

Transport is a critical and dynamic process of tourism, which facilitates physical movement to points of interest [22–24]. Transportation affects the accessibility to the tourist destination, the distance travelled, and the comfort of the trip [25,26]. The authors in [27] develop a genetic algorithm (GA) to solve the TTDP that included multimodal transport and real traffic parameters and time constraints.

In [14] the influencing factors of the tour route choices of tourists are analyzed by means of a questionnaire survey. Moreover, tour routes multiobjective optimization functions are prompted for the tour route design with the aim of maximise the user satisfaction with the minimum tour distance. The authors in [17] analyze the planning of a multipleday tour by considering a set of PoIs, a visiting duration, and a set of multiple opening and closing times per day combined with the trip constraints of the tourist. The authors in [28] apply an evolutionary algorithm to solve the tour planning problem in time-dependent urban areas. Gavalas et al. [29] develop a tool for tourist itineraries that considered the departure time and the mode of transport on the tourist route. Wu et al. [30] develop a mathematical model to consider the selection of transport modes, the travel budget, and the maximum travel times. Zheng et al. [31] design a multi-objective model of one-day urban tourist routes, taking into account the transport modes and the complexity of urban tourism transport systems, as congestion, and the transport needs of tourists. Zhang et al. [32] develop a model for the construction of itineraries in scenic routes considering the modes of transport. The authors in [33] analyze the environmental implications of tourist itineraries by creating groups of tourists that use a single mode of transportation (i.e., taxis). Some works study the use of electric vehicles (EV) for the generation of more environmentally friendly tourist itineraries, such as [34–36].

Other works focus on planning trips for tourist group [33,37], that consider the individual preferences of each tourist. The authors in [38] develop a model for the route design problem for various cycle-tourists. The model consider the preferences of each tourist who incorporates different benefits on the same route. Finally, the authors in [33] develop a route planning model that considers multiple days, urban tourism, PoI categories, and heterogeneous preferences for a group of tourists that maximise profit and minimises travel time, distance, and cost.

1.3. Contribution of the Paper

From the analysis of the above reported studies, the OP is not suitable in case the PoIs need to be selected and exchanged among different itineraries, like outward and return paths of a one day trip, because of time constraint. In this case, it is necessary to implement a multi-level algorithm to be able to consult the tourist on any relocation of PoIs in the tour. To these aims, we applied the Travel Salesman Problem (TSP) [39–41] method that involves finding the shortest route through n nodes that begins and ends at the same city and visits every node. The TSP is among the best-known combinatorial optimization problems and has been intensely studied by researchers in various research fields.

In this paper, we aim to determine the optimal itinerary for the one-day tourist, maximizing the number of PoIs to be visited in the outward and return parth, and at the same time minimizing the travel times taking into account the tourist preferences and hard time constraints. The idea is to allow the tourist to select the preferred PoIs to be visited on the first part of the day, i.e., in the outward trip, and on the second part of the day, i.e., in the return trip, respectively. We formulate our optimization problem on the basis of Graph theory, TSP and multi-level algorithms. We model the city PoIs network on the basis of the graph theory, where the nodes represent the various attractions (PoI) of the city and two separate graphs are derived. The tourist can select the PoIs of the starting graph to be visited with high priority in the outward and return journeys, respectively. In our application the tourist is part of the multi-algorithms approach interacting with it and taking decisions for one-day tourist. The proposed approach plans the tourist itinerary, minizing travel times based on the TSP algorithm, taking into account the priority list of PoIs and the decisions of the user. The TSP is used in this paper since it allows to consider a first itinerary solution including all the PoIs of the city that is refined by the multi-level algorithms interacting with the tourists. In detail, compared with the analyzed works, this paper presents the following novelties:

- an innovative multi-level algorithm approach is proposed to determine the optimal roundtrip path: the outward and return journeys are specified and customized, minimizing the total travel time, including the visiting time of each PoI.
- the number of attractions to be visited is maximized and is splitted between the outward and return path in order to improve the visiting experience on the basis of user preferences.
- the tourist is seen as an active and informed user who directly interacts with the system for the optimal planning of both the outward and return journeys, not only providing initial inputs and preferences but taking decisions at intermediate stages of the approach.

The rest of the paper is organized as follows: Section 2 describes the one-day tourist itinerary planning problem; Section 3 presents the Multi-level algorithm approach for the itinerary planning while Section 4 provides the analysis of the algorithms performance and complexity; Section 5 demonstrates the effectiveness of the proposed approach by a case study focusing on the cruise tourist in the city of Bari and Section 6 provides the conclusions and future works perspectives.

2. The One-Day Tourist Itinerary Planning Problem

The one-day tourist, having reached a stage of his journey through the airplane, train, car or cruise ship etc., wonders how to spend at best his/her time in the city in a short time period (e.g., one day).

Due to the limited time, it is therefore necessary to pay attention to the organization of the visits and excursions. The tourist can opt for a tour pre-organized by the operator or he/she can plan it on his/her own. In the first case, one of the advantages concerns the mere observation of the predefined roadmap to visit the city, without any worries. This case, on the other hand, does not always satisfy the personal interests of the individual tourist who must follow the visiting group and, in addition, can not personally manage the route and the stops. In the second case, however, the tourist has more freedom of choice but he/she must plan independently the trip in a city and respect the departure times that are mandatory. Instead of relying on the tours organized by the company, sometimes with unsatisfactory results, the tourist by use only a smartphone can select the preferred attractions.

Today, there are numerous online travel planning systems that allow to automatically generate a selection and routing plan to visit PoIs that suit the tourist's personal interests [42]. These systems implement various functionalities that aim to satisfy different profiles of tourist interest [43,44]. Therefore, considering a tourist discovering the city, in addition to walking through its most famous streets, he/she wants to head, for example, to a restaurant near an attraction to have lunch and taste the typical dishes of the place, and then resume the tour and return back. For instance, by simply accessing an app from the smartphone, he/she can set the time available to carry out the tour from a starting point to a restaurant and the time to return back. The visiting times must also include the stop times for activities such as take photos in front of a monument, go shopping, visit a museum and so on. The tourist can also select the preferred PoIs to be visited on the first part of the day, i.e., in the outward trip, and on the second part of the day, i.e., in the return trip, respectively. In addition, the tourist can also indicate the PoIs that are less important and that can be deleted by the roundtrip in case of time unavailability.

Then, let us describe an example in order to present the addressed problem. Firstly, the following assumptions are made:

- the tourist is an active user who wants to interact with the application in order to customize the daily roundtrip;
- the PoIs of the city are initialized by the application;
- the tourist indicates the starting and destination PoIs, the travel modes and time preferences as well as the PoIs to be visited with high and secondary priority, in the first and second part of the day, respectively.

Let us consider the case of a cruise tourist who wants to visit the city in one day, without loss of generality. When arriving at the port, the tourist needs to have a plan for the daily tour. In particular, he/she needs to decide which PoIs to visit based on the available time and in which order, also making a priority list to be sure to visit the most important ones. There can be also the necessity to specify which PoIs to be visited in the first part of the day, that is usually lightful and more appropriate to visit outdoor spaces like parks, before to have a lunch, usually in a typical restaurant. The tour for the second part of the day, starts after lunch allowing to complete the city visit going towards the final destination point, i.e., the port. Of course, an application is needed to help the tourist at planning the less time consuming roundtrip, respecting the preferences. In our scenario, the application initializes the PoIs network and shows to the user the map of the city PoIs with related information, including traveling times among each PoI couple based on transport means. Different trip solutions can be provided by the application based on the user choice regarding the stop time at each PoI and preferred way of transport: (1) fastest, (2) by foot, (3) by metro/bus. The tourist is also asked to indicate the starting and destination points of the roundtrip, that are different from the origin/final point (i.e., the port), as well as the time deadline for the roundtrip. In addition, the tourist is asked to decide which PoIs to be visited in the first and second part of the day, indicating the priority and the desired time to dedicate to the visit. On this basis, the application try to generate the customized outward and return tours of the day by applying the heuristic procedure presented in Section 3. If the deadline time both for the outward and return trips are respected, the heuristic procedure investigates the addition of secondary importance PoIs and generates the final roundtrip itineraries. On the contrary, if the deadline time of the outward and/or the return tours is violated, the heuristic procedure can exchange PoIs between the first and the second tour. In case some feasible solutions are determined, i.e., the deadlines of the outward and return trips are satisfied, the procedure delegates to the user the choice of a solution from a list created by the application. Afterwards, the heuristic procedure determines the final customized outward and return tours of the day including possible addition of secondary PoIs. Finally, if the deadline time of the outward and/or the return tours is violated and no feasible solution is achievable, a PoI deletion procedure must be implemented in order to respect the deadline travel time. Let us summarize the necessary input and output information and data of the proposed itinerary planning application:

Initial inputs from the user:

- starting and ending PoI;
- deadline time both for the outward and return trips;
- stop times;
- priority list of PoI;
- list of PoIs both for the outward and return trips;
- preferred mode of travel;

Moreover, other inputs are required to the user from the application while running in order to refine the roundtrip customization as described in detail in Section 3.

Real time inputs from the user:

- preferred itinerary from a list of feasible solutions determined by the proposed automatic procedure.

Outputs by the application:

- outward and return paths;
- outward and return travel times.

3. The Multi-Level Algorithm Approach for Tourist Itinerary Planning

In this Section, we want to present an adequate solution to the problem of the one-day tourist, whose goal is to visit the greatest number of attractions and carry out activities of his own liking, respecting the times available for visiting. First of all the city PoI network needs to be modeled in order to connect all the PoIs and decide the best itinerary. We apply the Graph theory [21,45] to model and study the PoIs network which in this paper is modeled as a weighted connected graph [46]. From each graph a path is determined ensuring that the tourist will visit only once those nodes representing the essential PoI: (i) the first path, called outward path, is from the source to the destination; (ii) the second path, called return path, is from the destination to the source.

In particular, the nodes of the graph represent the city attractions. In addition, the weight of an arc connecting two nodes represents the travel time between two attractions. More in detail, the proposed approach uses two graphs G_o and G_r that are built considering the following tourist inputs: the starting and ending PoIs of the outward path (they correspond to the ending and starting PoIs of the return path), the other preferred PoIs to visit during the outward and return path, the preferred transport mode. The starting PoI (ending PoI) of the outward path is represented by the source node (destination node) v_s (v_d) as shown in Figure 1. Moreover, the starting PoI is the place that the tourist firstly reaches after leaving the airport, port or station that are respresented in Figure 1 with the origin node V_{origin}. The origin node is not included in the set of nodes of G_o and G_r. Finally, each arc of graphs is weighted by the travel time between two PoIs and the time depends by the preferred transport mode chosen by the tourist.

The proposed approach to solve the tourist problem is based on a multi-level algorithm approach [39]. The proposed Algorithms are modeled by means of UML diagrams. UML is a standard highly recognized language widely used to visually describe software programs and algorithms [47]. More specifically, there is the main algorithm, so called Algorithm 1, that is responsible for the data initialization and for the determination of the initial paths. Moreover, Algorithm 1 makes use of two sub-algorithms to find an optimal planning of the itinerary based on the tourist needs in term of time and places of interest, by applying the TSP algorithm. The TSP is about a traveling man who wants to visit only once each PoI of the list returning to the initial PoI through the least cost route. The TSP is suitable to be modeled through a graph in which the nodes are the PoI and each arc connects a couple of PoI (i, j) including a travel cost from i to j. The total lenght of a journey is given by the sum of the arc weights included in the round-trip of the traveler. In order to formulate the generic version of the asymmetric TSP, the following binary variables are needed:

$$x_{ij} = \begin{cases} 1 & \text{if arc (i,j) is in the tour} \quad i,j \in \{1,\ldots,m\} \\ 0 & \text{otherwise} \end{cases} \tag{1}$$

with m total number of PoIs. Now, according to the Dantzig–Fulkerson–Johnson formulation the TSP can be formalized as the following integer linear programming problem:

$$min \sum_{i=1}^{m} \sum_{j=1,j\neq i}^{m} c_{ij} x_{ij}$$

$$s.t.$$

$$\sum_{i=1,i\neq j}^{m} x_{ij} = 1 \quad j = 1,\ldots,m \tag{2a}$$

$$\sum_{j=1,j\neq i}^{m} x_{ij} = 1 \quad i = 1,\ldots,m \tag{2b}$$

$$\sum_{i\in K} \sum_{j\in K,j\neq i} x_{ij} \leq |K| - 1 \quad \forall K \subset \{1,\ldots,m\}, |K| \geq 2 \tag{2c}$$

with $c_{ij} > 0 \ \forall i,j \in \{1,\ldots,m\}, i \neq j$ time cost to travel from i to j, K nonempty subset of the set of m PoIs and $m(m-1)$ number of binary variables. In particular, constraints (2c) ensures that no subset K can generate sub-tours, i.e., only a single tour will be generated. In order to obtain the symmetric version of the TSP it is necessary to have $c_{ij} = c_{ji} \ \forall i,j \in \{1,\ldots,m\}, i \neq j$. It holds that the number of variables in the symmetric TSP is halves with respect to the asymmetric TSP. In this paper, we consider the symmetric TSP inside the proposed heuristic approach modeled with an undirect graph to find the optimal travel times and paths associated to the outward and return tours.

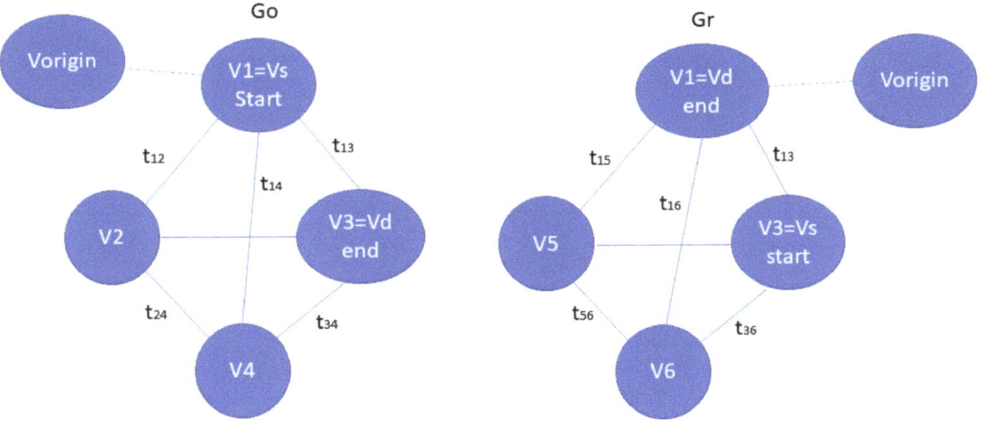

Figure 1. G_o and G_r graph examples.

3.1. The Proposed Heuristic Approach

The proposed approach starts by Algorithm 1 described by the UML diagram in Figure 2 that is the upper level Algorithm that executes two phases: (1) initialization phase; (2) itinerary planning phase.

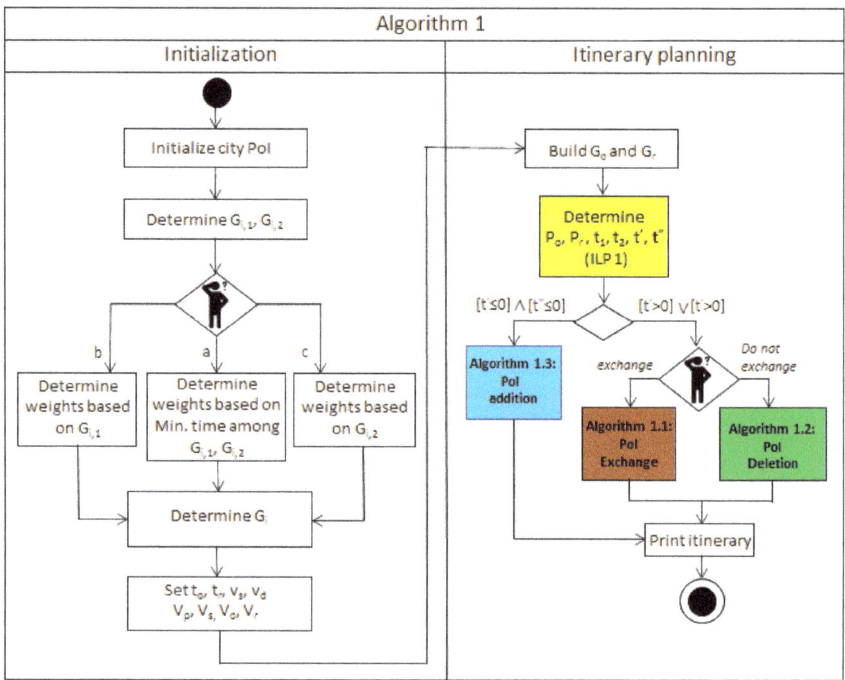

Figure 2. The UML diagram of Algorithm 1: data initialization and itinerary planning.

3.1.1. Initialization Phase

The first phase of the algorithm concerns the initialization of data and it is divided into two parts. In the first part, the algorithm, by knowing the city map, creates the initial graphs of the city tourist attractions: a tourist attraction is associated with a node and all the nodes are connected to each other through indirect arcs. For each pair of nodes the time needed to go from one attraction to another is specified, on the basis of the transport means, with a weight associated with the arc that connects the nodes. In addition, the times to reach each attraction starting from the origin PoI and vice versa are also provided. In particular, two weighted graphs are initially considered, respectively named $G_{i,1}$ and $G_{i,2}$, composed by the same nodes and arcs, i.e., $V_{i,1} = V_{i,2}$ and $E_{i,1} = E_{i,2}$, where each node represents a PoI of the city, each arc indicates the connection between two nodes and the arc weight represent respectively the travel times by foot in $G_{i,1}$ and by bus/metro in $G_{i,2}$. Considering two nodes a and b of $G_{i,2}$, here we assume that the trip from a to b is performed mainly by bus and/or metro, with the possibility that a short segment of this trip must be traveled by foot. Moreover, for each node of $G_{i,1}$ and $G_{i,2}$ we are assuming that the travel time to go from the origin PoI to the node is equal to the travel time to go from the node to the origin PoI. After that, the graph $G_i = \{V_i, E_i\}$ is defined composed by the same nodes and arcs of $G_{i,1}$ and $G_{i,2}$. On each arc of G_i the weight is set among the following three possibilities, based on the user preferences:

(a) time preference: the weight of the arc is given by the minimum travel time among the corresponding ones of $G_{i,1}$ and $G_{i,2}$;

(b) foot preference: the weight of the arc is given by the travel time by foot on the corresponding arc of $G_{i,1}$;

(c) bus/metro preference: the weight of the arc is given by the travel time by bus/metro on the corresponding arc of $G_{i,2}$.

In addition, the resulting graph G_i also keeps track of the transport means used on each trip segment, i.e., by foot or by bus/metro.

In the second part of the initialization phase, the tourist sets the following preferences:

- t_o, time available for the outward trip and t_r, time available for the return trip;
- t_s, stop time at each node $v \in V_i$;
- define set $V_p \subseteq V_i$ of nodes of high priority and set $V_s \subseteq V_i$ of nodes of secondary priority with $V_p \cap V_s = \varnothing$;
- v_s source node and v_d destination node, among the nodes $v \in V_p$;
- define set $V_o \subset V_p$ and $V_r \subset V_p$, $V_o \cap V_r = \{v_d, v_s\}$, of the nodes to be visited on the outward and return journeys, respectively.

3.1.2. Itinerary Planning Phase

On the basis of the input from the initialization phase, the Algorithm 1 proceeds with the construction of two separate graphs G_o and G_r to determine the *outward* and *return* paths, respectively. The graph G_o and G_r are composed by the nodes $v \in V_o$ and $v \in V_r$, respectively. Moreover, the graphs G_o and G_r are of order N_o (cardinality of V_o) and N_r (cardinality of V_r), respectively, and are implemented through the adjacency matrices. Since the graphs are not oriented, connected and complete, the adjacency matrices are symmetric with a null diagonal. We solve the TSP (2) for the graphs G_o and G_r, respectively, in order to find the minimum path P_o and P_r and the associated travel time cost t_1 and t_2. Now, let us define the following integer linear programming problem ILP1 in order to maximize the available travel times for the outward and return paths:

$$F(\lambda) = \max \lambda$$

$$s.t.$$

$$
\begin{cases}
\lambda \leq t' = t_1' - t_o & \text{(3a)} \\
\lambda \leq t'' = t_1'' - t_r & \text{(3b)} \\
t_1' = t_1 + t_{stop}(P_o) + y_1 * t_{stop,s} + y_2 * t_{stop,e} + t_p & \text{(3c)} \\
t_2'' = t_2 + t_{stop}(P_r) + y_3 * t_{stop,s} + y_4 * t_{stop,e} + t_p & \text{(3d)} \\
y_1 + y_3 = 1 & \text{(3e)} \\
y_2 + y_4 = 1 & \text{(3f)} \\
\lambda \in \mathcal{R} & \text{(3g)} \\
y_1, y_2, y_3, y_4 \in \{0,1\} & \text{(3h)}
\end{cases}
$$

where λ is the real decision variable that has to be maximized in order to maximize the difference between the effective travel times t_1' and t_2'' and the respective available times t_o and t_r. Let us specify that for the outward path P_o, the source PoI is set as the initial node while the destination PoI as the final node. On the contrary, for the return journey P_r the destination PoI is set as the initial node and the source PoI as the final node. Moreover, with constraints (3c) and (3d) ILP1 takes into account the following variables: the sum of the stop time period at each PoI of P_o and P_r, respectively called $t_{stop}(P_o)$ and $t_{stop}(P_r)$; the time period t_p to reach the origin PoI from the final point; the stop time period for visiting the source and destination nodes denoted respectively by $t_{stop,s}$ and $t_{stop,e}$. In particular, constraints, (3c), (3d), (3e) and (3f) are introduced to ensure that the source and end PoI can be visited only one time respectively on the outward or on the return journey. More precisely, constraint (3e) states that if the tourist stops for visiting the source PoI on the outward path, he/she will not repeat the visit on the return path: on the return path, the tourist will just pass through the source PoI without stopping there. The same statement of (3e) is done for the end/destination PoI by applying constraint (3f).

The resulting paths P_o and P_r are not definitive and a further analysis is required to satisfy the tourist preferences.

The travel times t'_1 and t''_2, must not exceed the time available for visiting t_o and t_r, respectively. Consequently, the following algorithms will manage the itinerary by adding and/or removing none, one or more PoI (nodes) from the path P_o and/or P_r so that the time constraints are respected. Now, considering that t_o and t_r are the available travel times to complete the outward and return journeys, respectively, two cases which needs to be managed can arise:

1. travel times exceed available times: $t' > 0$ OR $t'' > 0$;
2. travel times do not exceed available times: $t' \leq 0$ AND $t'' \leq 0$;

In particular, the management of case 1 is performed by Algorithm 1.1 and Algorithm 1.2, respectively described by the UML diagrams in Figures 3 and 4, while case 2 is managed through Algorithm 1.3 described by the UML diagram of Figure 5 in the following. At the end of Algorithm 1.1 and Algorithm 1.3, Algorithm 1 displays the final itinerary to the tourist.

Algorithm 1.1: PoI Exchange Procedure

In case 1, it is necessary to manipulate the outward and return paths, P_o and P_r, in order to respect the time constraints, to avoid delay in the origin PoI. An attempt is made to keep all the PoI of high priority by exchanging nodes between those selected for the outward and the return journeys. To this aim, a node belonging to the outward graph G_o is exchanged with a node belonging to the return graph G_r. Once the exchange has been made, the Algorithm 1.1 determines the new paths P_o and P_r and the travel time t_1 and t_2 by applying ILP1. Afterwards, it checks if the times t' and t'' are positive or negative and one of the two cases can occur, as highlighted in Figure 3.

If case 1 occurs, Algorithm 1.1 updates the table FS of feasible solutions, i.e., records the paths P_o and P_r obtained by feasible nodes exchange. Afterwards, all the possible nodes exchange are tried between the two graphs (see Figure 3) and all the feasible solutions are recorded in Table FS. If case 2 occurs the Algorithm 1.1 ignores the obtained solution beacuse it is not feasible and proceeds with the node exchange procedure until other combinations are no longer possible.

After all possible nodes exchange have been made, the Algorithm 1.1 checks if suitable solutions have been found. If table FS is not empty, Algorithm 1.1 asks the tourist to indicate one of the solution in table FS and Algorithm 1.1 goes to Algorithm 1.3. On the contrary, if table FS is empty, i.e., no feasible node exchange are possible, Algorithm 1.1 goes to Algorithm 1.2 to start the node deleting procedure.

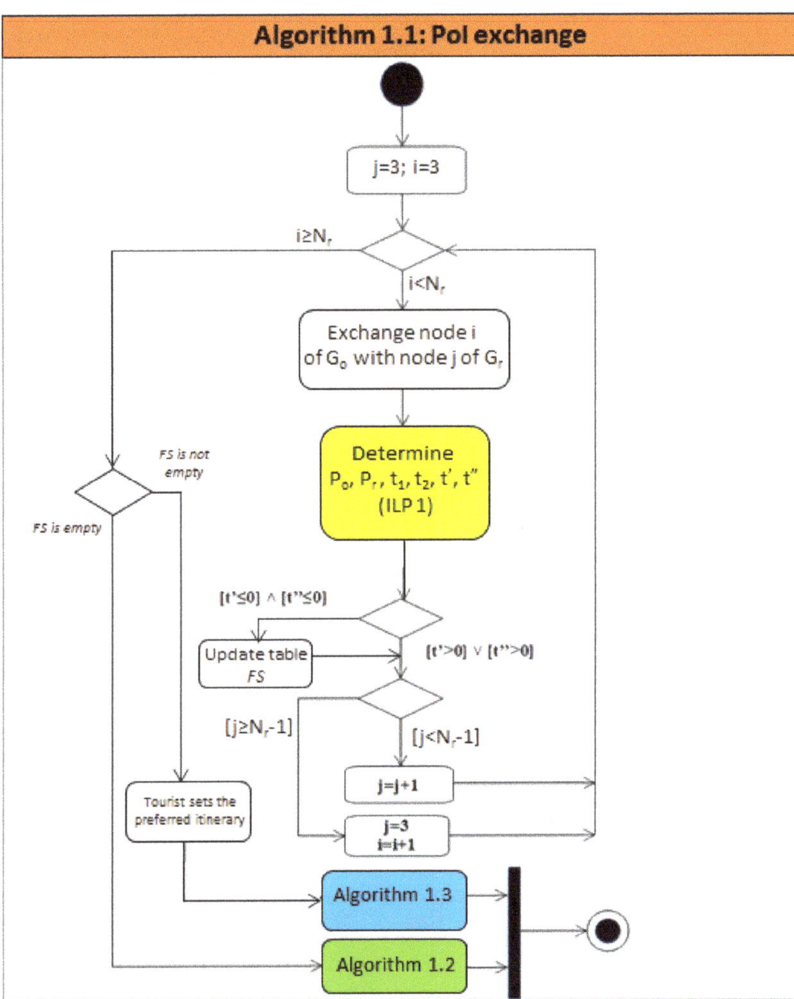

Figure 3. The UML diagram for the node exchange procedure.

Algorithm 1.2: PoI Deletion Procedure

Algorithm 1.2 starts the node deleting procedure (see UML diagram of Figure 4). If $t' > 0$, a node is eliminated from G_o and, in case $t'' > 0$, a node is simultaneously deleted from G_r. After that, Algorithm 1.2 computes P_o and t_1, P_r, t_2 and, in particular, t' and t'' by applying ILP1. Then, the algorithm checks if the time constraints on t' and t'' are satisfied. These steps are repeated iteratively by Algorithm 1.2 until time constraints are not respected or no more node can be deleted. The Algorithm 1.2 displays an error message in case, after all possible nodes of G_o or G_r have been deleted, it still holds $t' > 0$ or $t'' > 0$, respectively. On the contrary, if $t' \leq 0$ and $t'' \leq 0$, the PoI elimination is not necessary anymore and the Algorithm 1.2 goes to Algorithm 1.3 where the possible addition of other points is evaluated.

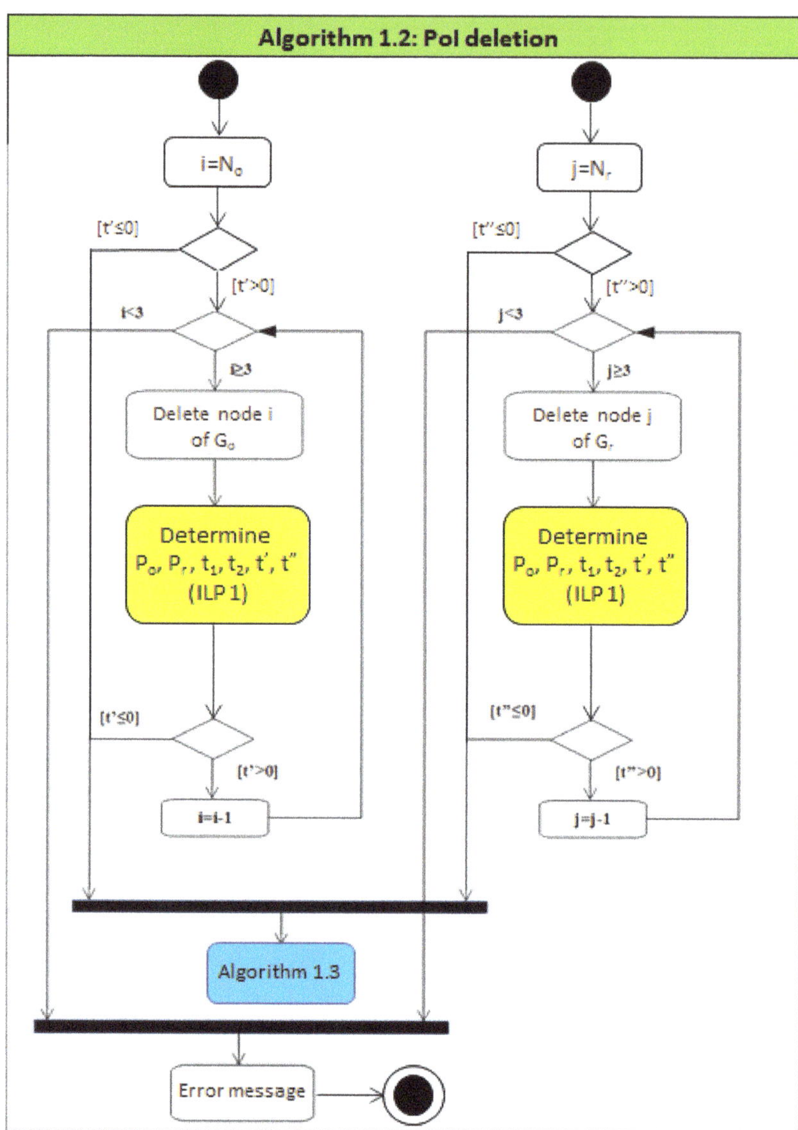

Figure 4. The UML diagram for the node deletion procedure.

Algorithm 1.3: PoI Addition Procedure

In the case $t' \leq 0$ and $t'' \leq 0$, it is reasonable to add one or more nodes to the paths P_o and P_r. Thus, a node $v \in V_s$ with secondary priority is temporarily added to G_o and G_r. Note that N_s is the cardinality of V_s. At this point, if condition $t' \leq 0$ and $t'' \leq 0$ is still verified the Algorithm 1.3 proceeds by adding another node $v \in V_s$, until no more nodes $v \in V_s$ can be added. On the other hand, if the addition of a node does not satisfy the time constraints, the added node is removed from the relative path and the Algorithm 1.3 checks whether it is possible to insert other nodes by repeating the operation until all nodes $v \in V_s$ are examined (see the UML diagram in Figure 5).

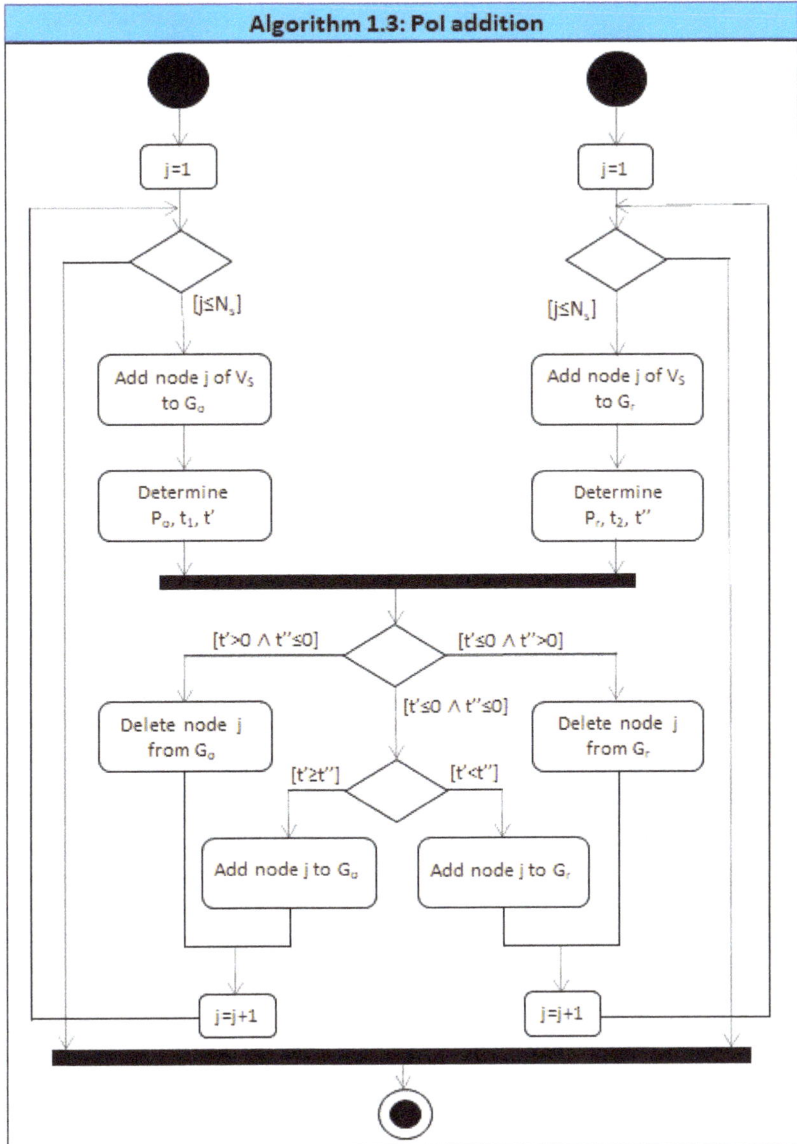

Figure 5. The UML diagram for the node addition procedure.

4. Complexity and Performance Analysis

In this section, the analysis of the complexity and performance of the proposed algorithms are provided. The following results describe the algorithms complexity:

- the Algorithm 1 requires the implementation of the ILP 1 problem including two TSP and 4 decision variables. Hence, considering a branch-bound approach, the complexity of Algorithm 1 is $O(2^4) + 2 * O(K) = O(K)$, with $O(K)$ TSP complexity;
- the Algorithms 1.1, 1.2 and 1.3 show complexity $O(N * K)$ since the ILP 1 problem is included in a N-dimensional "while" loop, where N is the number of PoI of graph G_i.

We can conclude that the complexity of the heuristic approach is $O(N * K)$. In order to be compliant with application time constraints, the TSP has been implemented using the Lin-Kernighan algorithm that often keeps its tours within 2% of the Held-Karp lower bound [48], then $K = O(N^{2.2})$. Therefore, our heuristic approach shows complexity $O(N^{3.2})$. Let us underline that the proposed application interacts with users to find the final best solution. Therefore, the time to complete the heuristic approach application and provide the final solution strongly depends on the velocity of the user given the necessary inputs to the application.

The performance of the proposed algorithms are validated with benchmark orienteering algorithms presented in [49]. In particular, the data set used in [49] and reported in [50] is used as benchmark (see Figure 6). Since, in the considered benchmark data set the PoI importance is defined by a score, we associate the priorities to the higher score values as reported in Figure 6. Note that in Figure 6, X and Y are the cartesian coordinates of the PoI and the PoIs distance is computed using the Euclidean distance formula [49]. Moreover, for comparison purpose, we assume that (i) the PoI V_d is selected as to minimize the total travel time, (ii) the Euclidean distance includes the stop time for visiting in our approach, (iii) the user preferences are randomly set in the instances simulation. Now, let us report the comparison results between the proposed heuristic and the D-algorithm and S-algorithm proposed by [49]. In particular, Figures 7–9 report respectively the comparison results considering data set of problem 1, 2 and 3, where Tmax represents the total travel time. Comparing the score and time values it can be concluded that the proposed approach performs much better than D-algorithm and little worse than the S-algorithm. However, in case only addition of PoI are needed by Algorithm 1.3, the proposed approach performs better than S-algorithm too. It is also remarked that the performance of Algorithm 1.2 can be further improved when a considerable number of PoIs must be deleted.

| | Problem 1 | | | | Problem 2 | | | | Problem 3 | | | |
Node	X	Y	Score	Priority	X	Y	Score	Priority	X	Y	Score	Priority
1	10.5	14.4	0		4.6	7.1	0		19.1	24.3	0	
2	18	15.9	10	x	5.7	11.4	20		12.6	24.9	20	
3	18.3	13.3	10	x	4.4	12.3	20		14.4	28.0	20	
4	16.5	9.3	10	x	2.8	14.3	30	x	16.9	28.1	20	
5	15.4	11	10	x	3.2	10.30	15		20.7	28.2	20	
6	14.9	13.2	5		3.5	9.8	15		12.5	26.6	20	
7	16.3	13.3	5		4.4	8.4	10		21.8	27.3	20	
8	16.4	17.8	5		7.8	11.0	20		12.5	22.6	20	
9	15	17.9	5		8.8	9.8	20		22.5	17.0	30	x
10	16.1	19.6	10	x	7.7	8.2	20		19.9	15.0	30	x
11	15.7	20.6	10	x	6.3	7.9	15		14.9	15.1	30	x
12	13.2	20.1	10	x	5.4	8.2	10		11.5	18.6	30	x
13	14.3	15.3	5		5.8	6.8	10		12.4	29.8	30	x
14	14	5.1	10	x	6.7	5.8	25	x	17.8	28.1	30	x
15	11.4	6.7	15	x	13.8	13.1	40	x	9.1	29.8	40	x
16	8.3	5	15	x	14.1	14.2	40	x	10.0	32.6	40	x
17	7.9	9.8	10	x	11.2	13.6	30	x	13.9	33.1	40	x
18	11.4	12	5		9.7	16.4	30	x	19.95	10.3	40	x
19	11.2	17.6	5		9.5	18.8	50	x	15.2	8.0	40	x
20	10.1	18.7	5		4.7	16.8	30	x	14.7	31.2	50	x
21	11.7	20.3	10	x	5.0	5.6	0		7.4	36.5	50	x
22	10.2	22.1	10	x					21.0	25.5	50	x
23	9.7	23.8	10	x					18.0	25.3	10	
24	10.1	26.4	15	x					19.5	24.7	10	
25	7.4	24	15	x					21.4	21.8	10	
26	8.2	19.9	15	x					16.0	21.4	10	
27	8.7	17.7	10	x					18.65	26.2	10	
28	8.9	13.6	10	x					17.9	28.9	10	
29	5.6	11.1	10	x					14.3	19.9	20	
30	4.9	18.9	10	x					17.0	19.0	20	
31	7.3	18.8	10	x					10.80	21.0	20	
32	11.2	14.1	0						15.7	23.7	10	
33									18.2	24.0	0	

Figure 6. The benchmark data set.

Tmax	D-algorithm		S-algorithm		Proposed algorithm		Time Optimality gap 1	Score Optimality gap 1	Time Optimality gap 2	Score Optimality gap 2
	Time	Score	Time	Score	Time	Score				
5	--	--	4.2	10	4.14	10	--	--	-1.4%	0.0%
10	--	--	7.0	15	6.87	15	--	--	-1.9%	0.0%
15	--	--	14.3	45	14.96	30	--	--	4.6%	-33.3%
20	19.99	40	19.6	65	19.99	40	0.0%	0.0%	2.0%	-38.5%
25	24.33	65	24.7	90	24.25	65	-0.3%	0.0%	-1.8%	-27.8%
30	28.94	80	28.8	110	27.41	80	-5.3%	0.0%	-4.8%	-27.3%
35	34.23	105	34.1	135	34.65	90	1.2%	-14.3%	1.6%	-33.3%
40	36.37	105	38.0	150	39.92	110	9.8%	4.8%	5.1%	-26.7%
46	45.33	130	44.5	175	45.82	140	1.1%	7.7%	3.0%	-20.0%
50	48.6	140	49.9	190	49.97	150	2.8%	7.1%	0.1%	-21.1%
55	51.96	160	54.8	205	54.81	190	5.5%	18.8%	0.0%	-7.3%
60	56.78	175	58.9	220	59.74	200	5.2%	14.3%	1.4%	-9.1%
65	63.91	200	63.9	240	64.58	225	1.0%	12.5%	1.1%	-6.3%
70	69.84	200	68.8	255	69.72	250	-0.2%	25.0%	1.3%	-2.0%
73	71.07	205	72.7	260	72.59	260	2.1%	26.8%	-0.2%	0.0%
75	73.35	210	74.7	270	74.62	265	1.7%	26.2%	-0.1%	-1.9%
80	78.17	220	79.0	275	79.83	275	2.1%	25.0%	1.1%	0.0%
85	82.41	235	82.2	280	81.79	285	-0.8%	21.3%	-0.5%	1.8%
						Average	**1.7%**	**11.7%**	**0.6%**	**-14.0%**

Figure 7. Algorithms comparison using data set of problem 1.

Tmax	D-algorithm		S-algorithm		Proposed algorithm		Time Optimality gap 1	Score Optimality gap 1	Time Optimality gap 2	Score Optimality gap 2
	Time	Score	Time	Score	Time	Score				
15	14.84	100	14.88	120	14.37	120	-3.2%	20.0%	-3.4%	0.0%
20	18.79	130	18.81	190	19.43	130	3.4%	0.0%	3.3%	-31.6%
23	18.79	130	22.75	205	22.26	190	18.5%	46.2%	-2.2%	-7.3%
25	24.84	145	24.89	230	24.13	230	-2.9%	58.6%	-3.1%	0.0%
27	26.00	160	24.89	230	26.76	165	2.9%	3.1%	7.5%	-28.3%
30	29.74	190	29.09	250	29.67	225	-0.2%	18.4%	2.0%	-10.0%
32	31.54	195	30.59	275	31.88	240	1.1%	23.1%	4.2%	-12.7%
35	34.17	200	34.25	315	34.91	265	2.2%	32.5%	1.9%	-15.9%
38	36.81	210	37.62	355	37.79	355	2.7%	69.0%	0.5%	0.0%
40	39.91	240	39.78	395	39.99	370	0.2%	54.2%	0.5%	-6.3%
45	44.37	270	44.28	430	44.44	450	0.2%	66.7%	0.4%	4.7%
						Average	**2.3%**	**35.6%**	**1.1%**	**-9.8%**

Figure 8. Algorithms comparison using data set of problem 2.

Tmax	D-algorithm		S-algorithm		Proposed algorithm		Time Optimality gap 1	Score Optimality gap 1	Time Optimality gap 2	Score Optimality gap 2
	Time	Score	Time	Score	Time	Score				
15	14.50	70	13.82	100	14.9	100	2.8%	42.9%	7.8%	0.0%
20	17.91	120	19.25	140	19.96	140	11.4%	16.7%	3.7%	0.0%
25	24.65	140	24.66	190	24.94	150	1.2%	7.1%	1.1%	-21.1%
30	27.50	180	29.60	240	29.84	190	8.5%	5.6%	0.8%	-20.8%
35	32.11	220	34.93	290	34.87	220	8.6%	0.0%	-0.2%	-24.1%
40	37.32	240	39.65	330	39.96	270	7.1%	12.5%	0.8%	-18.2%
45	44.39	280	44.04	370	44.32	350	-0.2%	25.0%	0.6%	-5.4%
50	49.03	310	48.35	410	49.98	420	1.9%	35.5%	3.4%	2.4%
55	54.73	340	54.31	450	54.92	450	0.3%	32.4%	1.1%	0.0%
60	57.05	350	59.07	500	59.81	490	4.8%	40.0%	1.3%	-2.0%
65	64.51	410	64.65	530	64.70	560	0.3%	36.6%	0.1%	5.7%
70	66.68	460	69.39	560	69.85	570	4.8%	23.9%	0.7%	1.8%
75	72.75	490	74.78	590	74.88	570	2.9%	16.3%	0.1%	-3.4%
80	75.47	510	79.80	640	79.98	610	6.0%	19.6%	0.2%	-4.7%
85	82.95	520	83.61	670	84.96	650	2.4%	25.0%	1.6%	-3.0%
90	86.30	580	89.07	690	89.23	720	3.4%	24.1%	0.2%	4.3%
95	92.42	610	94.40	720	94.48	770	2.2%	26.2%	0.1%	6.9%
100	98.35	640	99.67	760	99.09	770	0.8%	20.3%	-0.6%	1.3%
105	103.48	660	104.55	770	104.89	790	1.4%	19.7%	0.3%	2.6%
110	109.72	680	107.97	790	106.19	800	-3.2%	17.6%	-1.6%	1.3%
						Average	**3.4%**	**22.4%**	**1.1%**	**-3.8%**

Figure 9. Algorithms comparison using data set of problem 3.

5. Case Study

This section presents a case study where the proposed multi-level algorithm approach is applied to solve the cruise tourist problem in the metropolitan port city of Bari. Bari is the capital city of Apulia Region and the second biggest city in southern Italy. Since the roman epoch, Bari became an important commercial center, during the Saracen domination. From 1071, it became a big maritime center and still today it is an important port hub of the Mediterranean sea.

In the initialization phase of Algorithm 1, all the main attractions of Bari are determined and represented by the adjacency matrix of the graph G_i as it shown in Figure 10. Note that the weight of arcs are decided on the basis of the time preference (a) described in Section 3, according to the user. For better understanding of Figure 10 let us consider two examples: "1mp" in the box from PoI 2 to PoI 1 means that the tourist should travel by foot for 1 min; "10m p+a" in the box from PoI 2 to PoI 7 means that the tourist should travel by foot and by bus for 7 min total. The tourist inputs are provided in the initialization phase of the mobile application, where the tourist inserts the preferences and constraints related to: (i) the maximum available time for the outward and return path; (ii) the PoIs to visit during the outward and return path; (iii) the preferred transport mode, i.e., by foot, by bus or the fastest way (see Figure 11). Moreover for each node of the PoI network, the tourist can edit the selection of transport mode according to his/her preferences (see Figure 12).

Let the Saint Nicolas Basilica (node 2) be the starting point and Lungomare Nazario Sauro (node 8) be the final point of the tourist itinerary. Note that for path P_o the starting point is node 2 and the final point is node 8. On the contrary, for P_r the starting point is node 8 and the final point is node 2. In particular, let us define the nodes of priority level 1 (maximum priority) $V_p = \{1, 2, 4, 5, 8, 9, 10\}$, and the secondary nodes with $V_s = \{3, 6, 7\}$. Furthermore, the PoI to be visited on the outward and return paths and the stop time at each PoI are also reported in Figures 10 and 13.

	1	2	3	4	5	6	7	8	9	10
	Museo Nicolaiano	Basilica San Nicola	Cattedrale San Sabino	Castello Svevo	fortino San Antonio	teatro Piccinni	via Sparano	lungomare Nazario Sauro	pinacoteca	parco Due Giugno
Museo Nicolaiano		1 m p	5 m p	7 m p	5 m p	11 m p	8 m p+a	12 m p+a	11 m p+a	18 m p+a
Basilica San Nicola	1 m p		4 m p	6 m p	5 m p	11 m p	10 m p+a	12 m p+a	12 m p+a	20 m p+a
Cattedrale San Sabino	5 m p	4 m p		4 m p	7 m p	6 m p	8 m p	13 m p+a	13 m p+a	21 m p+a
Castello Svevo	7 m p	6 m p	4 m p		9 m p	4 m p	10 m p	14 m p+a	13 m p+a	23 m p+a
fortino San Antonio	5 m p	5 m p	7 m p	9 m p		8 m p	11 m p	12 m p+a	11 m p+a	18 m p+a
teatro Piccinni	11 m p	11 m p	6 m p	4 m p	8 m p		7 m p	13 m p+a	11 m p+a	19 m p+a
via Sparano	8 m p+a	10 m p+a	8 m p	10 m p	11 m p	7 m p		14 m p+a	12 m p+a	18 m p+a
lungomare Nazario	12 m p+a	12 m p+a	13 m p+a	14 m p+a	12 m p+a	13 m p+a	11 m p+a		3 m p	16 m p+a
pinacoteca	11 m p+a	12 m p+a	13 m p+a	13 m p+a	11 m p+a	11 m p+a	12 m p+a	3 m p		16 m p+a
parco Due Giugno	18 m p+a	20 m p+a	21 m p+a	23 m p+a	18 m p+a	19 m p+a	18 m p+a	16 m p+a	16 m p+a	

Figure 10. Adjacency matrix of graph G_i.

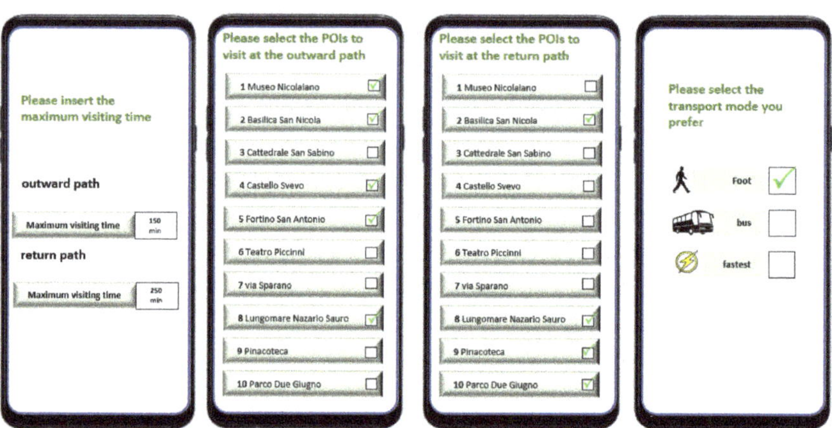

Figure 11. The GUI of the user preferences.

Figure 12. The GUI showing the travel times in the PoIs network from museo Nicolaiano. Preferred transport mode can be edit.

	1. Museo Nicolaiano	2. Basilica san Nicola	3. Cattedrale San Sabino	4. Castello Svevo	5. Muraglia e Fortino Sant'Antonio	6. Teatro Piccinni	7. Via Sparano	8. Lungomare Nazario Sauro	9. Pinacoteca Corrado Giaquinto	10. Parco due Giugno
stop time	60 min	30 min	30 min	50 min	10 min	60 min	120 min	40 min	60 min	10 min

Figure 13. Stop time at each PoI.

The Algorithm 1 creates the two graphs G_o and G_r by using only the nodes of G_i of priority level 1 (maximum priority) and calculates the travel times t_1 and t_2, as it is reported in Figure 14. It implies $V_o = \{1, 2, 4, 5, 8\}$ and $V_r = \{2, 8, 9, 10\}$. The travel times t_1 and t_2 do not include the time needed to go from the port to the first stop of the tour, i.e., node 2, and vice versa that is equal to 10 min. In Figure 15 the vector of the times to reach the port from each node is shown.

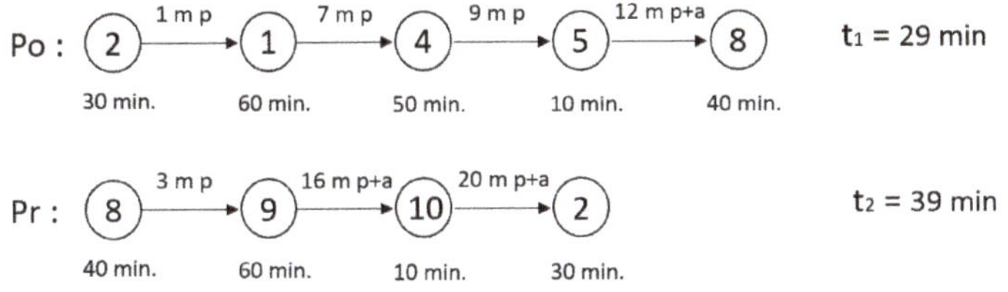

Figure 14. The Graphs G_o and G_r and the travel times t_1 and t_2.

	1. Museo Nicolaiano	2. Basilica san Nicola	3. Cattedrale San Sabino	4. Castello Svevo	5. Muraglia e Fortino Sant'Antonio	6. Teatro Piccinni	7. Via Sparano	8. Lungomare Nazario Sauro	9. Pinacoteca Corrado Giaquinto	10. Parco due Giugno
travel time to the port	9 min P	10 min P	12 min P	10 min A	11 min A	11 min A	15 min A	15 min A	15 min A	20 min A

Figure 15. Travel times from each node of G_i to the port.

Let us set $t_o = 150$ min and $t_r = 250$ min. Since $t' > 0$, as $t'_1 = 159$ (obtained by the sum of the travel and stop times from one node to another in the outward path, except the stop times of node 2 and 8) and $t''_2 = 189$ (obtained by the sum of the travel and stop times from one node to another in the return path, considering also the stop times of node 2 and 8), this solution is not feasible.

Therefore, Algorithm 1 goes to Algorithm 1.1 that tries to exchange the nodes between graph G_o and G_r in order to obtain all the feasible solutions that are stored in FS. In the table FS, two solutions are stored which are obtained by the following two nodes exchanging: (i) node 4 of G_o with node 10 of G_r, (ii) node 1 of G_o with node 10 of G_r. In Figure 16, P_o and P_r of solution (i) obtained by solving ILP1 are shown. In this case $t' < 0$ and $t'' < 0$, since $t'_1 = 130$ and $t''_2 = 212$ (node 2 and 8 are visited during the return path).

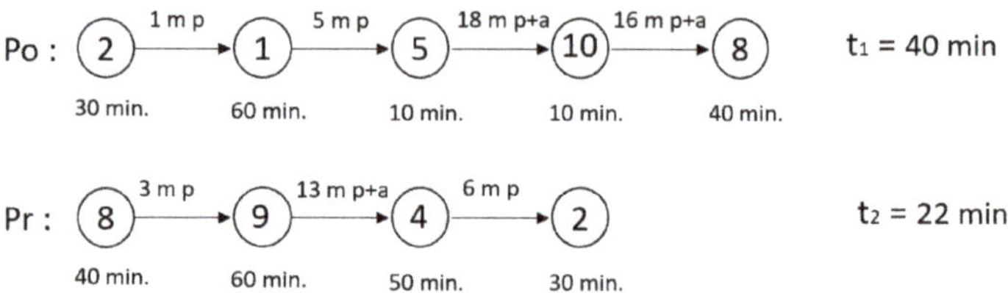

Figure 16. Algorithm 1.1 exchanges node 4 of G_o with node 10 of G_r

In Figure 17, P_o and P_r of solution (ii) obtained by solving ILP1 are shown. In particular, in this case $t' < 0$ and $t'' < 0$, since $t'_1 = 129$ (node 2 and 8 are visited during the outward path) and $t''_2 = 215$.

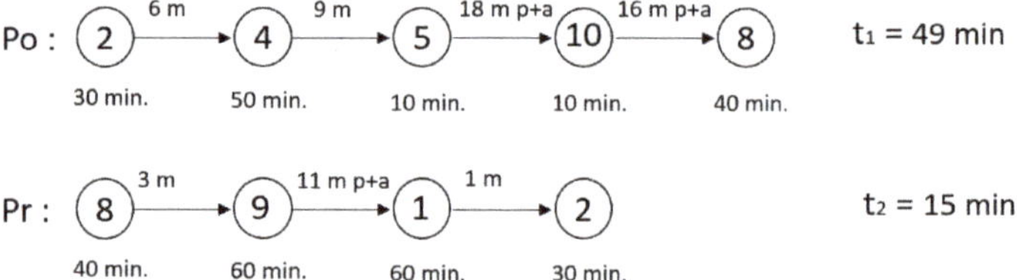

Figure 17. Algorithm 1.1 exchanges node 1 of G_o with node 10 of G_r

The two resulting feasible solutions are shown to the user that can select the preferred one in the mobile application (see Figure 18). Let us suppose that the tourist selects path (i). Therefore, Algorithm 1.1 goes to Algorithm 1.3 that tries to add nodes of priority 2 both on the outward and on the return paths. Since it implies that $t' > 0$, node 3 cannot be added to the route and it is removed. The adding of node 3 is checked for the outward path (see Figure 19). In this case, there is time available since $t'' < 0$ and $t''_2 = 244$. Therefore, node 3 with priority 2 is added to G_r. Moreover, since there are no other nodes of G_i to be added, the outward journey does not change.

Figure 18. The GUI of the feasible solutions.

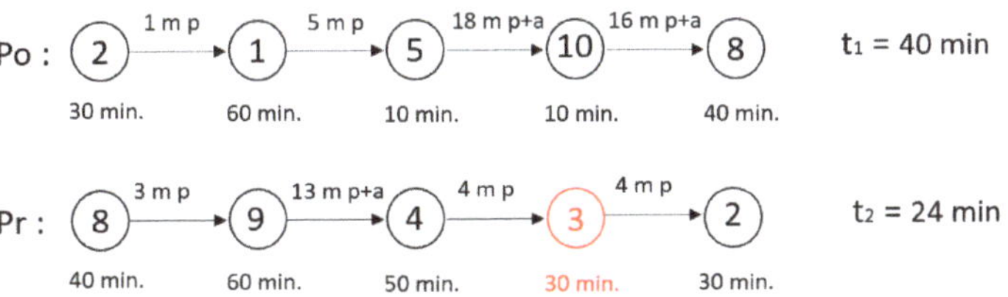

Figure 19. The outward path after adding node 3 to G_r.

Figures 20 and 21 depicts the GUI of the mobile application showing the final outward and return paths, respectively, connecting the identified PoIs.

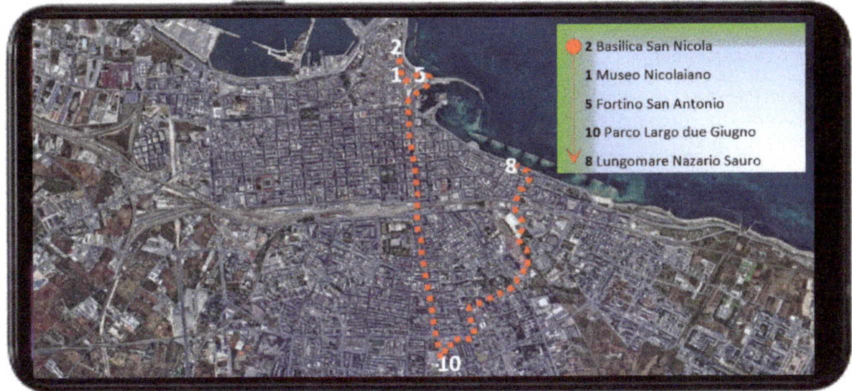

Figure 20. The GUI of the outward path.

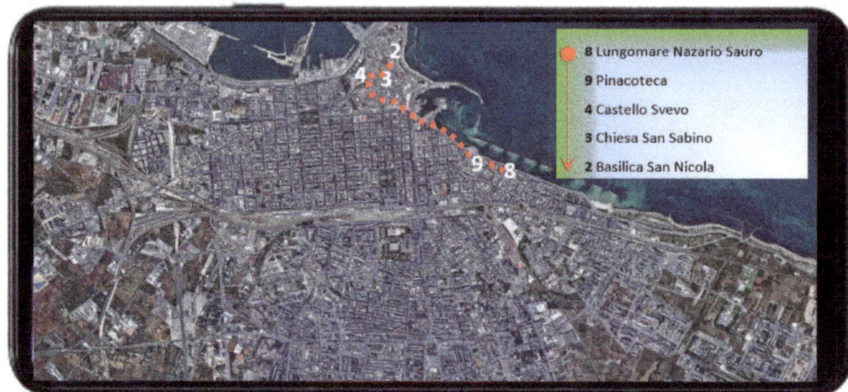

Figure 21. The GUI of the return path.

The procedure ends and the tourist can get the sequence of attractions in both trips with the relative means of transport. In the proposed case study, the mathematical calculations to obtain the final solution are performed in about 2 seconds.

The presented application demonstrates the effectiveness of the proposed heuristic approach in planning the itinerary for the one-day tourist, customizing the outward and return paths of the roundtrip in the city of Bari. In particular, the application can manage the map of city PoIs, showing the travel means and times for each PoI pair, decided based on the user preferences. The tourist preferences can initially be input through specific app pages like presented in Figures 11 and 12, where it is possible to decide which PoIs to be visited in the outward and return paths, respectively, and the preferred traveling mode. In particular, in the proposed case study it is evident how the Algorithm 1.1 recovers the initial unfeasible solution providing to the tourist alternative feasible trips including all the high priority PoIs. After the user choice, made as in the app page in Figure 20, Algorithm 1.3 further improves the solution adding node 3 belonging to secondary priority list, given one more PoI to be visited.

Let us remark that the obtained solution by applying the proposed procedure can be sub-optimal because of the human intervention. Indeed, with respect to a classical orienteering problem, the user subdivides the PoIs between outward and return trip and can choose a feasible itinerary according to the preferences affecting the real time procedure. Nevertheless, even if the obtained solution can be non optimal, it is surely customized based on the user preferences. Moreover, in Section 4, we enlighten that the proposed heuristic procedure shows better performances than other algorithms such as D-algorithm and S-algorithm [49] in some specific cases.

6. Conclusions

This paper is aimed at providing a tool to help the one-day tourist in the difficult choice to plan an itinerary in a city. Indeed, a tourist often renounces relying on professionals of the sector who offer a service although complete, often pre-packaged and not taking into account her/his passions and preferences.

The proposed approach builds a network of points of interest (PoIs), proposing the city attractions but leaving the choice of the PoIs to be visited by the tourist. Based on the obtained graph, a multi-algorithms approach is provided to determine the optimal itinerary. The tourist is part of the multi-algorithms approach interacting with it and taking decisions.

An Integer Linear Programming (ILP) problem is introduced to find the optimal outward and return paths of the touristic itinerary and the multi-algorithms strategy is used to maximize the number of PoIs to be visited in the paths. Moreover, a case study demonstrates the approach efficiency and the steps of the procedure.

Finally, an app prototype has been developed and several use cases are being tested to iron out bugs before writing the final code. Future works will focus on useful developments of the application: integrating the possibility of making reservations at hotels, purchasing entrance tickets for the various sites and promoting sustainable mobility by providing simple but complete and updated information on how to get around the city (bus/metro time schedule, opening/closing hours of attractions, etc.); locating bike-sharing stations and bicycle parking; considering the cost as additional objective function to be minimized. Furthermore, Algorithm 1.2 will be object of further studies in order to improve its performance.

Author Contributions: Conceptualization, A.M.M.; methodology, M.R.; software, A.R.; validation, A.M.M., M.R. and A.R.; formal analysis, A.R.; investigation, M.R.; resources, A.M.M.; data curation, A.R.; writing–original draft preparation, A.M.M. and M.R.; writing–review and editing, M.R. and A.R.; visualization, A.R.; supervision, M.R.; project administration, A.M.M.; funding acquisition, A.M.M. All authors have read and agreed to the published version of the manuscript.

Funding: This research received no external funding.

Institutional Review Board Statement: Not applicable.

Informed Consent Statement: Not applicable.

Data Availability Statement: Not applicable.

Conflicts of Interest: The authors declare no conflict of interest.

Abbreviations

The following abbreviations are used in this manuscript:

TSP	Traveling Salesman Problem
PoI	Point of Interest
UML	Unified Modeling Language
ILP	Integer Linear Programming

References

1. De Falco, I.; Scafuri, U.; Tarantino, E. Optimizing personalized touristic itineraries by a multiobjective evolutionary algorithm. *Int. J. Inf. Technol. Decis. Mak.* **2016**, *15*, 1269–1312. [CrossRef]
2. Angskun, T.; Angskun, J. A travel planning optimization under energy and time constraints. In Proceedings of the 2009 International Conference on Information and Multimedia Technology, Jeju, Korea, 16–18 December 2009; pp. 131–134.
3. Diosteanu, A.; Cotfas, L.A.; Smeureanu, A.; Dumitrescu, S.D. Natural language processing applied in itinerary recommender systems. In Proceedings of the 10th WSEAS International Conference on Applied Computer and Applied Computational Science, Venice, Italy, 8–10 March 2011; World Scientific and Engineering Academy and Society (WSEAS): Athens, Greece, 2011; pp. 260–265.
4. Vansteenwegen, P.; Souffriau, W.; Berghe, G.V.; Van Oudheusden, D. Iterated local search for the team orienteering problem with time windows. *Comput. Oper. Res.* **2009**, *36*, 3281–3290. [CrossRef]
5. Souffiau, W.; Maervoet, J.; Vansteenwegen, P.; Berghe, G.V.; Van Oudheusden, D. A mobile tourist decision support system for small footprint devices. In Proceedings of the International Work-Conference on Artificial Neural Networks, Salamanca, Spain, 10–12 June 2009; pp. 1248–1255.
6. Booth, J.; Sistla, P.; Wolfson, O.; Cruz, I.F. A data model for trip planning in multimodal transportation systems. In Proceedings of the 12th International Conference on Extending Database Technology: Advances in Database Technology, Saint Petersburg, Russia, 24–26 March 2009; pp. 994–1005.
7. Navabpour, S.; Ghoraie, L.S.; Malayeri, A.A.; Chen, J.; Lu, J. An intelligent traveling service based on SOA. In Proceedings of the 2008 IEEE Congress on Services-Part I, Honolulu, HI, USA, 6–11 July 2008; pp. 191–198.
8. André, P.; Wilson, M.L.; Owens, A.; Smith, D.A. Journey planning based on user needs. In Proceedings of the CHI'07 Extended Abstracts on Human Factors in Computing Systems, San Jose, CA, USA, 28 April–3 May 2007; pp. 2025–2030.
9. Gonzalez, H.; Han, J.; Li, X.; Myslinska, M.; Sondag, J.P. Adaptive fastest path computation on a road network: A traffic mining approach. In Proceedings of the 33rd International Conference on Very Large Data Bases, VLDB 2007, Vienna, Austria, 23–27 September 2007; Association for Computing Machinery, Inc.: New York, NY, USA, 2007; pp. 794–805.
10. Garcia, A.; Linaza, M.T.; Arbelaitz, O.; Vansteenwegen, P. *Intelligent Routing System for a Personalised Electronic Tourist Guide*; ENTER: Amsterdam, The Netherlands, 2009; pp. 185–197.

11. Gavalas, D.; Konstantopoulos, C.; Mastakas, K.; Pantziou, G. A survey on algorithmic approaches for solving tourist trip design problems. *J. Heuristics* **2014**, *20*, 291–328. [CrossRef]

12. Gunawan, A.; Lau, H.C.; Vansteenwegen, P. Orienteering problem: A survey of recent variants, solution approaches and applications. *Eur. J. Oper. Res.* **2016**, *255*, 315–332. [CrossRef]

13. Kai, W. Operational Research Problems in Tour Itinerary Design and Optimization. *Tour. Sci.* **2004**, *1*. Available online: https://en.cnki.com.cn/Article_en/CJFDTotal-LUYX200401008.htm (accessed on 10 September 2021).

14. Han, Y.; Guan, H.; Duan, J. Tour route multiobjective optimization design based on the tourist satisfaction. *Discret. Dyn. Nat. Soc.* **2014**, *2014*, 603494. [CrossRef]

15. Kenteris, M.; Gavalas, D.; Pantziou, G.; Konstantopoulos, C. Near-optimal personalized daily itineraries for a mobile tourist guide. In Proceedings of the IEEE symposium on Computers and Communications, Riccione, Italy, 22–25 June 2010; pp. 862–864.

16. Cotfas, L.A. Collaborative itinerary recommender systems. *Acad. Econ. Stud. Econ. Inform.* **2011**, *11*, 191.

17. Vansteenwegen, P.; Souffriau, W.; Berghe, G.V.; Van Oudheusden, D. The city trip planner: An expert system for tourists. *Expert Syst. Appl.* **2011**, *38*, 6540–6546. [CrossRef]

18. Evans, J.R. *Graph Theory and Networks*; Industrial Engineering-New York Basel-Marcel Dekker Incorporated: New York, NY, USA, 1996; Volume 20, pp. 271–306.

19. Gavalas, D.; Kasapakis, V.; Konstantopoulos, C.; Pantziou, G.; Vathis, N. Scenic route planning for tourists. *Pers. Ubiquitous Comput.* **2017**, *21*, 137–155. [CrossRef]

20. Geng, J.; Ye, Q.; Wu, D.; Yang, H. Research of the evaluation and optimization of tourism route based on graph theory in Tibet. *Areal Res. Dev.* **2011**, *30*, 104–109.

21. Chen, J.; Tang, D. Tour routes optimization based on Graph theory for improving instruction services in scenic spots. In Proceedings of the ICSSSM11, Tianjin, China, 25–27 June 2011; pp. 1–3.

22. Le-Klaehn, D.T.; Hall, C.M. Tourist use of public transport at destinations—A review. *Curr. Issues Tour.* **2015**, *18*, 785–803. [CrossRef]

23. Van Truong, N.; Shimizu, T. The effect of transportation on tourism promotion: Literature review on application of the Computable General Equilibrium (CGE) Model. *Transp. Res. Procedia* **2017**, *25*, 3096–3115. [CrossRef]

24. Tussyadiah, I. A review of research into automation in tourism: Launching the Annals of Tourism Research Curated Collection on Artificial Intelligence and Robotics in Tourism. *Ann. Tour. Res.* **2020**, *81*, 102883. [CrossRef]

25. Gavalas, D.; Konstantopoulos, C.; Mastakas, K.; Pantziou, G.; Vathis, N. Heuristics for the time dependent team orienteering problem: Application to tourist route planning. *Comput. Oper. Res.* **2015**, *62*, 36–50. [CrossRef]

26. Pellegrini, A.; Scagnolari, S. The relationship between length of stay and land transportation mode in the tourism sector: A discrete–continuous framework applied to Swiss data. *Tour. Econ.* **2021**, *27*, 243–259. [CrossRef]

27. Abbaspour, R.; Samadzadegan, F. Itinerary planning in multimodal urban transportation network. *J. Appl. Sci.* **2009**, *9*, 1898–1906. [CrossRef]

28. Abbaspour, R.A.; Samadzadegan, F. Time-dependent personal tour planning and scheduling in metropolises. *Expert Syst. Appl.* **2011**, *38*, 12439–12452. [CrossRef]

29. Gavalas, D.; Kasapakis, V.; Konstantopoulos, C.; Pantziou, G.; Vathis, N.; Zaroliagis, C. The eCOMPASS multimodal tourist tour planner. *Expert Syst. Appl.* **2015**, *42*, 7303–7316. [CrossRef]

30. Wu, X.; Guan, H.; Han, Y.; Ma, J. A tour route planning model for tourism experience utility maximization. *Adv. Mech. Eng.* **2017**, *9*, 1687814017732309. [CrossRef]

31. Zheng, W.; Liao, Z.; Lin, Z. Navigating through the complex transport system: A heuristic approach for city tourism recommendation. *Tour. Manag.* **2020**, *81*, 104162. [CrossRef]

32. Zhang, R.; Liu, Z.; Feng, X. A novel flexible shuttle vehicle scheduling problem in scenic areas: Task-divided graph-based formulation and algorithm. *Comput. Ind. Eng.* **2021**, *156*, 107295. [CrossRef]

33. Kargar, M.; Lin, Z. A socially motivating and environmentally friendly tour recommendation framework for tourist groups. *Expert Syst. Appl.* **2021**, *180*, 115083. [CrossRef]

34. Wang, Y.W.; Lin, C.C.; Lee, T.J. Electric vehicle tour planning. *Transp. Res. Part D Transp. Environ.* **2018**, *63*, 121–136. [CrossRef]

35. Karbowska-Chilinska, J.; Chociej, K. Optimization of multistage tourist route for electric vehicle. In *Computer Science On-Line Conference*; Springer: Cham, Switzerland, 2018; pp. 186–196.

36. Karbowska-Chilinska, J.; Chociej, K. Genetic Algorithm for Generation Multistage Tourist Route of Electrical Vehicle. In Proceedings of the International Conference on Computer Information Systems and Industrial Management, Bialystok, Poland, 16–18 October 2020; pp. 366–376.

37. Sylejmani, K.; Dorn, J.; Musliu, N. Planning the trip itinerary for tourist groups. *Inf. Technol. Tour.* **2017**, *17*, 275–314. [CrossRef]

38. Malucelli, F.; Giovannini, A.; Nonato, M. Designing single origin-destination itineraries for several classes of cycle-tourists. *Transp. Res. Procedia* **2015**, *10*, 413–422. [CrossRef]

39. Korte, B.; Vygen, J. The traveling salesman problem. In *Combinatorial Optimization*; Springer: Cham, Switzerland, 2012; pp. 557–592.

40. Miller, C.E.; Tucker, A.W.; Zemlin, R.A. Integer Programming Formulation of Traveling Salesman Problems. *J. ACM* **1960**, *7*, 326–329. [CrossRef]

41. Abbatecola, L.; Fanti, M.P.; Mangini, A.M.; Ukovich, W. A decision support approach for postal delivery and waste collection services. *IEEE Trans. Autom. Sci. Eng.* **2016**, *13*, 1458–1470. [CrossRef]

42. Silvestri, B.; Rinaldi, A.; Berardi, A.; Roccotelli, M.; Acquaviva, S.; Fanti, M.P. A Serious Game Approach for the Electro-Mobility Sector. In Proceedings of the 2019 IEEE International Conference on Systems, Man and Cybernetics (SMC), Bari, Italy, 6–9 October 2019; pp. 674–679.

43. Souffriau, W.; Vansteenwegen, P. Tourist trip planning functionalities: State–of–the–art and future. In *Proceedings of the International Conference on Web Engineering*; Springer: Berlin/Heidelberg, Germany, 2010; pp. 474–485.

44. Sylejmani, K.; Dika, A. A survey on tourist trip planning systems. *Int. J. Arts Sci.* **2011**, *4*, 13.

45. Bender, E.A.; Williamson, S.G. *Lists, Decisions and Graphs*; Williamson, S.G., Ed.; University of California: San Diego, CA, USA, 2010.

46. Costantino, N.; Dotoli, M.; Falagario, M.; Fanti, M.P.; Mangini, A.M.; Sciancalepore, F.; Ukovich, W. A fuzzy programming approach for the strategic design of distribution networks. In Proceedings of the 2011 IEEE International Conference on Automation Science and Engineering, Trieste, Italy, 24–27 August 2011; pp. 66–71.

47. Pender, T. *UML Bible*; John Wiley & Sons: Hoboken, NJ, USA, 2003.

48. Nilsson, C. Heuristics for the traveling salesman problem. *Linkop. Univ.* **2003**, *38*, 85–89.

49. Tsiligirides, T. Heuristic methods applied to orienteering. *J. Oper. Res. Soc.* **1984**, *35*, 797–809. [CrossRef]

50. The Orienteering Problem: Test Instances Benchmark Dataset. Available online: https://www.mech.kuleuven.be/en/cib/op (accessed on 10 September 2021).

Article

Long Term Household Electricity Demand Forecasting Based on RNN-GBRT Model and a Novel Energy Theft Detection Method

Santanu Kumar Dash [1], Michele Roccotelli [2],*, Rasmi Ranjan Khansama [3], Maria Pia Fanti [2] and Agostino Marcello Mangini [2]

[1]   TIFAC-CORE, Vellore Institute of Technology (VIT), Vellore 632014, India; santanu4129@gmail.com or santanukumar.dash@vit.ac.in

[2]   Department of Electrical and Information Engineering, Politecnico di Bari, 70126 Bari, Italy; mariapia.fanti@poliba.it (M.P.F.); agostinomarcello.mangini@poliba.it (A.M.M.)

[3]   Department of Computer Science, C. V. Raman Global University, Bhubaneswar 752054, India; rasmiranjankhansama@gmail.com

*   Correspondence: michele.roccotelli@poliba.it

Citation: Dash, S.K.; Roccotelli, M.; Khansama, R.R.; Fanti, M.P.; Mangini, A.M. Long Term Household Electricity Demand Forecasting Based on RNN-GBRT Model and a Novel Energy Theft Detection Method. *Appl. Sci.* **2021**, *11*, 8612. https://doi.org/10.3390/app11188612

Academic Editor: Luis Hernández-Callejo

Received: 30 June 2021
Accepted: 13 September 2021
Published: 16 September 2021

Abstract: The long-term electricity demand forecast of the consumer utilization is essential for the energy provider to analyze the future demand and for the accurate management of demand response. Forecasting the consumer electricity demand with efficient and accurate strategies will help the energy provider to optimally plan generation points, such as solar and wind, and produce energy accordingly to reduce the rate of depletion. Various demand forecasting models have been developed and implemented in the literature. However, an efficient and accurate forecasting model is required to study the daily consumption of the consumers from their historical data and forecast the necessary energy demand from the consumer's side. The proposed recurrent neural network gradient boosting regression tree (RNN-GBRT) forecasting technique allows one to reduce the demand for electricity by studying the daily usage pattern of consumers, which would significantly help to cope with the accurate evaluation. The efficiency of the proposed forecasting model is compared with various conventional models. In addition, by the utilization of power consumption data, power theft detection in the distribution line is monitored to avoid financial losses by the utility provider. This paper also deals with the consumer's energy analysis, useful in tracking the data consistency to detect any kind of abnormal and sudden change in the meter reading, thereby distinguishing the tampering of meters and power theft. Indeed, power theft is an important issue to be addressed particularly in developing and economically lagging countries, such as India. The results obtained by the proposed methodology have been analyzed and discussed to validate their efficacy.

Keywords: time series analysis; energy demand forecast; ARIMA; hybrid model; power theft

1. Introduction

With the current increase in global warming, the focus of energy dependency has moved towards renewable energy sources (RESs), which seemingly have zero emission of greenhouse gases. As the percentage of carbon footprint rises with the use of traditional sources of energy, such as coal, the utilization of solar or hydro energy helps in reducing carbon footprints, providing a green energy alternative [1,2]. In order to ensure a pollution free ecosystem, we must move towards the utilization of RESs. Hence, a proper investment in RESs is essential [1–3].

RESs are integrated to the existing grid infrastructure to satisfy the energy demand of consumers, reducing the need for power from the main grid [1]. At the same time, storage devices are essential to optimize the use of renewable energy by storing energy when available and supplying it to consumers according to their requirement. Hence, for an optimal use of the energy from RESs, it is important to predict the energy demand of customers based on historical measurements. However, the evaluation of energy demand from the consumer load side cannot be efficiently performed through any of the present energy meter reading techniques in India [4,5]. Many researchers have focused on various power measuring devices [6–8]. These literature studies point out that the dynamic consumer electricity requirement has become a challenge for the power grid and sudden inflation in power demand and future energy requirement by the various categories of load cannot be efficiently predicted.

To overcome this limitation, many forecasting techniques to be applied by using smart metering and control systems are proposed [9,10]. In this context, efficient strategies to address the current challenges should aim at analyzing the consumer's energy requirements in the past, forecasting the demand and finally generating and producing the energy [9–11]. Demand forecasting using time series can be useful for informing the control unit regarding the energy over time to be produced in the future [12–14]. This can significantly help to cope with the problem of managing the generation from different sources, such as RESs, produce and distribute power more efficiently according to demand, and lessen the fee of depletion.

In addition, to improve energy distribution efficiency and reduce financial losses, power theft detection has become necessary to reduce energy loss at the consumer as well as generation point. Due to the intensity of economic and financial losses it provides, power theft is a punishable offence across the globe [14–16]. Indeed, power theft has enhanced the financial losses of both the supply units and the consumers in many countries. As an example, the Indian government is losing its financial integrity and is continuously working on this issue to produce the required devices and systems for proper grid operation and distribution. Nevertheless, the Indian government has, as of yet, failed to eliminate the power theft practice. Hence, to curb such illegal practices, a proper theft tracking scheme is necessary [17,18]. Although many electricity meter schemes are introduced by researchers in the literature, an efficient solution for detecting the illegal power theft is desired. The work [19] deals with the internal power larceny issues by constantly tracking the data consistency to detect any kind of abnormal and sudden change in the meter reading. The societal impact of the work lies in the fact that the utilities can distinguish the source of theft and the real consumers. Ultimately, the Indian government can limit energy losses by curbing the illegal practices of power felonies, which would favor the financial and economic stability of the country [20,21].

In this context, the contribution of the paper is twofold. The first contribution is analyzing and forecasting the energy demand of the domestic household. The daily consumption of a household is predicted using and comparing various forecasting models and the energy demand for the next few days over a specific period of time is determined. The proposed RNN-GBRT hybrid model shows more accurate forecasting performance than other models. Thereby, creating awareness among the consumers and the utilities with regard to their energy usage is the major focus. The second contribution is towards identifying the power theft by constantly tracking the data consistency to detect any kind of abnormal and sudden change in the energy meter reading.

The rest of the paper is organized as follows. Section 2 presents the literature survey, wherein various papers are analyzed with respect to the proposed research work. Additionally, some of the limitations of the papers are highlighted. Section 3 presents the algorithms for demand forecasting and power theft detection, while their implementation is described in Section 4, which presents various simulation results and demonstrates the superiority of the proposed prediction model and the efficacy of the power theft strategy. Finally, Section 5 concludes the work and highlights the future scope.

2. Related Works

The issues related to variation in load demand at the consumer end have raised serious concerns for the grid utility provider. Therefore, analyzing the future energy requirement of the domestic end-user is essential both for short term and long term load demand management. Focusing on the raised problems, researchers have proposed different load demand forecasting models to resolve the issues related to dynamic consumer load demand. In this section, we have analyzed related works and their approaches for the load demand forecasting.

In this context, to analyze the consumer behavior and electricity consumption over time, ref. [10] proposes an innovative strategy based on cluster analysis with K-means algorithm. The authors have identified the time variable separated groups of individual electricity consumption patterns, which will help in predicting the consumer electricity requirements. Considering the climate change effect and utilization of clean energy resources for power generation, the long term load demand prediction and demand side management has been addressed in [11] for the Taiwan government. The photovoltaic energy generation prediction and its demand have been analyzed by the authors of [12] by the application of R programming to estimate its efficiency for a Photovoltaic (PV) based microgrid. For environmental and economic security, energy demand management and forecast are essential to satisfy the future energy grid needs [13]. Therefore, future energy demand prediction by the utilization of conventional and advanced intelligent methods has been discussed in [13,14] based on time series models and soft computing-based prediction models. In addition, energy demand prediction has also been studied for short term period applications. In particular, short term load forecasting is presented in [15,16] by studying residential behavior learning, explaining the decay of radial basis function (RBF) neural networks (DRNNs), and demonstrating that deep-learning forecasting framework-based models have higher efficiency than the traditional forecasting strategies. The utilization of a fuzzy logic regression method for short term load forecasting is presented in [17], using previous three-year data with high accuracy. For the intraday electricity demand prediction in highly developed countries from Europe has been discussed in [22]. This paper has focused on the shot term a day ahead prediction based on principal component analysis of the daily demand profiles. The short term load demand prediction by the application of deep learning network-based CNN and multi-layer bi-directional LSTM prediction method for a household has been analyzed in [23]. This methodology has shown its effectiveness in achieving the lowest value of root mean square error and mean square error for an individual household. The utilization of the LSTM Network based on artificial neural networks to predict the long term electricity demand in Poland is presented in [24]. This method focused on cost management and has been adopted for strategic plan development in electric power system of Poland. As the deep learning approach has become popular among researchers for developing energy demand forecasting models, in [25], deep learning approaches including deep neural networks and long short term memory have been considered. This work extensively verified the efficiency of deep neural network-based prediction methods under different consumer electricity patterns. Bio-inspired algorithmic models have also been adopted to develop efficient energy forecasting models. In this context, [26,27] have investigated the utilization of autoregressive neural network based on genetic algorithm and particle swarm optimization for development of energy prediction models. The adopted hybrid method shows its effectiveness in energy prediction compared with other bio inspired algorithms. Moreover, the Autoregressive integrated moving average (ARIMA) model is a popular methodology for load demand evaluation and forecasting [28], and hybrid ARIMA strategies have been studied and implemented to achieve higher efficiency in [28–31]. In particular, for a Brazilian northeast company [28], the energy demand based on natural gas has been studied and predicted by applying an ARIMA-ANN based hybrid methodology. Furthermore, for the prediction of a single day ahead energy price, ARIMA and Holt-Winters forecasting models have been implemented in [29], achieving results that overcome those of the traditional models. Authors in [30] have proposed adaptive online ensemble learning with a recurrent neural network (RNN)-ARIMA-based hybrid model to reach the target of load demand prediction. Similarly, an ARIMA-GBRT (gradient boosting regression tree)-based hybrid model has also been developed and implemented in [31] to simulate and forecast the electrical energy consumption of residential buildings. The GBRT technique is used in combination with

ARIMA in the proposed model to improve the forecast performance with respect to existing prediction models. The comparison of forecasting models is performed based on standard performance indicators and demonstrates the superiority of the proposed ARIMA-GBRT model.

In addition, other contributions have proposed methods to identify the energy theft detection in power grids. In this context, the authors in [32,33] have studied the load profile of various customers to expose the degree of nontechnical losses. The study on load profile and load demand provided satisfactory data to analyze the possibilities of energy theft at the distribution point. In [33,34], a solution to the fraud activity by implementing the technique of decision tree is discussed. The authors of [35] used a time series neural network to identify the source of power tampering activity. Moreover, a machine learning algorithm allows the comparison of various local households to determine the number of honest customers and the rate of illegal consumers. In [36], the authors applied various classification models on regular energy expenditure data and encoded data. A comparison of accuracies of the adopted models is also presented. In addition, contributions [37–41] have studied how the power theft phenomenon affects the consumers as well as the power grid.

The analysis of the related literature reveals that there is the lack of a unique methodology to predict energy consumption, especially for detecting illegal use of energy, and it does not emerge a suitable and effective strategy to be applied in this context. Therefore, the authors of this paper are motivated to analyze the performance of various demand forecasting techniques and propose an efficient method for energy demand prediction. In addition, in order to protect the domestic household from power theft, a unique methodology is adopted and analyzed.

3. Demand Forecasting Models and Power Theft Detection Strategy

In this section, the proposed architecture for the energy demand forecasting and power theft detection is presented in Figure 1, highlighting the main components and functionalities. In particular, the conventional forecasting models for demand forecasting in domestic household are the focus of this section. Existing models, as well as the proposed hybrid RNN-GBRT forecasting strategy, are presented and analyzed. The considered dataset for models training and testing is taken from a household in Sceaux (Paris) and includes a total of 727 daily energy consumption values from 16 December 2008 to 20 December 2017. Moreover, a new strategy is presented to provide an effective solution to the power theft detection problem.

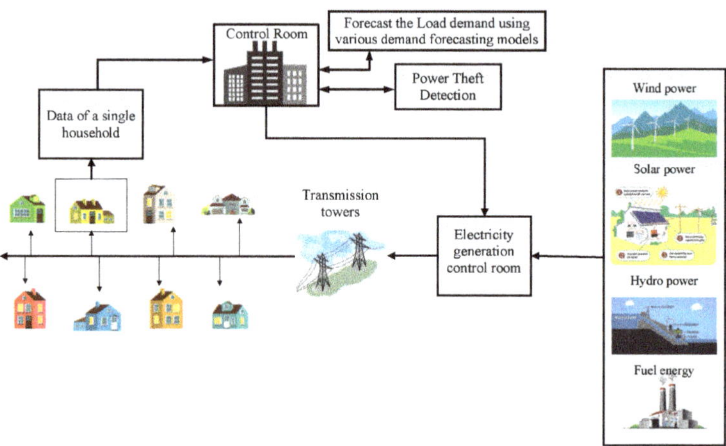

Figure 1. Architecture of the energy demand forecasting system.

3.1. ARIMA Model

Autoregressive integrated moving average (ARIMA) is a time series model designed to predict future points of an event with regard to its historic time series data [27,28]. The ARIMA model is generally considered with stationary data, wherein an initial differencing is applied several times to nonstationary data to attain a stationary time series. The AR part of ARIMA specifies that the generation variable repeats with respect to its lag values. The MA part of ARIMA specifies that the generation variable repeats with respect to its prior forecast errors obtained continuously. The first part of ARIMA indicates that the data values are restored by the difference value between original value and previous value. The aim of each of these features of ARIMA is to make the model fit the data. The model being used for prediction is an ARIMA model given by ARIMA $(p, d, q)_s$, where 'p' denotes the number of autoregressive terms, and 'd' represents number of seasonal differences required for stationarity. The 'q' is the number of lagged forecast errors in the prediction errors, and 's' is the number of lagged forecast errors in the periods per season (generally 12 in the present case). The structure of ARIMA model is shown in Figure 2.

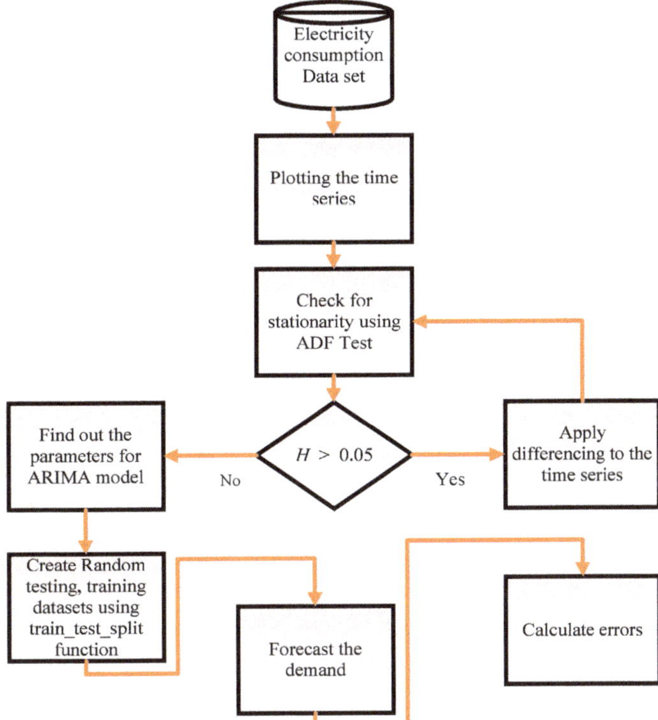

Figure 2. Architectural view of the ARIMA model.

The mathematical representations of AR and MA models can be given as a pure Auto Regressive (AR) model, where Z_t is only dependent on its own lag. In an AR model, Z_t is a function of the 'lag components of Z_t'. Hence, the mathematical representation of AR model becomes:

$$Z_t = \alpha + \beta_1 Z_{t-1} + \beta_2 Z_{t-2} + \ldots\ldots \beta_n Z_{t-n} \in_t + \ldots\ldots + \phi_1 \in_{t-1} + \phi_2 \in_{t-2} + \ldots + \phi_n \in_{t-n} \tag{1}$$

The architecture of an ARIMA model works on the basis of a time series analysis. Time series analysis (TSA) is a method to determine the futuristic trend of an event with a view of its past trend. The technique is based on the assumption that the future trend will hold similar to the historical trend. TSA focuses on two aspects, which are the identification of the nature of event (with respect to the series of observations) and the forecasting thereof.

Electricity demand forecasting by the utilization of ARIMA model consists of the following steps:

Step 1: *Collection of dataset:* Forecasting is always triggered by a set of values.

Step 2: *Observation of series:* The obtained series of data are plotted, and the pattern of series is observed. The underlying characteristics, such as trend, seasonality, and noise, are identified. Various break points and elevated points of the series are also observed. This is a crucial part of TSA, wherein the data, with respect to its time series, is thoroughly analyzed; various statistical components of the series are discovered.

Step 3: *Stationarity:* When the statistical properties of a series, such as mean, variance, autocorrelation, etc., are constant over time, then such a time series is stationary. The augmented Dickey–Fuller (ADF) test is the test for stationarity, where the time series must have a 'H' parameter less than 0.05 to be stationary [28].

Step 4: *Extraction of the optimal model parameters:* This can be done by identifying the Autocorrelation function (ACF) and partial autocorrelation function (PACF) plots. The autocorrelation function is the plot of autocorrelation of a time series by lag; partial autocorrelation is the relationship among the prior time observations of time series. In this step, all the parameters are checked, and the most appropriate models are selected.

Step 5: *Fitting Model:* Optimal model parameters are followed by the ARIMA model, which is fit to learn the series pattern. ARIMA(7,1,6) is the prediction model.

Step 6: *Prediction:* After fitting the model, the event is predicted. In the present research paper, the predicted test data is obtained and compared with the original dataset.

Step 7: *Determination of accuracy:* Finally, the various erroneous parameters are considered to determine the efficiency of the model. Various error parameters are tabulated in the results section. Generally, a time series data (S_m) is a combination of seasonality (L_m), trend (T_m), and noise (N_m) components, which can be determined as $S_m = L_m + T_m + N_m$, known as an additive model. This additive model does not show good performance for the household electricity consumption data prediction due to its dependence on seasonality.

Therefore, a multiplicative model, given as $S_m = L_m \times T_m \times N_m$, is preferable. The multiplicative model can be transformed into an additive model with the introduction of logarithms, given as:

$$\log S_m = L_m + \log T_m + \log N_m \tag{2}$$

3.2. Support Vector Regression Forecasting Model

Support vector network is a branch of machine learning that analyzes data with respect to operational learning techniques. Support vector regression (SVR) uses the principles of support vector machine (SVM), except the fact that SVR adjusts the prediction function with the threshold error. SVR tries to minimize the generalization error so as to achieve generalized performance. On the contrary, most of the other regression techniques try to decrease the observed error between the forecasted value and the original value. SVR is the most common application of SVM. SVR can be applied for time series prediction, financial forecasting, estimation of challenging engineering tasks, etc. SVR is classified into linear, polynomial, and rbf kernel. Due to its high accuracy, the rbf kernel model is considered in this paper.

In this paper, ε-SVR was implemented, and the value of ε was set as 0.2. The regularization parameter 'C' and the gamma 'γ' were defined through grid search.

The train test split function was used, with random classification of 70% of data as train set and using the remaining 30% as test set. The train set was fitted using rbf kernel, and forecasting of the data was achieved. The forecasted data was compared with the test set to verify the result, and mean square error (MSE), mean absolute error (MAE), and root mean squared error (RMSE) [35,36] indices were calculated. The following steps were implemented for the SVR model forecasting, as also shown in Figure 3:

Step 1: Create Testing and training datasets with the train_test_split function. Here, we divided it into 70% training dataset, 30% testing dataset.

Step 2: Build the support vector regression model using SVR function for RBF kernel with appropriate parameters for the training dataset.

Step 3: Forecast the consumption values for the testing dataset.

Step 4: Plot the actual and forecasted values of the testing dataset.

Step 5: Calculate MSE, MAE, and RMSE.

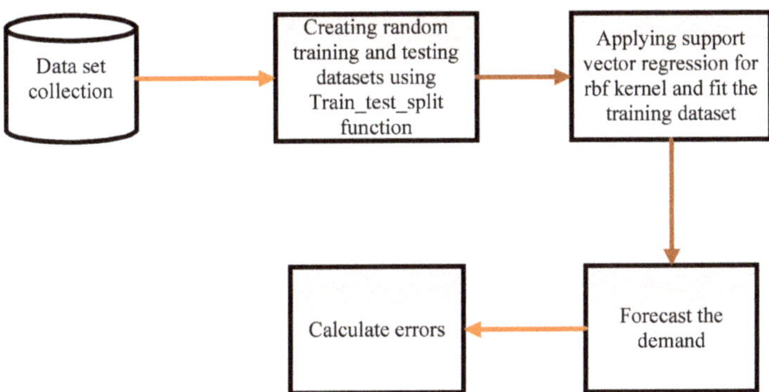

Figure 3. Architecture of support vector regression.

3.3. Linear Regression Forecasting Model

Linear regression is a concept derived from statistics. It is a linear technique for modelling the correlation between a dependent variable and one or more independent variables. If one independent variable is considered, then the process is a simple linear regression. On the contrary, if several independent variables are considered, then the process is a multiple linear regression. Unlike multivariate analysis, which entirely focuses on joint probability, linear regression focuses on conditional probability.

Linear regression is extensively used for the study of practical applications, since it is easy to fit the models that have linear dependence with its historic data. Hence, linear regression has greater significance in forecasting applications.

The simple linear regression model is typically formulated as $y = a + bx$; 'y' is the output, 'x' is the input, 'b' is the input coefficient, and 'a' is a constant. In case of multiple inputs, such as $x1, x2, x3$, the model representation is $y = a + b1x1 + b2x2 + b3x3$.

Additionally, in this case, 70% of data are randomly considered as train data, and the remaining 30% are considered as test data, using the train_test_split function. Finally, the energy demand with respect to the test data is predicted and compared with the original data. To determine the accuracy of the model, MSE, MAE, and RMSE are calculated. The architecture of linear regression forecasting is presented in Figure 4. The model implementation is done according to the listed steps:

Step 1: Create testing and training datasets with the train_test_split function. Here, we divided it into 70% training dataset, 30% testing dataset.

Step 2: Build the linear regression model using a linear regression function with appropriate parameters for the training dataset.

Step 3: Forecast the consumption values for the testing dataset.

Step 4: Plot the actual and forecasted values of the testing dataset.

Step 5: Calculate MSE, MAE, and RMSE.

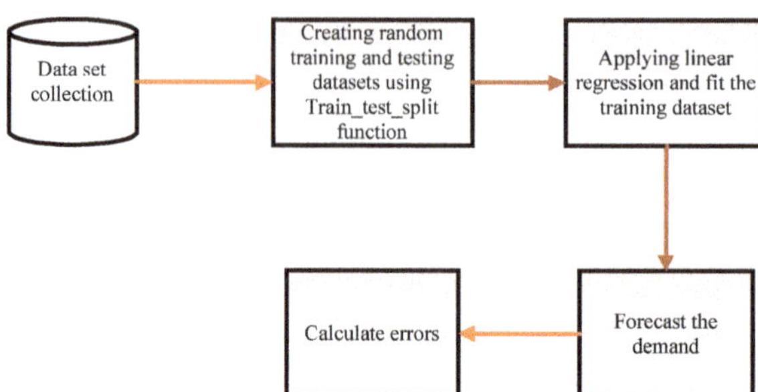

Figure 4. Architecture of linear regression.

3.4. Long Short Term Memory Model

The LSTM is a sort of RNN that is capable of remembering information for a significantly long period of time. In contrast to basic neural networks, where each node is characterized by a single activation function, each node in LSTM is employed as a memory cell that may store other information. LSTMs, in particular, have their own cell state, an input gate, an output gate and a forget gate. The cell remembers values over arbitrary time intervals, and the three gates regulate the flow of information into and out of the cell. These characteristic allow LSTMs address the problem of disappearing gradients from prior time-steps [23–25]. In our application, there are three LSTM layers, each with 40 units, a *tanh* activation function, and a drop out value of 0.15. The input sequence length is defined as 20.

Step 1: Create testing and training datasets with the train _test_split function. Here, we divided it into 70% training dataset, 30% testing dataset.

Step 2: Build the linear regression model using long short term memory model with predefined parameters for the training dataset.

Step 3: Forecast the consumption values for the testing dataset.

Step 4: Plot the actual and forecasted values of the testing dataset.

Step 5: Calculate MSE, MAE, and RMSE.

3.5. Recurrent Neural Networks Model

RNNs are networks that contain loops, which allow information to endure. They are utilized to model data that changes over time [30]. The data is fed into the network one by one, and the network's nodes save their current state at one time step and utilize it to influence the following time step. RNNs exploit the temporal information in the input data, and for this reason, are more suited to manage time series data. The ability of a RNN stands in using recurrent connections between neurons and can generally be described by the following equation [31]:

$$x_t = \begin{cases} 0, \text{if}(t=0) \\ \phi(x_{t-1}, a_t), \text{otherwise} \end{cases} \tag{3}$$

3.6. Proposed RNN-GBRT Hybrid Model

An RNN's goal is to forecast the next step in a sequence of observations in relation to the previous phases in the series [31]. In order to predict future trends, RNN makes use of consecutive observations and learns from previous phases. Data must be remembered throughout the early phases while estimating the following moves. The hidden layers in RNN serve as internal storage for the information obtained during the previous phases of the sequential data processing.

The GBRT algorithm, which is a mix of the CART (classification and regression trees) and GB (gradient boosting) algorithms, is also considered [31]. It is noted that the CART outperforms most artificial intelligence models in terms of prediction, since it can simulate nonlinear interactions without having previous knowledge of the probability distribution of variables. Inspired by the contribution of [30,31] we propose the hybrid model RNN-GBRT in order to exploit the advantages of the two methods and obtain better forecasting performances. In GBRT, the current iteration's model reduces the previous iteration's residuals. At each iteration, it builds a new regression tree to reduce residuals with the gradient descent of the objective function. In the proposed hybrid model, RNN-GBRT, the generated series after RNN forms the training examples for GBRT. Generally, the performance of GBRT depends on learning rate and the total number of regression trees. In this paper, the value of the learning rate is set from 0.1 to 0.3, and the total number of regression trees is set from 20 to 150.

3.7. Power Theft Detection Algorithm

Energy is the fundamental resource to make possible every application in domestic, commercial, and industrial environments. The electric grid refers to the combination of transformers, transmission lines, substations, and other components that make energy delivery possible, from the source layout to the field of work in each sector. The complexity of energy generation and distribution systems leads to the necessity of managing and solving several possible issues and challenges. In this context, one of the most significant issues to be addressed is power theft. Particularly in India, power theft is a serious issue that the country has dealt with for many years. No effective solutions have yet been found. A recent survey by the Central Electricity Authority suggests that over 27 percent of the total produced energy from various sources is lost due to the illegal practice of power felony [36]. Consequently, this affects over 5 percent of the country's GDP (gross domestic product). To address the issue, a new scheme is proposed in this paper for analyzing consumer energy and thereby detecting the source of theft, according to the procedure shown in Figure 5. To detect the source of power theft, the developed system is trained with recent data of energy consumption of the past 90 days from the available dataset. The system starts evaluating the mean of the collected data. The statistical mean of the given data is the sum of all the energy consumption values divided by its frequency, which is 90 in our case. The calculated mean 'A' is then stored in the memory manager of the proposed system. Afterwards, the system starts estimating the standard deviation of the data fed.

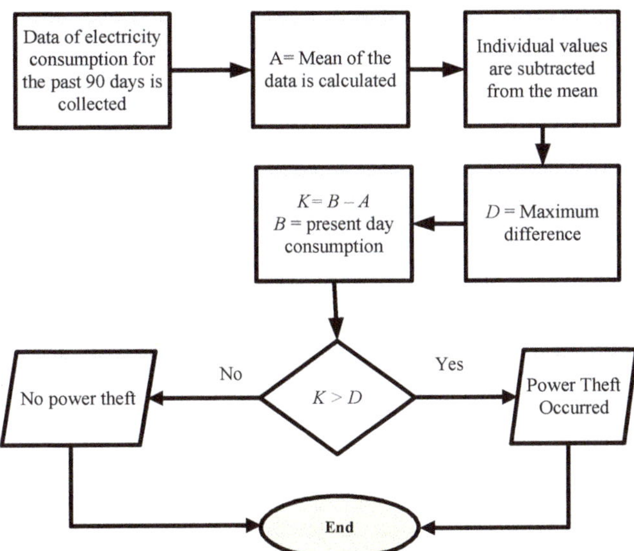

Figure 5. Flowchart of power theft detection procedure.

Standard deviation is defined as the individual differences of the data values with its mean. The purpose of standard deviation in our system is to define the degree of deviation between the adjacent data points. All the odd values of standard deviation are recorded. Among the recorded values of standard deviation, the values of maximum and minimum deviation are considered 'd' and 'D', respectively. Now, for testing the developed system, we consider the current day energy expenditure. In particular, the current energy expenditure 'B' is subtracted with the recorded mean value of the trained data 'A'. The obtained value is denoted by 'K'.

The final step of the theft detection scheme compares K and D values. If the K value is greater than the D value, then it can be concluded that power theft is occurring. The intensity of power theft can also be determined from the magnitude of calculated difference 'D'.

4. Implementation of Energy Forecasting and Power Theft Detection Models

Let us recall that the available dataset from Sceaux, Paris (France), is randomly classified into train and test datasets. The train dataset consists of 70% of total data, and the test dataset consists of the remaining 30% of total 727 values. The daily energy consumption values were forecasted by fitting ARIMA parameters (7,1,6). The household energy demand forecasting results, obtained by applying the ARIMA, linear regression, SVR, LSTM, and RNN models, are, respectively, presented in Figures 6–10.

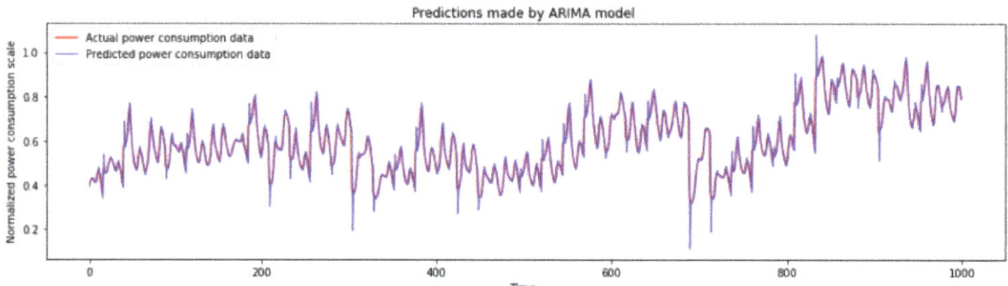

Figure 6. Forecast result of ARIMA model with respect to test dataset over time (day).

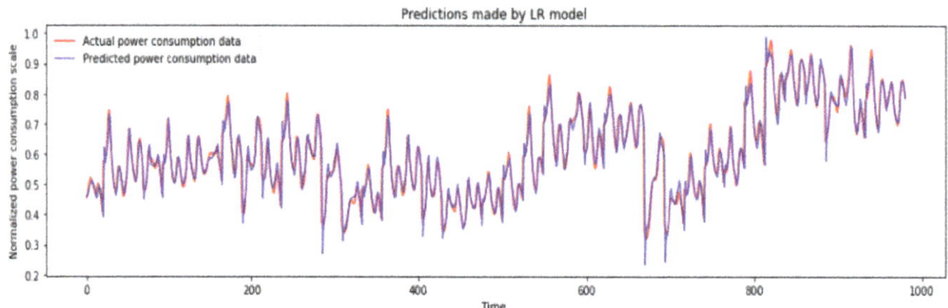

Figure 7. Forecast result of linear regression with respect to test dataset over time (day).

For better prediction performances, the authors propose the hybrid RNN-GBRT model, as described in Section 3.6. From the simulation results, it is evident that the RNN-GBRT model shows better prediction performance than other conventional methods, as presented in Figure 11.

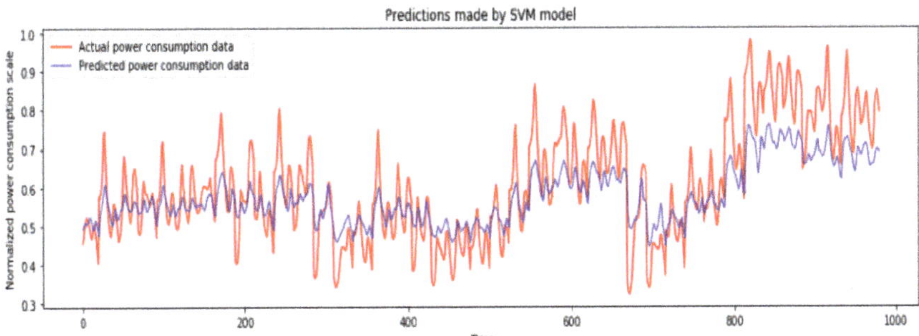

Figure 8. Forecast result of support vector model with respect to test dataset over time (day).

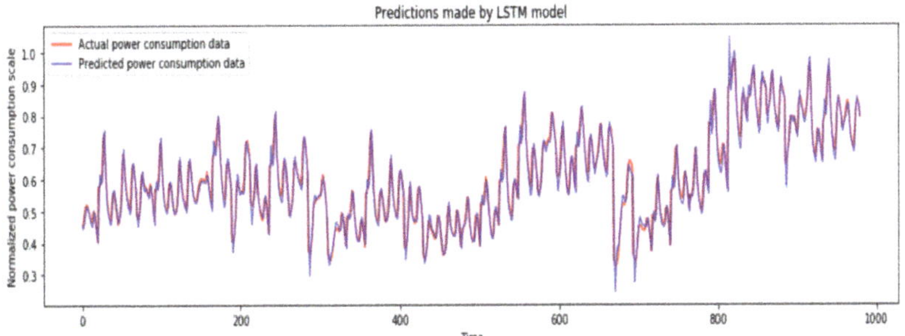

Figure 9. Forecast result of LSTM model with respect to test dataset over time (day).

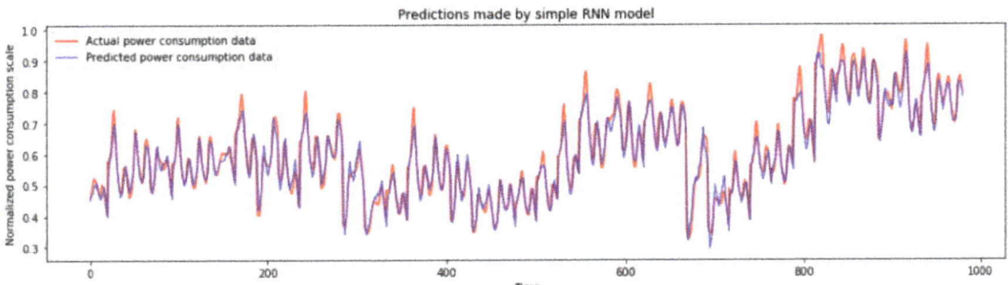

Figure 10. Forecast result of simple-RNN with respect to test dataset over time (day).

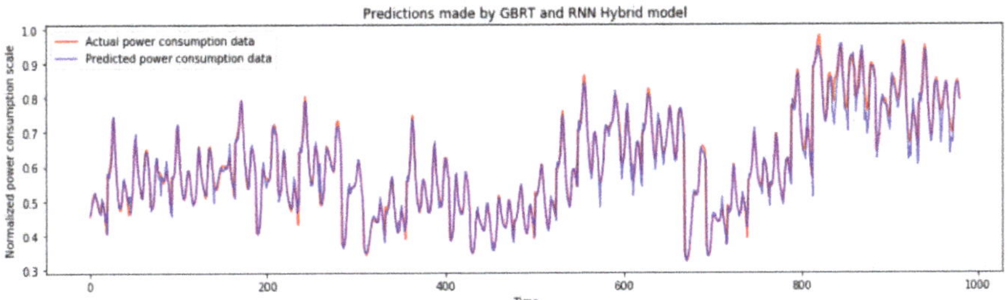

Figure 11. Forecast result of GBRT-RNN with respect to test dataset over time (day).

4.1. Comparative Analysis Based on Error Indices Calculations

In order to compare the prediction accuracy of the analyzed forecasting models, error indices can be used. With this aim, in this paper, three error indices are used to compare the efficiency of the analyzed prediction models. The first error index is the MSE [30], which is the mean squared difference between the estimated value and original value of the prediction technique. The second index is the MAE [31], which is the difference between the most similar observations of the model. Finally, the RMSE index [31] is calculated as the standard deviation of the residuals, representing the dispersion of residuals in the series. The comparison of the error values is presented in Figure 12. As is evident, the proposed RNN-GBRT performs better than other forecasting methods.

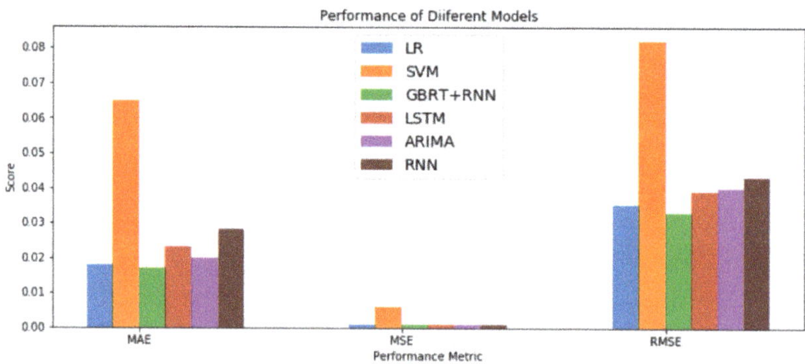

Figure 12. Comparison of Errors for accuracy analysis.

4.2. Power Theft Detection Results Analysis

The proposed scheme to detect the illegal practice of power theft is simulated in this section in order to show its effectiveness.

The range of theft detection is shown in Figure 13, where hourly data samples are evaluated, and no power theft occurs. The figure shows the threshold limit to detect power theft as 7.5 kW mean difference, as proposed in [34,35]. If the power consumption crosses the mentioned threshold, then users will receive a message regarding power theft that is occurring in their connection line. The utilization of a conventional distribution network and energy meters in the city of Vellore in India leads to the possibility of power theft events. Therefore, in order to provide the power theft case, the authors refer to a 727 day data set from Vellore city. In particular, Figure 14 shows the case of power theft detection where two power theft periods are highlighted.

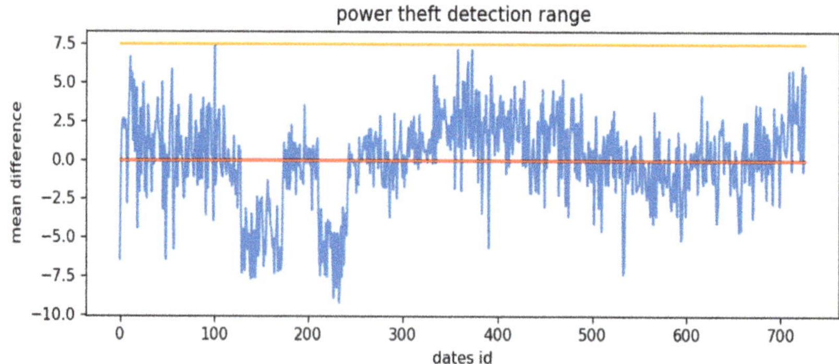

Figure 13. The optimal range of power theft (kW), no power theft detected.

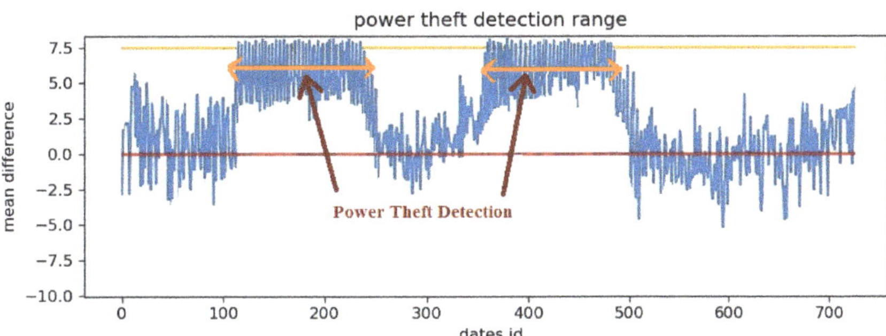

Figure 14. The case of power theft detection.

5. Conclusions

The exhaustible and inexhaustible sources of energy influence the economic growth of a country. The energy demand is a determinant of various functions, such as individual income, market structure, economic structure, lifestyle of individuals, and population change. The world may experience numerous challenges if there is uncertainty in energy supply in the near and far future. To determine the economic stability of a country, sustainable management of energy is needed. Therefore, the prediction of energy demand is of utmost importance for the uniform allocation of available resources relating to industrial production, healthcare, agriculture, population, accessibility of water, education, and quality of life. By forecasting the demand, we can accommodate the power generated using the available storage facilities. Incorporating this feature in the power grid will help in maintaining a balance between all the power sources. Therefore, this paper has proposed the hybrid RNN-GBRT model for forecasting the load demand, validating its efficiency. In particular, the comparative analysis among all the considered forecasting models is presented based on three error indices to evaluate the performance of the proposed hybrid model for energy forecasting.

Another important issue addressed in this paper is the power theft that can affect the quality of the energy distribution service and cause economic losses. In most of the sectors of energy distribution, a medium to excessive rate of larceny and medium to low rate of detection exist despite numerous technologies. The intensity of theft differs among several parts of the country. However, the detection and punishment of illegal consumers are extremely challenging tasks. Tracing power theft at the root can prove invaluable to the government's power sector. The proposed algorithm for individual household consumption tracking will be extremely helpful in notifying anomalies both to the user, who is paying the extra amount in his/her power bill, and to the government, which is losing power and money.

Future works will propose techniques to address the problem of forecasting the energy demand in a network of householders in a smart district context, managing different users and detecting possible power theft or grid malfunctioning within the district.

Author Contributions: Conceptualization, methodology, formal analysis, and writing—original draft preparation, S.K.D., R.R.K., M.R.; writing—review and editing, S.K.D. and M.R.; visualization, M.P.F.; supervision, M.R., M.P.F. and A.M.M. All authors have read and agreed to the published version of the manuscript.

Funding: This research received no external funding.

Institutional Review Board Statement: Not applicable.

Informed Consent Statement: Not applicable.

Data Availability Statement: Not applicable.

Conflicts of Interest: The authors declare no conflict of interest.

References

1. Sadorsky, P. Renewable energy consumption and income in emerging economies. *Energy Policy* **2009**, *37*, 4021–4028. [CrossRef]
2. Zou, C.; Zhao, Q.; Zhang, G.; Xiong, B. Energy revolution: From a fossil energy era to a new energy era. *Nat. Gas Ind. B* **2016**, *3*, 1–11. [CrossRef]
3. Wen, Y. *The Making of an Economic Superpower-Unlocking China's Secret of Rapid Industrialization*; World Scientific: Singapore, 2015; pp. 1–180.
4. Ghosh, S.; Manna, D.; Chatterjee, A.; Chatterjee, D. Remote Appliance Load Monitoring and Identification in a Modern Residential System With Smart Meter Data. *IEEE Sens. J.* **2021**, *21*, 5082–5090. [CrossRef]
5. Chakraborty, S.; Das, S. Application of Smart Meters in High Impedance Fault Detection on Distribution Systems. *IEEE Trans. Smart Grid* **2019**, *10*, 3465–3473. [CrossRef]
6. Haben, S.; Singleton, C.; Grindrod, P. Analysis and clustering of residential customers' energy behavioral demand using smart meter data. *IEEE Trans. Smart Grid* **2016**, *7*, 136–144. [CrossRef]
7. Palensky, P.; Dietrich, D. Demand side management: Demand response, intelligent energy systems, and smart loads. *IEEE Trans. Ind. Inform.* **2011**, *7*, 381–388. [CrossRef]
8. Tom, R.J.; Sankaranarayanan, S. IoT based SCADA Integrated with Fog for Power Distribution Automation. In Proceedings of the 12th Iberian Conference on Information Systems and Technologies, Lisbon, Portugal, 21–24 June 2017; pp. 1772–1775.
9. Sankar, V.J.; Hareesh, V.; Nair, M.G. Integration of Demand Response with Prioritized Load Optimization for Multiple Homes. In Proceedings of the 2017 International Conference on Technological Advancements in Power and Energy (TAP Energy), Kollam, India, 21–23 December 2017; pp. 1–6.
10. Cerquitelli, T.; Chicco, G.; Di Corso, E.; Ventura, F.; Montesano, G.; Del Pizzo, A.; González, A.M.; Sobrino, E.M. Discovering Electricity Consumption over Time for Residential Consumers through Cluster Analysis. In Proceedings of the International Conference on Development and Application Systems, Suceava, Romania, 24–26 May 2018; pp. 164–169.
11. Wang, C.-K.; Lee, C.-M.; Hong, Y.-R.; Cheng, K. Assessment of Energy Transition Policy in Taiwan—A View of Sustainable Development Perspectives. *Energies* **2021**, *14*, 4402. [CrossRef]
12. Vink, K.; Ankyu, E.; Kikuchi, Y. Long-Term Forecasting Potential of Photo-Voltaic Electricity Generation and Demand Using R. *Appl. Sci.* **2020**, *10*, 4462. [CrossRef]
13. Suganthi, L.; Samuel, A.A. Energy models for demand forecasting—A review. *Renew. Sustain. Energy Rev.* **2012**, *16*, 1223–1240. [CrossRef]
14. Singh, A.K.; Khatoon, S. An Overview of Electricity Demand Forecasting Techniques. In Proceedings of the 2012 Emerging Trends in Electrical, Instrumentation and Communication Engineering Conference, Uttar Pradesh, India, 6–7 April 2012.
15. Cecati, C.; Kolbusz, J.; Różycki, P.; Siano, P.; Wilamowski, B.M. A novel rbf training algorithm for short-term electric load forecasting and comparative studies. *IEEE Trans. Ind. Electron.* **2015**, *62*, 6519–6529. [CrossRef]
16. Kong, W.; Dong, Z.Y.; Hill, D.J.; Luo, F.; Xu, Y. Short-term residential load forecasting based on resident behaviour learning. *IEEE Trans. Power Syst.* **2018**, *33*, 1087–1088. [CrossRef]
17. Song, K.B.; Baek, Y.S.; Hong, D.H.; Jang, G. Short-term load forecasting for the holidays using fuzzy linear regression method. *IEEE Trans. Power Syst.* **2005**, *20*, 96–101. [CrossRef]
18. Ceperic, E.; Ceperic, V.; Baric, A. A Strategy for Short-Term Load Forecasting by Support Vector Regression Machines. *IEEE Trans. Power Syst.* **2013**, *28*, 4356–4364. [CrossRef]
19. Amjady, N. Short-term hourly load forecasting using time-series modeling with peak load estimation capability. *IEEE Trans. Power Syst.* **2001**, *16*, 498–505. [CrossRef]
20. Mohsenian-Rad, A.H.; Leon-Garcia, A. Optimal residential load control with price prediction in real-time electricity pricing environments. *IEEE Trans. Smart Grid* **2010**, *1*, 120–133. [CrossRef]
21. Ugurlu, U.; Oksuz, I.; Tas, O. Electrical price forecasting using recurrent neural networks. *Energy* **2018**, *11*, 1255.
22. Taylor, J.W.; McSharry, P.E. Short-term load forecasting methods: An evaluation based on European data. *IEEE Trans. Power Syst.* **2007**, *22*, 2213–2219. [CrossRef]
23. Ullah, F.U.M.; Ullah, A.; Haq, I.U.; Rho, S.; Baik, S.W. Short-term prediction of residential power energy consumption via CNN and multi-layer bi-directional LSTM networks. *IEEE Access* **2020**, *8*, 123369–123380. [CrossRef]
24. Manowska, A. Using the LSTM Network to Forecast the Demand for Electricity in Poland. *Appl. Sci.* **2020**, *10*, 8455. [CrossRef]
25. Son, N.; Yang, S.; Na, J. Deep Neural Network and Long Short-Term Memory for Electric Power Load Forecasting. *Appl. Sci.* **2020**, *10*, 6489. [CrossRef]
26. Le Cam, M.; Daoud, A.; Zmeureanu, R. Forecasting electric demand of supply fan using data mining techniques. *Energy* **2016**, *101*, 541–557. [CrossRef]
27. Yu, S.-W.; Zhu, K.-J. A hybrid procedure for energy demand forecasting in China. *Energy* **2012**, *37*, 396–404. [CrossRef]
28. Cardoso, C.A.V.; Cruz, G.L. Forecasting Natural Gas Consumption using ARIMA Models and Artificial Neural Networks. *IEEE Lat. Am. Trans.* **2016**, *14*, 2233–2238. [CrossRef]
29. Bissing, D.; Klein, M.T.; Chinnathambi, R.A.; Selvaraj, D.F.; Ranganathan, P. A Hybrid Regression Model for Day-Ahead Energy Price Forecasting. *IEEE Access* **2019**, *7*, 36833–36842. [CrossRef]
30. Jagait, R.K.; Fekri, M.N.; Grolinger, K.; Mir, S. Load Forecasting Under Concept Drift: Online Ensemble Learning With Recurrent Neural Network and ARIMA. *IEEE Access* **2021**, *9*, 98992–99008. [CrossRef]

31. Nie, P.; Roccotelli, M.; Fanti, M.P.; Ming, Z.; Li, Z. Prediction of home energy consumption based on gradient boosting regression tree. *Energy Rep.* **2021**, *7*, 1246–1255. [CrossRef]
32. Glauner, P.; Boechat, A.; Dolberg, L.; State, R.; Bettinger, F.; Rangoni, Y.; Duarte, D. Large-scale detection of non-technical losses in imbalanced data sets. Presented at the Innovative Smart Grid Technologies Conference (ISGT), 2016 IEEE Power & Energy Society, Minneapolis, MN, USA, 6–9 September 2016; pp. 1–5.
33. Glauner, P.; Meira, J.A.; Valtchev, P.; State, R.; Bettinger, F. The Challenge of Non-Technical Loss Detection Using Artificial Intelligence: A Survey. *Int. J. Comput. Intell. Syst.* **2017**, *10*, 760–775. [CrossRef]
34. Depuru, S.S.S.R.; Wang, L.; Devabhaktuni, V. Electricity theft: Overview, issues, prevention and a smart meter based approach to control theft. *Energy Policy* **2011**, *39*, 1007–1015. [CrossRef]
35. Richardson, C.; Race, N.; Smith, P. A privacy preserving approach to energy theft detection in smart grids. In Proceedings of the 2016 IEEE International Smart Cities Conference, Trento, Italy, 12 September 2016.
36. Salinas, S.A.; Li, P. Privacy-preserving energy theft detection in microgrids: A state estimation approach. *IEEE Trans. Power Syst.* **2016**, *31*, 883–894. [CrossRef]
37. Luan, W.; Wang, G.; Yu, Y.; Lin, J.; Zhang, W.; Liu, Q. Energy theft detection via integrated distribution state estimation based on AMI and SCADA measurements. In Proceedings of the International Conference on Electric Utility Deregulation and Restructuring and Power Technologies, Changsha, China, 26–29 November 2015; pp. 751–756.
38. Su, C.L.; Lee, W.H.; Wen, C.K. Electricity theft detection in low voltage networks with smart meters using state estimation. In Proceedings of the IEEE International Conference on Industrial Technology, Taipei, Taiwan, 14–17 March 2016; pp. 493–498.
39. Huang, S.C.; Lo, Y.L.; Lu, C.N. Non-technical loss detection using state estimation and analysis of variance. *IEEE Trans. Power Syst.* **2013**, *28*, 2959–2966. [CrossRef]
40. Maurya, A.; Akyurek, A.S.; Aksanli, B.; Rosing, T.S. Rosing, Time series clustering for data analysis in Smart Grid. In Proceedings of the 2016 IEEE International Conference on Smart Grid Communications, Sydney, Australia, 6–9 November 2016; pp. 606–611.
41. Fanti, M.P.; Mangini, A.M.; Roccotelli, M.; Nolich, M.; Ukovich, W. Modeling Virtual Sensors for Electric Vehicles Charge Services. In Proceedings of the 2018 IEEE International Conference on Systems, Man, and Cybernetics (SMC), Miyazaki, Japan, 7–10 October 2018; pp. 3853–3858.

applied sciences

Article

Design and Experimental Validation of Power Electric Vehicle Emulator for Testing Electric Vehicle Supply Equipment (EVSE) with Vehicle-to-Grid (V2G) Capability

Eduardo García-Martínez [1,*], Jesús Muñoz-Cruzado-Alba [1], José F. Sanz-Osorio [2] and Juan Manuel Perié [1]

[1] Fundación CIRCE, Parque Empresarial Dinamiza, Avenida Ranillas Edificio 3D, 1ª Planta, 50018 Zaragoza, Spain; jmunoz@fcirce.es (J.M.-C.-A.); jmperie@fcirce.es (J.M.P.)
[2] Instituto Universitario de Investigación CIRCE, Universidad de Zaragoza, Edificio CIRCE, Campus Río Ebro, C/Mariano Esquillor Gómez 15, 50018 Zaragoza, Spain; jfsanz@unizar.es
[*] Correspondence: egarcia@fcirce.es

Abstract: Nowadays, the global decarbonization and electrification of the world's energy demands have led to the quick adoption of Electric Vehicle (EV) technology. Therefore, there is an urgent need to provide a wide network of fast Vehicle-to-Grid (V2G) charging stations to support the forecast demand and to enable enough autonomy of such devices. Accordingly, V2G charging stations must be prepared to work properly with every manufacturer and to provide reliable designs and validation processes. In this way, the development of power electric vehicle emulators with V2G capability is critical to enable such development. The paper presents a complete design of a power electric vehicle emulator, as well as an experimental testbench to validate the behaviour of the proposal.

Keywords: electric vehicle; vehicle-to-grid; vehicle-to-building; electric vehicle emulator; electric vehicle supply equipment; smart grid

Citation: García-Martínez, E.; Muñoz-Cruzado-Alba, J.; Sanz-Osorio, J.F.; Perié, J.M. Design and Experimental Validation of Power Electric Vehicle Emulator for Testing Electric Vehicle Supply Equipment (EVSE) with Vehicle-to-Grid (V2G) Capability. *Appl. Sci.* **2021**, *11*, 11496. https://doi.org/10.3390/app112311496

Academic Editor: Michele Roccotelli

Received: 19 October 2021
Accepted: 30 November 2021
Published: 4 December 2021

Publisher's Note: MDPI stays neutral with regard to jurisdictional claims in published maps and institutional affiliations.

1. Introduction

The development of the Smart Grid is contributing to the integration of renewable energy into the electric grid, which guides the needed decarbonization of energy consumption to deal with climate change and the depletion of fossil fuels. However, this development is increasing the complexity of the electric grid [1], adding new power system elements which must provide reliable operation in all kinds of different situations. One of these new systems is the Electric Vehicle Supply Equipment (EVSE), which is required for the grid integration of the Electric Vehicle (EV), including Battery Electric Vehicle (BEV) and Plug-in Hybrid Electric Vehicle (PHEV). Thanks to the Vehicle-to-Grid (V2G) functionality, the EVSE can work both as a load or as a generation source, using the energy of the EV battery available for secondary purposes, such as peak shaving, load frequency control, demand response or the management of renewable energy surplus [2,3].

It is expected than in 2030, more than the 90% of the EVSE will be private [4], while over the 80% of EV charging currently takes place at home [5]. Accordingly, the main domains for V2G functionality will be Vehicle-to-Building (V2B) or Vehicle-to-Home (V2H), whether the building is connected to the grid or isolated, where several advantages has been proved [6]. In order to use this functionality, EVSE must be integrated into the building Energy Managment System (EMS), which will handle the charging or discharging according to the needs of the building [7,8]. Therefore, to ensure the proper behaviour of these new systems, the research of suitable test systems can establish the Smart Grid development [9,10].

There are several test system techniques in the literature, but the ones that exchange real power with the Hardware-Under-Test (HUT) give the most accurate results, due to the fact that they can probe the full system. A very promising technique to test the full system

Appl. Sci. **2021**, *11*, 11496. https://doi.org/10.3390/app112311496 https://www.mdpi.com/journal/applsci

is the Power Hardware-In-the-Loop (PHIL) technique, which has the best trade-off between test fidelity and test coverage [11]. However, in applications like testing EV chargers, in which the number of tests carried out can be very high and they have a very defined functionality, the PHIL technique can be replaced by a power test bed, or also known as machine emulator [12].

Electric Vehicle Emulators EVE are powerful tools to develop and test EV charging stations. They have been used for a while for testing EVSE communication [13], unidirectional power [14,15] or unidirectional power and communications [16,17]. Besides, the requirements needed for a test bench based on PHIL for testing and verification of EV and EVSE are defined in [18]. However, bidirectional vehicles and chargers have a better impact on the stability of the Smart Grid, offering expanded flexibility services [5]. Therefore, the testing of these systems is an important step in the development of the present and future power electric grid.

This paper presents the details of the design and development for manufacturing the electric vehicle emulator for testing V2G chargers, with power factor grid correction functionality. The paper is organized as follows. Firstly, Section 2 analyses the main needs of the Electric Vehicle Emulator (EVE). The design of the system is described in Section 3, explaining every developed component. Then, in Section 4 the complete test system and some of the main results are shown. Finally, the conclusions are drawn in Section 5.

2. EV Emulator Needs

The EV battery needs a bidirectional power electronics system to charge and discharge its energy to the electric grid. However, current on-board EV chargers are only unidirectional and are not able to give energy to the grid. The main reason is that, as a rule of thumb, the unidirectional topologies can get higher power densities (W/m^3) and more specific power (W/kg) than the bidirectional ones. Therefore, the V2G functionality is only available in DC standards, because in this case, the bidirectional power electronics converter is located in the EVSE, where the specific power index is not especially important.

Table 1 shows the current status of the DC chargers standards. Among these standards, CHAdeMO [19] is the first and most used standard with V2G capability [20]. There have been five updates of the protocol in order to include the different necessities of the new EV and their uses. Consequently, an electric vehicle emulator compatible with this standard will cover most of V2G EVSE in the market.

Table 1. Current status of the different DC chargers standards [20].

Standard	CHAdeMO	GB/T	CCS Type 1	CCS Type 2	Tesla	ChaoJi
Maximum Voltage (V)	1000	750	600	900	500	1500
Maximum Current (A)	400	250	400	400	631	600
Maximum Power (kW)	400	185	200	350	250	900
Communication Protocol	CAN	CAN	PLC	PLC	CAN	CAN
V2X Function	Yes	No	No	No	Unknown	Yes
Start year	2009	2013	2014	2013	2012	2020

Accordingly, the EVE also needs a bi-directional power electronics system to test both charge and discharge. If the emulator takes energy from the same point of common coupling as the V2G charger, the electric consumption during the test is only the sum of EV emulator and V2G charger losses, which saves in general more than 90% of the test energy. Furthermore, if the power factor is close to 1, the test can be done in facilities with lower electric power supply, which also saves money and allows testing in several places. This is important for testing unidirectional chargers, especially old EVSE [21], since the Active Front End (AFE) can be a topology with no control of the reactive power consumed during the charging state. Therefore, an EVE with a four-quadrants AFE will be able to test different types of EVSE, ensuring low apparent power consumption during the complete test.

The end user of EVE should be laboratories which need to check the integration of the EV in a specific electric grid; for instance to test stability, time response, compatibility, etc. However, it could also be interesting for EVSE manufacturers to check the behaviour of their developments and for maintenance purposes. Therefore, extra functionality to debug the correct behaviour of the EVSE will be desired.

3. EV Emulator Design

3.1. Overview

A block diagram of the complete EV emulator system proposed with the HUT EV charger connection is shown in Figure 1.

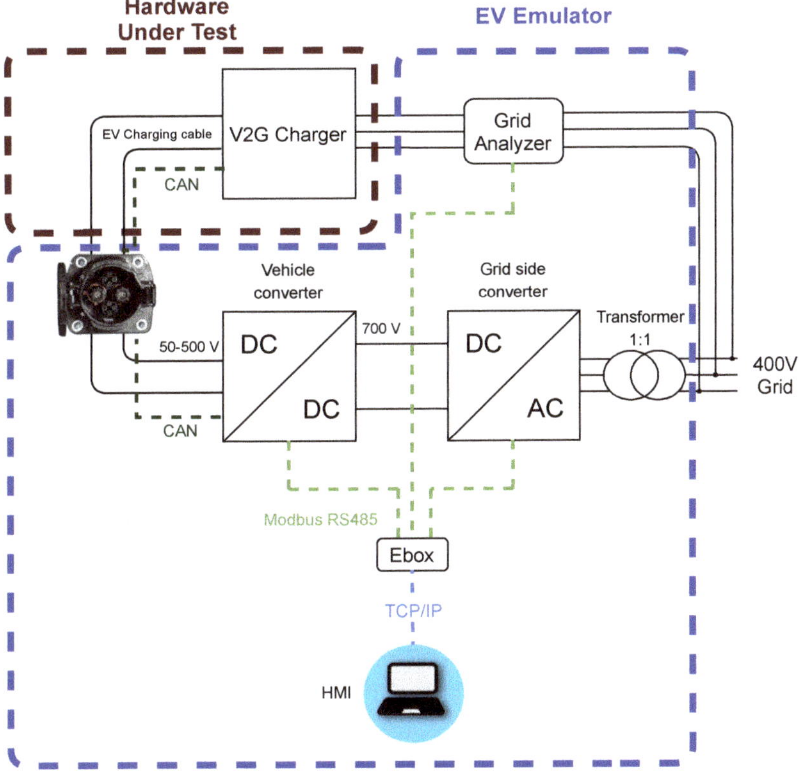

Figure 1. General EVE testbench block diagram, pointing out the main power subsystems of the EVE (blue) and the HUT (red), as well as the communication interfaces defined between them (dark and light green).

The power electronics system is built up of one AC/DC grid side converter and one DC/DC vehicle converter, whose specifications are listed in Table 2. The two components communicate through an embedded low cost gateway, which is called Energy Box (EBox) [22], via Modbus RS485. The EBox also communicates via Modbus RS485 with the grid analyzer to measure in real time the active and reactive power of the HUT and with the Human–Machine Interface (HMI) via TCP/IP. Furthermore, a CHAdeMO protocol communication via CAN has been implemented in the DC/DC, which allows the emulator to interact with HUT, setting the power limits and the desired current during the test. In order to have galvanic isolation in the whole system, a three-phase transformer $\curlywedge/\Delta$ is connected between the AFE and the electric grid at 400 Vrms and 50 Hz.

Table 2. Main electrical parameters of the designed EVE.

EV Emulator	
Nominal Power	50 kW
Efficiency	>96%
Internal DC-link voltage	700 V
Switching frequency	20 kHz
Control frequency	20 kHz
Input	
Nominal RMS phase voltage	230 V
Nominal RMS phase current	80 A
I ripple	1% Inom
Nominal frequency	50 Hz
Output	
Output voltage range	50–500 V
Output current range	−100 A to 100 A
Maximum voltage ripple	<2% Vmax

3.2. DC/DC Converter

3.2.1. Hardware Design

Figure 2 shows the schematic of the DC/DC converter. A three-branches in parallel bridge topology to generate the battery output voltage emulation has been designed. Switching frequency of the IGBTs (SEMIKRON SEMIX302GB12E4s) are 20 kHz, working in the non audible spectrum. A carrier phase-shift scheme is adopted to make the equivalent switching frequency up to 60 kHz, reducing the output voltage ripple, and also the selection of IGBTs with less current capability but better switching efficiency.

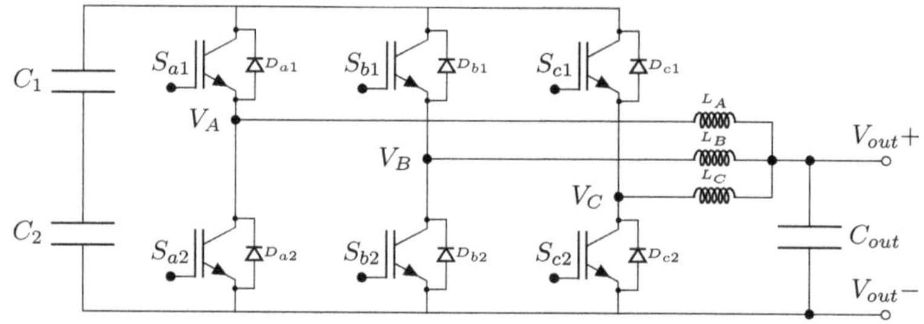

Figure 2. Topology of the DC/DC converter of the proposed EVE: a three-branches two-level in parallel bridge topology.

The output filter is an LC filter, with three coils connected in parallel to the output capacitor. It is a second-order low-pass filter with a resonance frequency ω_{res} given by Equation (1):

$$\omega_{res} = \sqrt{\frac{3}{LC_{out}}} \tag{1}$$

The resonance frequency of the filter needs to be placed at least one tenth of the switching frequency in order to have a sufficient rejection of the switching components. To avoid resonance problems of the filter, ω_{res} is also placed lower than the control frequency of the system. It allows the control to compensate the resonance current implementing a virtual resistance ($R_{virtual}$), which is placed in series with each inductor. This control method improves the overall efficiency of the system, avoiding physical resistance in the filter to dampen the resonance. However, due to the lower ω_{res}, the overall dynamic is reduced but is still enough to guarantee the stability and fidelity of the test. In order to decide the $R_{virtual}$, Figure 3 shows the bode plot of the LC filter (Equation (2)) with different resistance values.

$$G_{DCfilter}(s) = \cfrac{1}{\cfrac{LC_{out}}{3}s^2 + \cfrac{R_{virtual}C_{out}}{3}s + 1} \tag{2}$$

Ideally, it can be seen in Figure 3 that with a higher R the system is damper. However, a high $R_{virtual}$ will also increase the measurement noise of the current, decreasing the steady state performance. A trade-off between dampening and performance has been chosen, selecting a $R_{virtual} = 1\ \Omega$. The frequency response of the filter with the selected resistance is shown in Figure 4, obtaining resonance gain close to 10 dB.

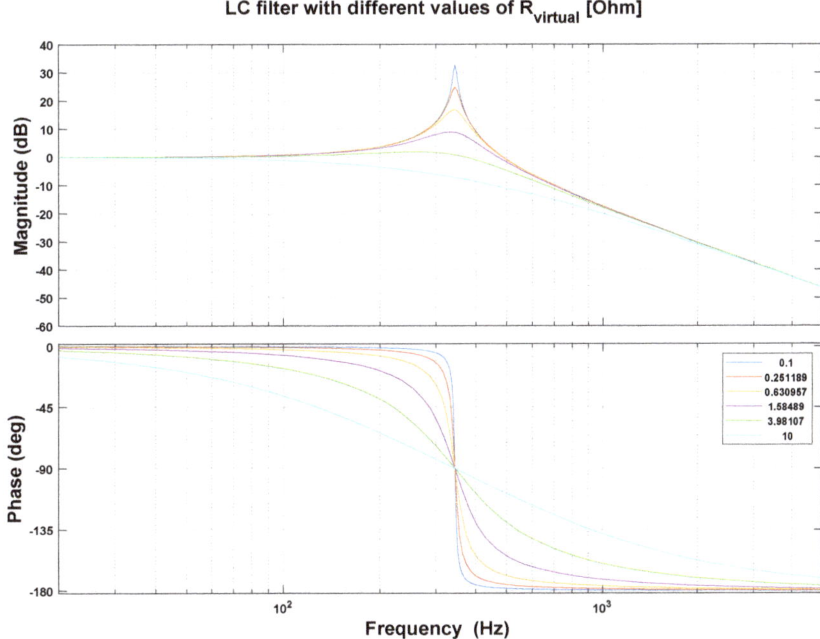

Figure 3. Sweep of the LC filter transfer function with different values of $R_{virtual}$: 0.1 Ω in blue, 0.251 Ω in red, 0.63 Ω in orange, 1.58 Ω in purple, 3.98 Ω in green, and 10.0 Ω in light blue.

3.2.2. Control Design

The block scheme control of the DC/DC converter is shown in Figure 5.

The current of every coil is measured and multiplied by the $R_{virtual}$, getting the emulation effect of a real resistance in the LC filter. A PI control has been chosen in order to get no voltage error in the steady state output capacitor voltage. The output of the PI regulator is subtracted in every branch by the previous calculation of the $R_{virtual}$ gain, and then divided by the input voltage V_{dc} to obtain the duty cycle in every branch. After that, a carrier phase-shift is implemented in order to obtain 120° phase in each branch PWM.

The gain of this integrator has been calculated to have a cut-off frequency to 16.7 Hz, which gives a step response of the system close to 3 ms. Figure 6 shows the step response of the system to a 40 V instantaneous set-point change.

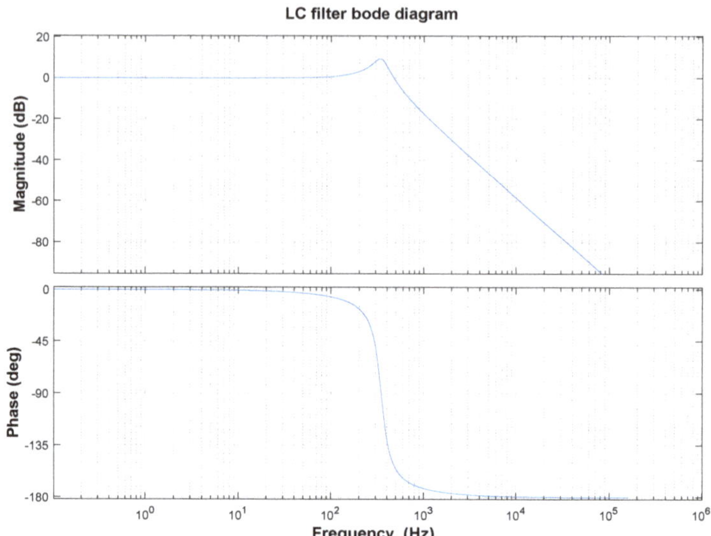

Figure 4. Bode diagram of the DC output filter with the selected $R_{virtual} = 1\ \Omega$.

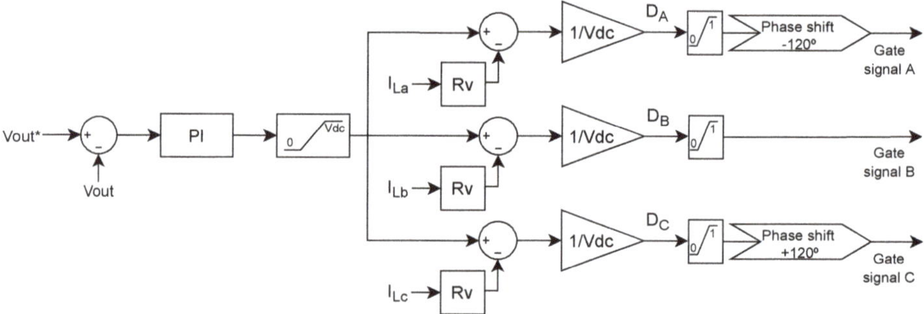

Figure 5. EVE output voltage control block diagram of the DC/DC converter. The carrier of the three gate signals has the same sample time in order to keep constant the phase between them.

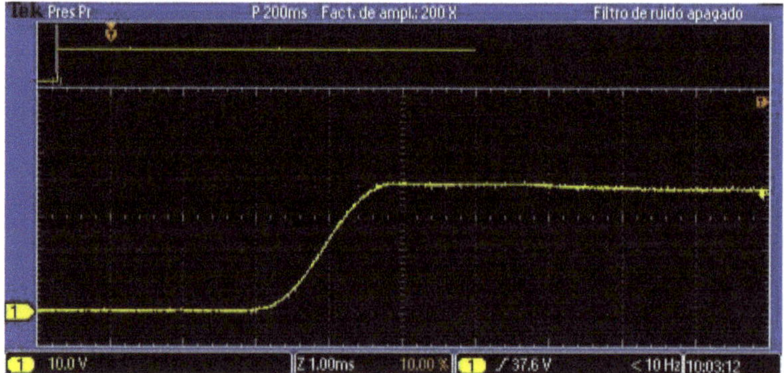

Figure 6. Output voltage response of the DC/DC converter to a step of 40 V in the set-point.

3.3. AC/DC Converter

3.3.1. Hardware Design

The AC/DC converter, used to exchange power with the electric grid, has been designed and manufactured with a three-level Neutral Point Clamped (NPC) topology with SEMIKRON SK150MLI066T IGBTs, as shown in Figure 7. In comparison with the two level converter, this topology reduces the voltage stress on the IGBTs and their switching losses, increasing the output voltage waveform quality.The attenuation of the current switching ripple is done by a third-order LCL grid filter, which is smaller and lower-priced than the first order L filter [23]. Furthermore, the AC/DC converter can operate in the four quadrants, compensating reactive power if needed, even when the DC/DC converter is not working. This functionality makes it possible to perform the complete set of tests in facilities consuming only the active power losses of the whole testbench. This is because the EVE can compensate whenever the power factor of the charger diverges from 1, especially at low charging levels.

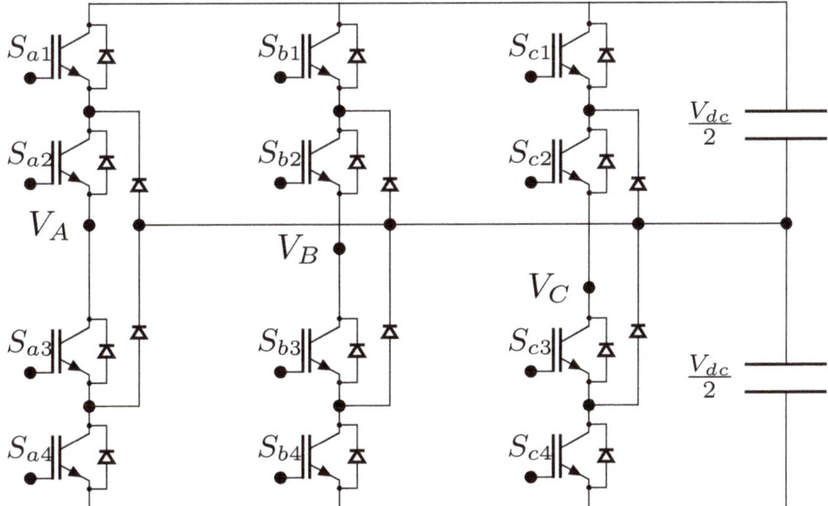

Figure 7. Topology of the AC/DC converter of the proposed EVE: a three-phase three-level NPC power converter.

3.3.2. Control Design

Figure 8 shows the general block scheme control of the AC/DC converter, which is based on [24,25]. The controller is divided into the following four layers:

- High-level controller: the higher level control is in charge of generating the appropriate active and reactive power references provided to the low level layers of the controller. The purpose is to lead the system to the desired goal: on the one hand, to maintain the desired DC voltage level by regulating the active power output with the voltage regulator box; on the other hand, to adjust the system's reactive power reference by means of providing the needed reactive power output.
- Middle-level controller: the middle-level controller is responsible for saturating the power references in order to guarantee that the system remains in its operation working range and does not exceed its limits. Therefore, safety features and operating constraints such as temperature and over-voltages are considered to evaluate whether the desired power objectives are reached or not. Finally, from the power references, the current references are set.

- Low-level controller: the low-level controller is divided into two parts: the current controller that determines the control actions needed to follow the control current references; and the duty control system, in charge of the converter's modulation technique.
- Hardware-level controller: the hardware level controls the power converter's drive system, translating the control signals to the physical pulses of the converter.

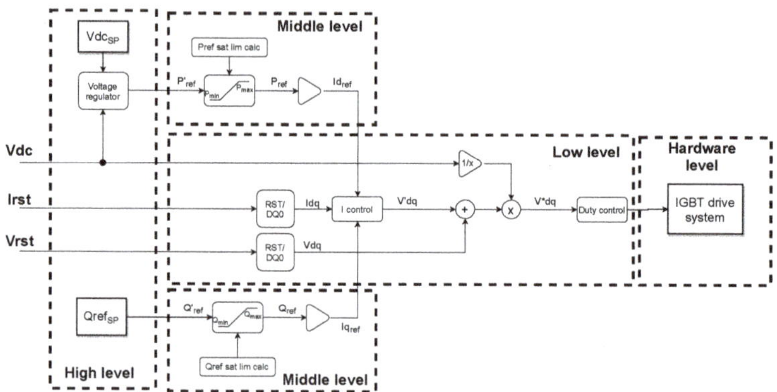

Figure 8. Control Scheme of the AC/DC converter of the proposed EVE pointing out four control levels: high, middle, low and hardware level.

3.4. EBox

3.4.1. Hardware Design

The EBox has been described previously in the literature [22]. It is a low cost gateway that uses a Raspberry Pi computing module to manage the high level power electronics system and to run the reactive control algorithm of the EVE. Furthermore, it can communicate with the HMI and the grid analyzer, which allows the gathering of the information of the complete system.

3.4.2. Control Design

To control and ensure the compensation of the EVSE reactive power consumption, the EBox communicates with a grid analyzer to know the reactive power value and send the calculated set-point to the emulator AFE, which generates the desired current. As shown in Figure 9, there are three different grid analyzer configurations: in the output of the V2G charger under test, in the Point of Common Coupling (PCC) of the facility or in both (V2G and PCC).

Depending on the location of the analyzer, the control algorithm implemented in the EBox has to be changed, achieving different performances:

- **V2G:** in this case, the grid analyzer can only measure the power consumed or returned by V2G charger. The control algorithm implemented in this context is an open loop control, which is shown in Figure 10a. This control loop is easy to implement and has a very good time response. However, it is not possible to determine the reactive power at the PCC, and problems such as an incorrect calibration or installation of the EVE could even increase the power consumption in the facility.
- **PCC:** the grid analyzer is located at the point where the facility is connected to the grid. Figure 10b shows the control algorithm, which is a closed loop control with a PI regulator to ensure zero error in steady-state operation. The problem is that the time response of this control is minimum 5–10 times the time step, so loads with an abrupt change of power can be over the limits for a few seconds.
- **V2G and PCC:** the two grid analyzers are installed, one in the V2G charger and the other one in the PCC. The control algorithm is shown in Figure 10c, which is based on the previous closed loop control with a PI regulator. However, in this case, the

reactive power of the charger is measured and directly compensated at the output of the regulator. In order to avoid the integration of the error produced by the charger, the derivative of this measure is compensated in the input of the regulator. In this way, the controller achieves a better time response capability. However, it has to be highlighted that the use of two grid analyzers increases close to 2% the final price of the solution.

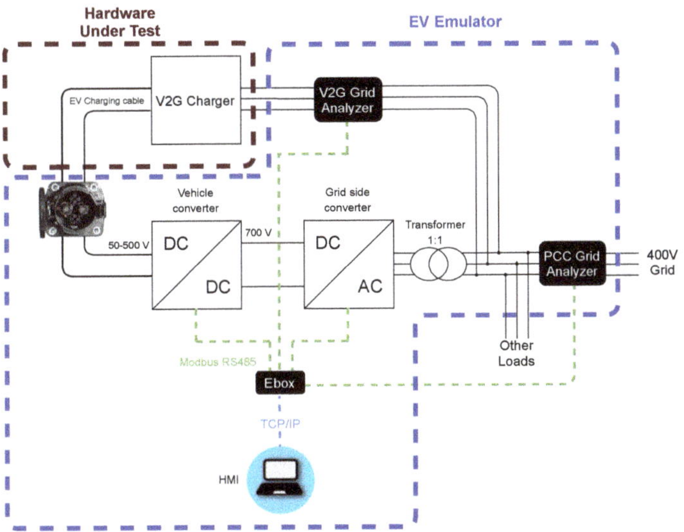

Figure 9. Representation of the two possible locations of the grid analyzer in order to compensate the reactive power injected by the EVE.

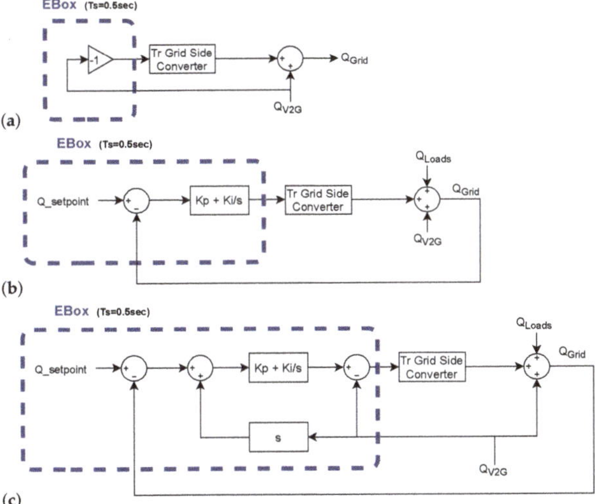

Figure 10. Three types of different reactive compensation control strategies: (**a**) Open loop control, measuring reactive power at the output of the V2G charger. (**b**) Closed loop control, measuring reactive power at the PCC of the facility. (**c**) Closed loop control with V2G compensation, measuring both V2G charger and the PCC of the facility.

Figure 11 shows the simulation in MATLAB/Simulink of the reactive power consumption of the three different types of control that have been shown in previous Figure 10. It can be seen that in Figure 11a without any external loads, the reactive consumption in the PCC with open control loop as well as with closed control loop with V2G compensation are the same, while with closed control loop the response is slower than the previous case. However, if there is reactive power consumption by an external load, Figure 11b shows different behaviour. On one hand, with open control loop the reactive power consumption at the PCC in steady-state is determined by the external loads. On the other hand, with closed control loop, the power at the PCC can be fully compensated by the EVE, with a better time response to the V2G power demand by compensating it using two grid analyzers.

Comparing the last two control modes in Figure 11b, it should be noticed that the reactive power error given by the abrupt change of the V2G power at time $t = 32$ s, is quickly balanced with the third type of reactive control strategy. The reason is that the compensation of the V2G power avoids the error integration in the PI control, improving the time response and performance of the reactive power control at the PCC.

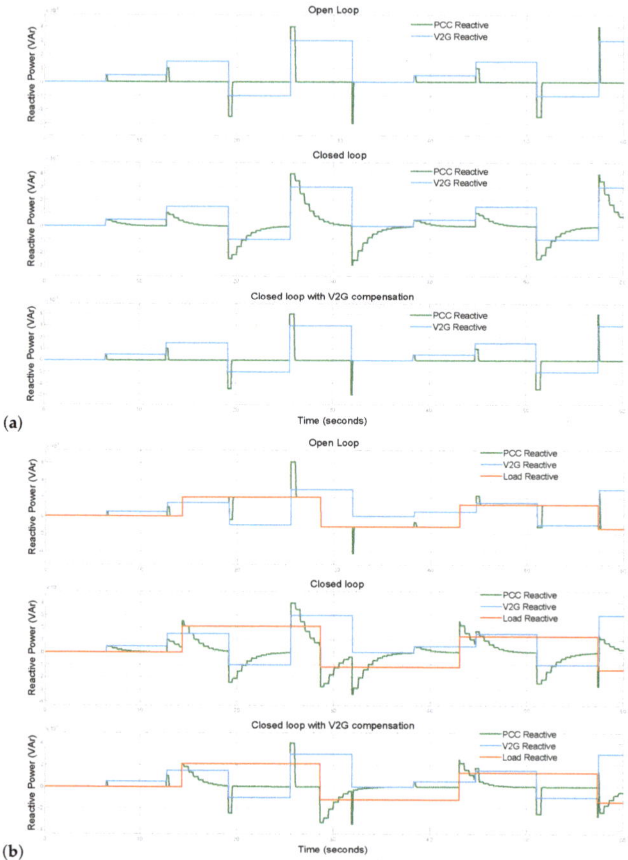

Figure 11. Simulation of the reactive power consumption using the same V2G consumption profile with abrupt consumption changes. Two scenarios have been considered: (**a**) without power consumption of the external loads; (**b**) with power consumption of the external loads. In every scenario, all three different control strategies have been tested: open loop, closed loop, and closed loop with V2G compensation.

4. Experimental Results

4.1. Test Description

The main equipment involved in the tests is shown in Figure 12. A V2G charger has been used as a HUT and it has been connected to the EVE through a CHAdeMO connection. The bottom box of the emulator is the AC/DC converter and the upper box is the DC/DC converter. Both of them are connected to the EBox, which is also connected to the grid analyzer (Circutor CVM-MINI) via RS-485 and to the HMI through TCP/IP. The transformer is also inside an enclosure for safety reasons.

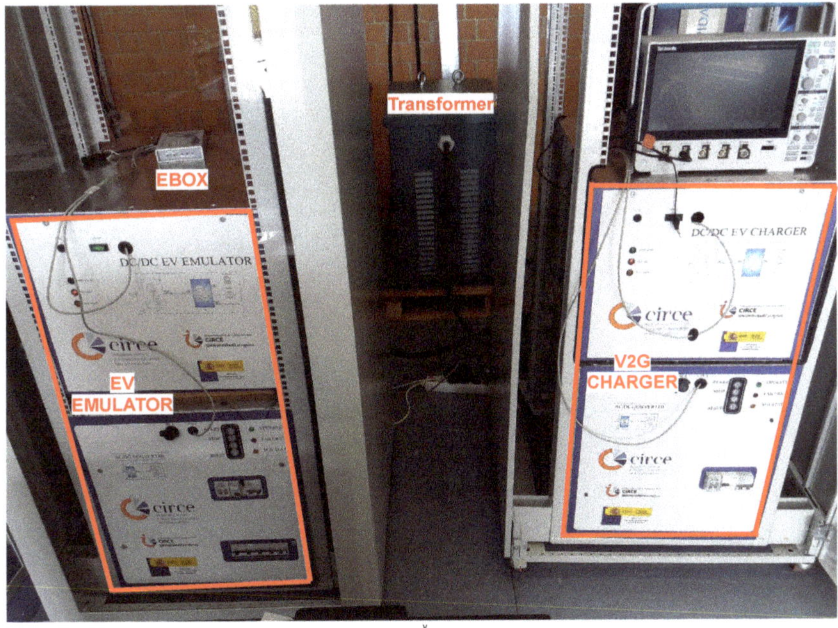

Figure 12. General overview of the testbench used for the power tests done with the proposed EVE. On the left, a 50 kW EVE together the EBox gateway; On the right the HUT system built up with a 50 kW V2G charger; and the power transformer in the centre.

The simplified sequence of the complete test is shown in Figure 13, where it is explained how the system works, and the main processes with their interactions that are running, which are needed to emulate the behaviour of an EV. First, the user has to start the EVE system through the HMI and wait until the DC/DC system is ready to initiate the charge/discharge operation in order to test the HUT accordingly. Then, the user has to plug-in the cable and begin the operation needed to launch the charging/discharging process of the HUT. The user can change the voltage and/or power manually at any time or load a script with the profile of EV charging/discharging, indicating every 0.5 s the required voltage and power via HMI. The test will be finished whenever: the user stops it via HMI, the EVE battery model determines that it has finished the charging/discharging process, or there is any error during the operation.

4.2. Manual Set-Point Adjustment

With the aim to test the stability of the emulator, the voltage response at different set-points has been analyzed. Figure 14 shows the manual change of current set-point from 65 to −65 A, with the emulator controlling the same set-point voltage at 300 V. It must be noticed that the slope of the current increment by the charger is 10 A/s, which is within the limits of the CHAdeMO standard [19]. Furthermore, the PI controller implemented in the emulator has a reduced velocity error during the transition of this current variation in the HUT.

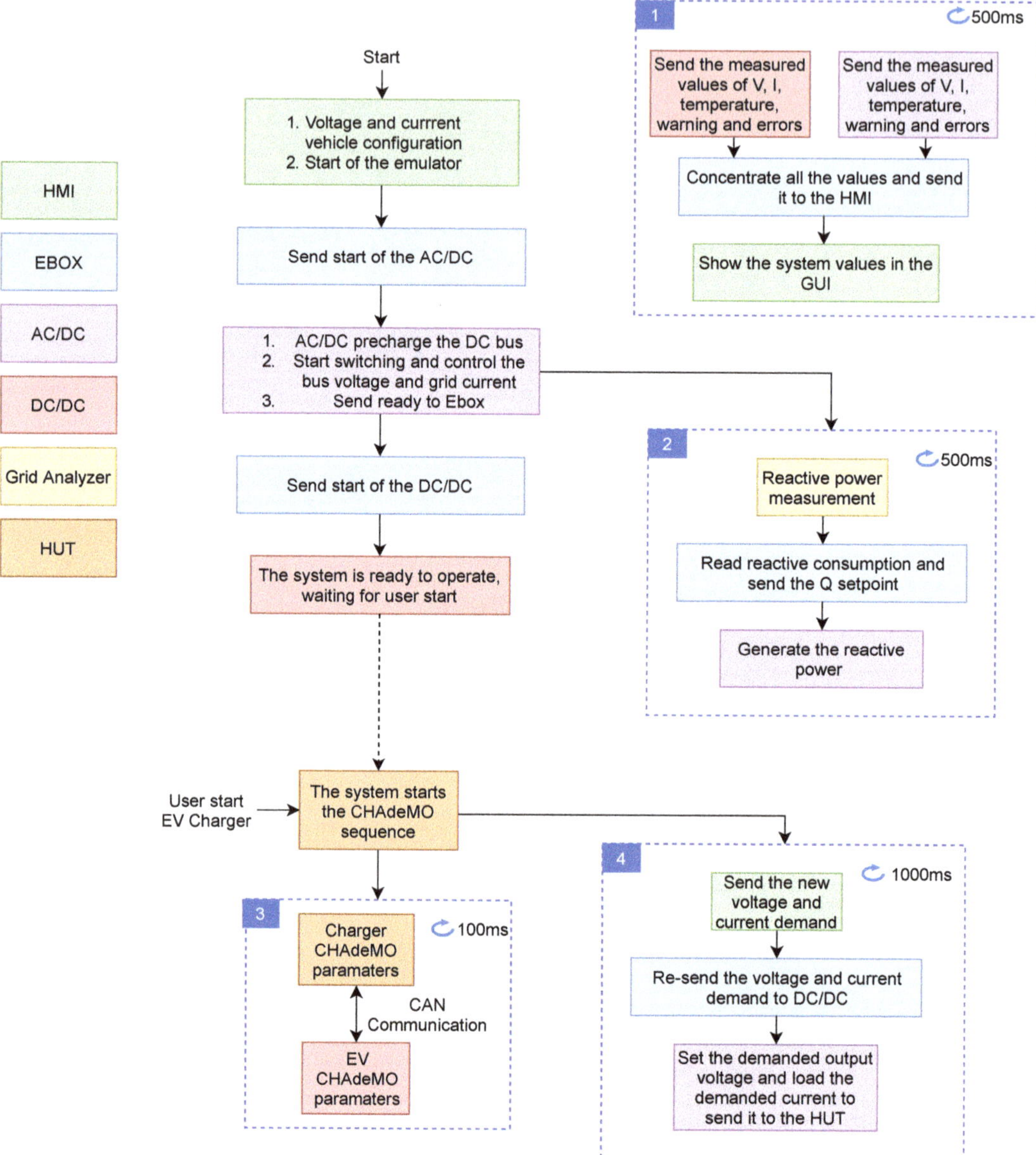

Figure 13. Simplified operation sequence diagram of the complete EVE test procedure.

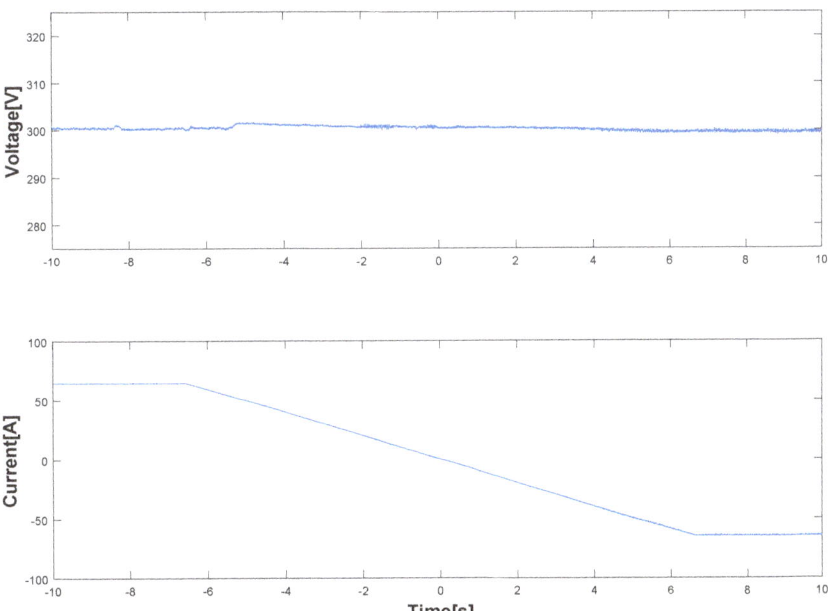

Figure 14. Experimental results of the variation of the current set-point from 65 A to −65 A at 300 V set-point. Voltage (**top**) and current (**bottom**) transient evolution is depicted in the figure.

Furthermore, it is possible to verify the behaviour of the system for grid management purposes. For example, thanks to this test, it has been verified that this charger could be used to perform frequency grid operation for current set-point changes up to 50 A [26]. Furthermore, the smooth transition between the two set-points, indicates that it is possible to use this EVSE with V2G capability for operations such as peak shaving or management of renewable energy surplus.

The variation of the set-point voltage from 300 V to 200 V is shown in Figure 15. These kinds of battery voltage fluctuations are abnormal in EV batteries, but they allow the user to know the stability and response of the charger to be tested. In this case, the emulator keeps the same set-point power, which produces an increment in the demanded current to the V2G charger. Depending on the time response of the chargers, maximum or minimum power peaks can appear during the voltage transition, which will be larger if the voltage transition is more abrupt.

4.3. Load an EV Battery Profile

In this case, a user defined EV battery charge profile has been defined. It consists of a charging process of 5 min, starting the voltage at 350 V with a demanded power of 12 kW. This demanded power decreases until reaching 6.5 kW and the voltage increases up to 358 V, the moment when the emulator sends a stop command to the charger through CHAdeMO communication. The evolution of the voltage and current measured at the emulator output is shown in Figure 16. At the beginning of the charging process, the CHAdeMO's isolation test procedure is performed by the charger, setting 500 V at the input of the emulator. Once the charger verifies that there is not any isolation problem, it closes the emulator power relay. From this point, the emulator control the output voltage and it sends the demanded power to the EVSE, which is defined in the loaded profile of the emulator. This profile can be modified in order to perform different user tests, changing current, voltage and time of the vehicle charge and executed as many times as needed. This feature allows the repeatability of the test, which makes it possible to compare and analyze the response of different HUTs.

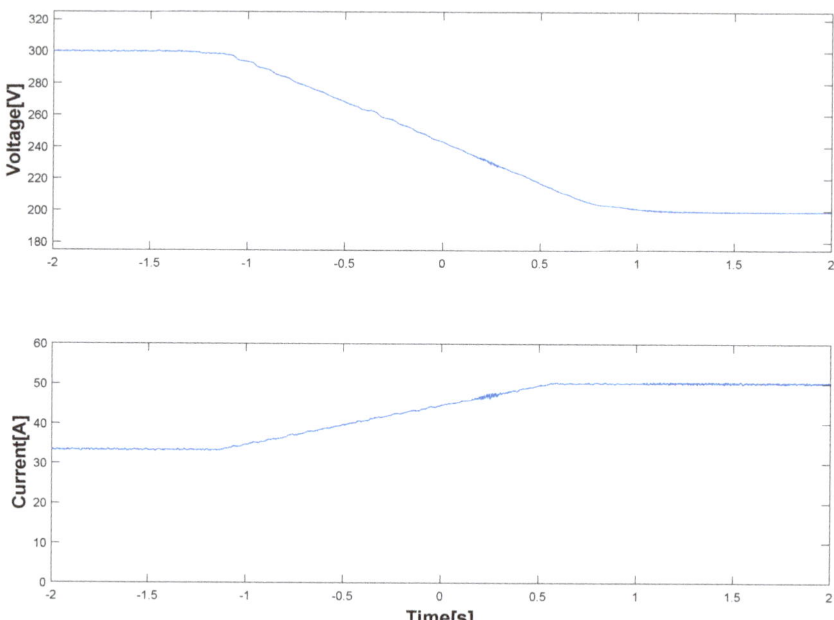

Figure 15. Experimental results of the variation of the voltage set-point from 300 V to 200 V at 10 kW set-point. Voltage (**top**) and current (**bottom**) transient evolution is depicted in the figure.

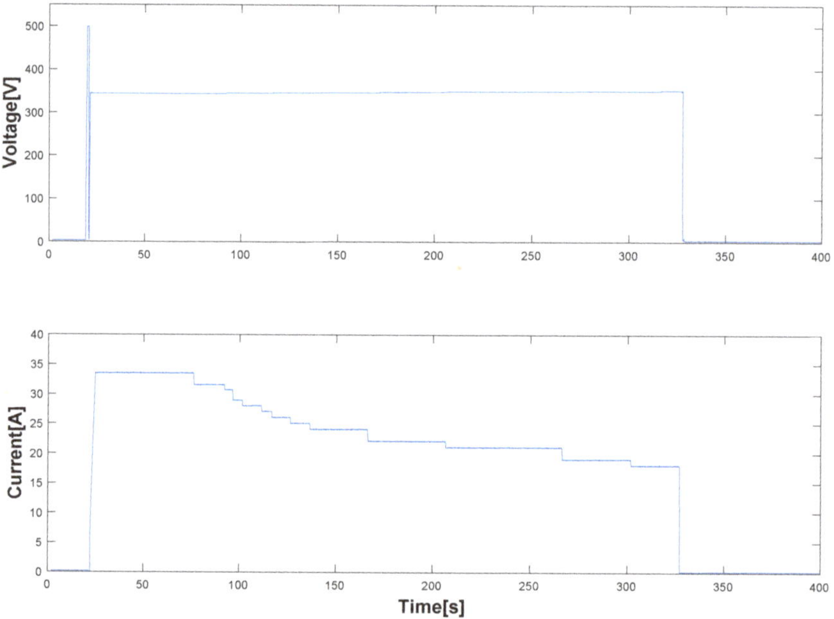

Figure 16. Emulation of an EV charging for 5 min. The upper graph is the output voltage of the charger and the lower graph is the charged current.

To test the reactive power compensation, the previous EV battery charging profile has been used, but also adding a profile of reactive power consumption to the charger. A profile with abrupt changes in the reactive consumption has been configured in order to verify the behaviour of the emulator. In this case, the grid analyzer has been placed at the output of the HUT, with an open-loop control implemented in the EBox. Figure 17 shows the results of the test, measured with a power analyzer data logger (Fluke 435-II) at the PCC of the facility. Firstly, the reactive power consumption, without compensating it by the emulator, has been measured by a grid analyzer. Secondly, the same test has been repeated and measured, but this time compensating the reactive power consumption by the emulator. The data from the two measurements have been downloaded and synchronized.

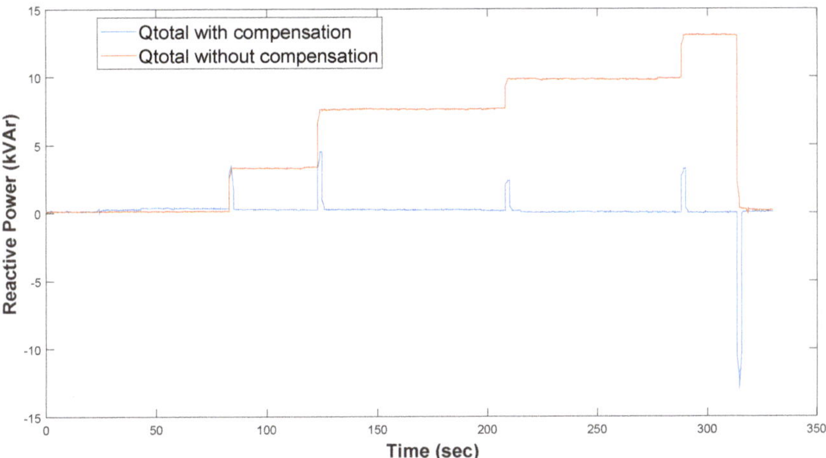

Figure 17. Reactive power consumption of the EV charging with (blue) and without (red) reactive power compensation, measured at the facility's PCC.

The steady reactive power consumption of the HUT is completely compensated by the emulator. However, if the HUT has quick step changes in the reactive power consumption in less than the execution time control of the EBox, it cannot be compensated properly. This behaviour can cause problems in the facility's PCC if the electric protection devices are less than the peak current occurred during the transition. To resolve this issue, it is possible to perform a power consumption characterization of the HUT at every charge/discharge current, and configure a charge/discharge profile in the emulator, including the reactive power that has to be generated by the emulator in order to compensate it.

5. Conclusions

EVs and V2G chargers have attracted a lot of attention for being considered as part of the solution to increase the share of renewable energy in the electric grid. Since most the EVSE are expected to be at home, these systems are an important tool towards the building EMS to achieve near-zero energy buildings. Therefore, the study and verification of the proper behaviour of these elements can smooth the path to the energy transition.

This paper has presented the design, test and results of a power EV emulator with V2G capability. The ability to emulate a real V2G EV to handle fast charge/discharge with an EV charger, consuming only the losses of the complete system, has been validated experimentally. In addition, for small and medium enterprises which only want to test EVSE and do not need the flexibility of a PHIL test bench, it is more economically affordable due to not requiring a fast real-time simulator and a high bandwidth power amplifier.

Future tests will consist of increasing the implemented EVSE protocol to CCS, due to the importance of this standard in Europe. Moreover, testing the AFE of the EVSE under test using the PHIL technique, allows us to verify the complete system: grid and vehicle side. Furthermore, thanks to the recirculation power of the EVE during test, a low power amplifier could be used to emulate the grid behaviour during the PHIL test. In order to test the fast dynamics behaviour of the DC side of a V2G charger, the implementation of a real-time battery model could be done in the EVE. Another future work will be the development of an EVE to test EVSEs which work on a DC grid.

Author Contributions: Conceptualization, E.G.-M., J.F.S.-O. and J.M.-C.-A.; methodology, E.G.-M., J.F.S.-O., J.M.-C.-A. and J.M.P.; software, E.G.-M. and J.M.P.; validation, E.G.-M.; investigation, E.G.-M., J.M.-C.-A. and J.M.P.; writing—original draft preparation, E.G.-M. and J.M.-C.-A.; writing—review and editing, E.G.-M., J.F.S.-O., J.M.-C.-A. and J.M.P.; visualization, E.G.-M. and J.M.-C.-A.; supervision, J.F.S.-O., J.M.-C.-A. and J.M.P. All authors have read and agreed to the published version of the manuscript.

Funding: This project has received funding from the European Union's Horizon 2020 research and innovation programme under Grant Agreement No 875683.

Institutional Review Board Statement: Not applicable.

Informed Consent Statement: Not applicable.

Data Availability Statement: Not applicable.

Conflicts of Interest: The authors declare no conflict of interest.

References

1. Monti, A.; Ponci, F. Power Grids of the Future: Why Smart Means Complex. In Proceedings of the Complexity in Engineering (COMPENG '10), Rome, Italy, 22–24 February 2010; pp. 7–11. [CrossRef]
2. Vahedipour-Dahraie, M.; Rashidizaheh-Kermani, H.; Najafi, H.R.; Anvari-Moghaddam, A.; Guerrero, J.M. Coordination of EVs Participation for Load Frequency Control in Isolated Microgrids. *Appl. Sci.* **2017**, *7*, 539. [CrossRef]
3. Habeeb, S.A.; Tostado-Véliz, M.; Hasanien, H.M.; Turky, R.A.; Meteab, W.K.; Jurado, F. DC Nanogrids for Integration of Demand Response and Electric Vehicle Charging Infrastructures: Appraisal, Optimal Scheduling and Analysis. *Electronics* **2021**, *10*, 2484. [CrossRef]
4. IEA. Global EV Outlook 2021. Paris. 2021. Available online: https://www.iea.org/reports/global-ev-outlook-2021 (accessed on 14 October 2021).
5. Jones, L.; Lucas-Healey, K.; Sturmberg, B.; Temby, H.; Islam, M. The A to Z of V2G: A Comprehensive Analysis of Vehicle-to-Grid Technology Worldwide. Realising Electric Vehicle to Grid Services Project. 2021. Available online: https://arena.gov.au/assets/2021/01/revs-the-a-to-z-of-v2g.pdf (accessed on 14 October 2021).
6. Tostado-Véliz, M.; León-Japa, R.; Jurado, F. Optimal electrification of off-grid smart homes considering flexible demand and vehicle-to-home capabilities. *Appl. Energy* **2021**, *298*, 117184. [CrossRef]
7. Odkhuu, N.; Lee, K.-B.; Ahmed, M.A.; Kim, Y.-C. Optimal Energy Management of V2B with RES and ESS for Peak Load Minimization. *Appl. Sci.* **2018**, *8*, 2125. [CrossRef]
8. Chen, J.; Zhang, Y.; Li, X.; Sun, B.; Liao, Q.; Tao, Y.; Wang, Z. Strategic integration of vehicle-to-home system with home distributed photovoltaic power generation in Shanghai. *Appl. Energy* **2018**, *263*, 114603. [CrossRef]
9. Strasser, T.; Pröstl Andrén, F.; Lauss, G.; Bründlinger, R.; Brunner, H.; Moyo, C.; Seitl, C.; Rohjans, S.; Lehnhoff, S.; Palensky, P.; et al. Towards holistic power distribution system validation and testing—An overview and discussion of different possibilities. *E I Elektrotechnik Inf.* **2017**, *134*, 71–77. [CrossRef]
10. Mylonas, E.; Tzanis, N.; Birbas, M.; Birbas, A. An Automatic Design Framework for Real-Time Power System Simulators Supporting Smart Grid Applications. *Electronics* **2020**, *9*, 299. [CrossRef]
11. García-Martínez, E.; Sanz, J.F.; Muñoz-Cruzado, J.; Perié, J.M. A Review of PHIL Testing for Smart Grids—Selection Guide, Classification and Online Database Analysis. *Electronics* **2020**, *9*, 382. [CrossRef]
12. Oettmeier, M.; Bartelt, R.; Heising, C.; Staudt, V.; Steimel, A.; Tietmeyer, S.; Bock, B.; Doerlemann, C. Machine emulator: Power-electronics based test equipment for testing high-power drive converters. In Proceedings of the 2010 International Conference on Optimization of Electrical and Electronic Equipment, Brasov, Romania, 20–22 May 2010; pp. 582–588. [CrossRef]
13. Anil, D.; Sivraj, P. Electric Vehicle Charging Communication Test-bed following CHAdeMO. In Proceedings of the 2020 International Conference on Computing, Communication and Networking Technologies (ICCCNT), Kharagpur, India, 1–3 July 2020; pp. 1–7. [CrossRef]
14. Jayawardana, I.; Ho, C.N.M.; Zhang, Y. A Comprehensive Study and Validation of a Power-HIL Testbed for Evaluating Grid-Connected EV Chargers. *IEEE J. Emerg. Sel. Top. Power Electron.* **2021**. [CrossRef]
15. De Herdt, L.; Shekhar, A.; Yu, Y.; Mouli, G.R.C.; Dong, J.; Bauer, P. Power Hardware-in-the-Loop Demonstrator for Electric Vehicle Charging in Distribution Grids. In Proceedings of the 2021 IEEE Transportation Electrification Conference & Expo (ITEC), Chicago, IL, USA, 21–25 June 2021; pp. 679–683. [CrossRef]

16. Cabeza,T.; Sanz, J.F.; Calavia, M.; Acerete, R.; Cascante, S. Fast charging emulation system for electric vehicles. In Proceedings of the 2013 World Electric Vehicle Symposium and Exhibition (EVS27), Barcelona, Spain, 17–20 November 2013; pp. 1–6. [CrossRef]
17. Ledinger, S.; Reihs, D.; Stahleder, D.; Lehfuss, F. Test Device for Electric Vehicle Grid Integration. In Proceedings of the 2018 IEEE International Conference on Environment and Electrical Engineering and 2018 IEEE Industrial and Commercial Power Systems Europe (EEEIC/I&CPS Europe), Palermo, Italy, 12–15 June 2018; pp. 1–5. [CrossRef]
18. Popov, A.; Tybel, M.; Schugt, M. Power hardware-in-the-loop test bench for tests and verification of EV and EVSE charging systems. In Proceedings of the 2014 IEEE International Electric Vehicle Conference (IEVC), Florence, Italy, 17–19 December 2014; pp. 1–8. [CrossRef]
19. CHAdeMO. Available online: https://www.chademo.com (accessed on 14 October 2021).
20. Wang, L.; Qin, Z.; Slangen, T.; Bauer, P.; van Wijk, T. Grid Impact of Electric Vehicle Fast Charging Stations: Trends, Standards, Issues and Mitigation Measures—An Overview. *IEEE Open J. Power Electron.* **2021**, *2*, 56–74. [CrossRef]
21. Krasselt, P.; Boßle, J.; Suriyah, M.R.; Leibfried, T. DC-Electric Vehicle Supply Equipment Operation Strategies for Enhanced Utility Grid Voltage Stability. *World Electr. Veh. J.* **2015**, *7*, 530–539. [CrossRef]
22. Andolšek, A.; Nemček, M.P.; Gómez, A.; Zocchi, A.; Bruna, J.; Oliván, M.A. Flexibility and optimization services validation in a microgrid. In Proceedings of the 2018 CIRED Ljubljana Workshop, Ljubljana, Slovenia, 7–8 June 2018. [CrossRef]
23. Azani, H.; Massoud, A.; Benbrahim, L.; Williams, B.W.; Holiday, D. An LCL filter-based grid-interfaced three-phase voltage source inverter: performance evaluation and stability analysis. In Proceedings of the 2014 IET International Conference on Power Electronics, Machines and Drives (PEMD 2014), Manchester, UK, 8–10 April 2014. [CrossRef]
24. Muñoz-Cruzado, J.; Villegas-Núñez, J.; Vite-Frías, J.A.; Carrasco-Solís, J.M.; Galván-Díez, E. New Low-Distortion Q–f Droop Plus Correlation Anti-Islanding Detection Method for Power Converters in Distributed Generation Systems. *IEEE Trans. Ind. Electron.* **2015**, *62*, 5072–5081. [CrossRef]
25. Muñoz-Cruzado-Alba, J.; Villegas-Núñez, J.; Vite-Frías, J.A.; Carrasco Solís, J.M. A New Fast Peak Current Controller for Transient Voltage Faults for Power Converters. *Energies* **2016**, *9*, 1. [CrossRef]
26. Martinenas, S.; Marinelli, M.; Andersen, P.B.; Træholt, C. Implementation and demonstration of grid frequency support by V2G enabled electric vehicle. In Proceedings of the 2014 49th International Universities Power Engineering Conference (UPEC), Cluj-Napoca, Romania, 2–5 September 2014. [CrossRef]

applied sciences

MDPI

Article

Inferring Long-Term Demand of Newly Established Stations for Expansion Areas in Bike Sharing System

Hsun-Ping Hsieh [1,*], Fandel Lin [2], Jiawei Jiang [1], Tzu-Ying Kuo [2] and Yu-En Chang [1]

[1] Department of Electrical Engineering, National Cheng Kung University, Tainan 70101, Taiwan; E24066218@gs.ncku.edu.tw (J.J.); N26090520@gs.ncku.edu.tw (Y.-E.C.)

[2] Institute of Computer and Communication Engineering, National Cheng Kung University, Tainan 70101, Taiwan; Q36084028@mail.ncku.edu.tw (F.L.); Q36074031@mail.ncku.edu.tw (T.-Y.K.)

* Correspondence: hphsieh@mail.ncku.edu.tw

Abstract: Research on flourishing public bike-sharing systems has been widely discussed in recent years. In these studies, many existing works focus on accurately predicting individual stations in a short time. This work, therefore, aims to predict long-term bike rental/drop-off demands at given bike station locations in the expansion areas. The real-world bike stations are mainly built-in batches for expansion areas. To address the problem, we propose LDA (Long-Term Demand Advisor), a framework to estimate the long-term characteristics of newly established stations. In LDA, several engineering strategies are proposed to extract discriminative and representative features for long-term demands. Moreover, for original and newly established stations, we propose several feature extraction methods and an algorithm to model the correlations between urban dynamics and long-term demands. Our work is the first to address the long-term demand of new stations, providing the government with a tool to pre-evaluate the bike flow of new stations before deployment; this can avoid wasting resources such as personnel expense or budget. We evaluate real-world data from New York City's bike-sharing system, and show that our LDA framework outperforms baseline approaches.

Keywords: bike sharing system; expansion areas; category clustering; batches prediction

Citation: Hsieh, H.-P.; Lin, F.; Jiang, J.; Kuo, T.-Y.; Chang, Y.-E. Inferring Long-Term Demand of Newly Established Stations for Expansion Areas in Bike Sharing System. *Appl. Sci.* **2021**, *11*, 6748. https://doi.org/10.3390/app11156748

Academic Editors: Agostino Marcello Mangini and Michele Roccotelli

Received: 12 June 2021
Accepted: 19 July 2021
Published: 22 July 2021

1. Introduction

A prominent sharing economy business model, the bike-sharing systems, has emerged in recent years as a popular way of public transportation [1]. For society, a bike-sharing system meets the theme of sustainable development because of convenience, lower prices, and environmental protection [2,3]. Consequently, many bike-sharing systems are being established to satisfy the need. One example of a bike-sharing system is Citi Bikes, with more than 85,000 active users [4].

Distributing a suitable bicycle network structure can not only connect the system of urban traffic and commuting but reduce the greenhouse effect. However, constructing unwanted stations in a bike-sharing system will cause environmental damage and resource waste. The framework presented in the paper aims to assist the government and planners in predicting bike demands at a macroscopic level in advance, i.e., evaluating and verifying whether new stations meet the needs of the public.

Research on bike-sharing systems has been widely studied in recent years. Some works [5–8] depend completely on station-based historical records and features, and their target is to make predictions for already established stations. The works of [9,10] aim to predict the demand in hours or only during rush hour. The work of [11] defines functional zones [12,13] and then predicts that the demand for bike expansion is the most relevant one to our work. Unfortunately, their mobility trip data in the expanded system is inapplicable for our long-term scenario, as it is also regarded as future data in the prediction stage. Different from previous works, we commit to long-term demand prediction, which is faced with two challenges. First, mobility and meteorology data used in previous works are unavailable in expansion areas, for example, taxi usages, temperature, wind speed, etc. Moreover, we cannot directly apply the methodology of existing works, which focus on short-term demand prediction for a single station, since they usually have enough training data but lack future events [5]. Second, the real-world bike stations are mainly built-in batches for expansion areas. However, the different geographical characteristics between regions make the prediction task hard.

To tackle these challenges, we propose a robust framework called LDA (Long-Term Demand Advisor) to predict long-term (e.g., six months) demand in newly established bike regions. Apart from the short-term prediction, which is highly affected by emergencies and other temporal factors [6,7], the proposed long-term prediction can not only reduce inaccuracies resulted from unpredictable social events or traffic accidents but also advise decision-makers on where to build new stations. This framework aims to provide governments with a preliminary estimation of the amount of bike usages in the following periods (e.g., half-year) in the new regions of a city, given merely the locations of the bike stations. Our contributions are as follows:

- To the best of our knowledge, this is the first work to predict long-term bike demand in batches for expansion areas.
- A G-clustering algorithm, a hierarchical POI clustering method to cluster POI categories, is proposed in this work, and it is shown to be effective. Experiments carried out on real-world datasets prove that our LDA framework outperforms baseline approaches.

2. Overview

We propose a robust framework called LDA (Long-Term Demand Advisor) to predict long-term (e.g., six months) demand in newly established bike regions. We first extract spatial and temporal features from multi-source open data, then apply our proposed G-clustering algorithm to measure the geographical characteristics and urban correlations in a city. The G-clustering algorithm takes the surrounding locations of the target candidate location into consideration to make a better prediction. Moreover, we extract the urban factors correlated with the long-term demand of sharing bikes, such as POIs (Point of Interests), road structure, and time. On the other hand, features from existing neighbor stations and future stations that have an overlapping operating period are also applied to new bike stations predictions since they will influence the number of demands and transit behaviors.

Our work focuses on long-term prediction, e.g., six months, since the short-term prediction (e.g., one month) is too difficult to predict and not worth studying in practice due to initially unstable environments. Moreover, the long-term effectiveness of stations seems worth investigating to aid in the government's decision and urban planning. For the reasons above, we consider that the predictions of no less than six months are relatively appropriate for urban decision-making. Figure 1 shows our proposed LDA framework, which consists of two major components: data preprocessing and batch prediction.

Data preprocessing. We first collect government open data and fetch others from Facebook Place API. We also record the latitude and the longitude of all bike stations. Next, we extract spatial features for each station, including nearby station features, seasons, number of POIs and number of check-ins, popular spots, number of intersections, and the length of bike routes based on the parameter r of the reachable station region. Finally, the proposed G-clustering algorithm is applied to cluster categories, and all of the extracted features are prepared to be fed into prediction models. Numerical data normalization, data cleaning, and missing data imputation are also applied to all features.

Batch prediction. We observe that new stations are sometimes constructed in batches in the real world. For example, the bike station deployment of New York from 2013 to 2017 can be mainly divided into four stages. Each stage contains at least 97 stations to be established in a newly expanded area. After data preprocessing, we split stations into original ones and the others in batches according to their month of establishment. From Batch 1 to Batch n ($n = 3$ for the NYC example) predictions, stations established before the corresponding period are set as training sets, and those in the period are testing sets. Finally, a strong prediction model can be applied to finish n batches of predictions.

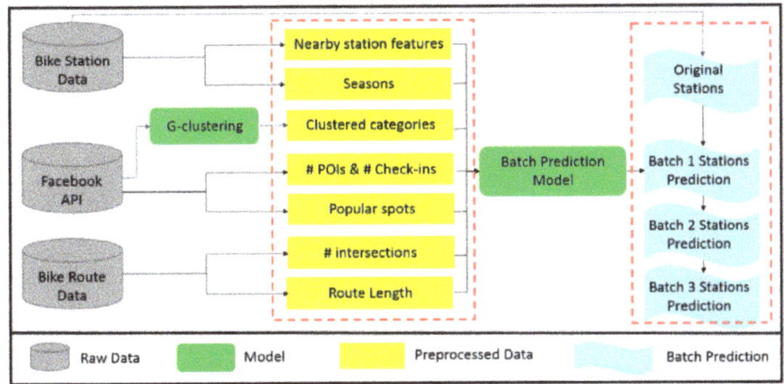

Figure 1. The overview of LDA Framework.

3. Methodology

In this section, we introduce (a) our proposed G-clustering algorithm, (b) extracted features correlated with rental/drop-off demand, and (c) demand prediction. We define notations used in this paper in Table 1. Problem definitions and our proposed framework are explained in Section 3.1.

Table 1. Notations used in this paper.

Notations	Descriptions
S	The station set $S = \{S_1, S_2, \ldots S_n\}$
C	The category set $C = \{CT_1, CT_2, \ldots CT_m\}$
POI	The POI set $POI = \{P_1, P_2, \ldots P_l\}$
R	The bike route set $R = \{R_1, R_2, \ldots R_k\}$
n, m, l, k	Number of stations/categories/POIs/bike routes
S_i	The feature set of ith station
$S_i \cdot \text{rent}$	Rental demand for S_i six months after the establishment
$S_i \cdot \text{drop}$	Drop-off demand for S_i six months after the establishment
$S_i \cdot \text{lat}$	Latitude of S_i
$S_i \cdot \text{long}$	Longitude of S_i
$S_i \cdot \text{date}$	The established date (e.g., operating date) of S_i
$CS\,(S_i, S_j)$	Cosine similarity between S_i and S_j
$P_{l \times m}$	Category matrix corresponding to POIs

3.1. Preliminary and Problem Definition

Definition 1. *Reachable Station Region. Considering how far a resident is willing to move and to get appropriate modeling of spatial factors, we define r as the radius of the farthest influencing area of a new station. In other words, when considering a location to build a new bike station, we propose to set a Euclidean distance r to extract the neighbor characteristics and features. Figure 2 gives an example. S_i is the target location, and we extract the density of our pre-defined POIs, which may be correlated with bike demands within the region.*

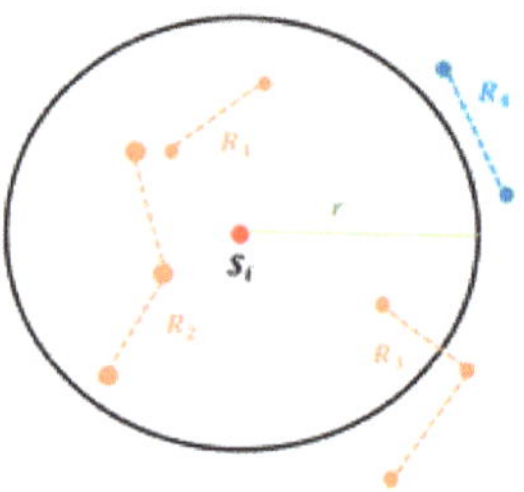

Figure 2. An illustration of the reachable region of station S_i and the corresponding bike routes.

Definition 2. *Nearby Stations. For the target location of a new station, we extract its top-k nearest stations whose establishment dates are earlier than the corresponding nearby stations. Three features of corresponding nearby stations are considered in our work: the difference of establishment dates, the number of cumulative demands, and the Euclidean distance between the target location and the nearby stations.*

Definition 3. *Bike Route Structure. We consider the road length of bike routes and the number of intersections in road structure as features to improve the demand prediction effectiveness. The reason that we consider the road length of bike routes is because a bike station might have a great demand in the long-term if its surrounding environment contains many bike routes, which are convenient for riders to travel by taking bikes. The high number of intersections might also indicate a traffic hub with significant human mobility, leading to increased potential bike flows.*

In Figure 3, there are three kinds of bike routes, and a bike route R_i is composed of multiple intersections (red points) and road segments (black dotted lines). Those route segments and intersections within the reachable station region of S_i are needed to be included. That is, the features extracted from R_1, R_2, and partial of R_3 in Figure 2 should be taken into consideration.

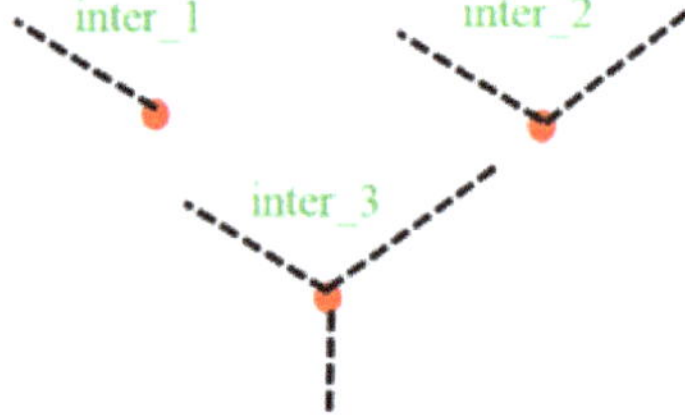

Figure 3. Examples of bike route intersection.

Definition 4. *Season. The period after building a station will span multiple seasons, and all of them should be considered since the commuting behavior of people will change with seasons. For each target station, we calculate how many months it will operate in each season. Spring is defined as the months from March to May, and the season changes every three months.*

Definition 5. *Category Vector P_i for Each POI. A POI P_i may have more than one corresponding category defined in Facebook Place API. Then, we define P_i as:*

$$P_i = \{p_{i,j}\} \tag{1}$$

where $p_{i,j} = 1$, if P_i belongs to CT_j; or 0, otherwise.
Where CT_j is the jth element in the category set defined by Facebook.

Problem Definition. *Rental/Drop-off demand prediction.* Given k new bike station locations $S_N = \{S_1, S_2, \ldots, S_k\}$, we want to predict the rental/drop-off demands of each station six months after its establishment; that is, S_i rent/S_i drop defined in Table 1.

3.2. G-Clustering

Since thousands of corresponding categories for *POI*s exist in certain regions, it is impractical to perform a one-to-one clustering for mapping a single category to a class. Therefore, we propose G-clustering to allocate categories into classes, where the characteristics of each category are similar to those of all the other categories in the same class. The G-clustering is inspired by the Gini coefficient [14], which is an index proposed by Corrado Gini to judge the fairness of annual income distribution according to the Lorenz curve. In order to apply the concept of the Lorenz Curve in our work, we modify the definition of it, which is illustrated in Figure 4. The Gini coefficient is equal to the area ratio between A and (A + B), and it is also equal to 2A since the sum area of A and B is 0.5.

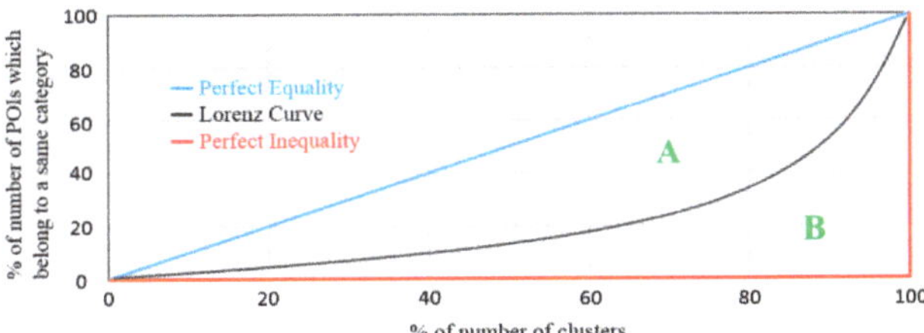

Figure 4. Lorenz Curve.

We apply this index to evaluate the distribution for each category in several regions clustered by geographical locations in the given problem space, and thus categories with similar distribution into the same clusters. The pseudocode for the G-clustering algorithm is depicted in Algorithm 1.

Algorithm 1 G-Clustering Algorithm

Input: *C, POI, P*;
Output: CC^D;
1. Cluster *POI* into D_1 clusters: $C_1, C_2, C_3 \ldots C_{D_1}$ by DBSCAN according to POI geographical locations;
2. Initialize $H^D = 0$;
3. **for** i = 1 : m **do**
4.   **for** j = 1: l **do**
5.   **if** $p_{j,i} == 1$ **then**
6. $h_{i,C_d} += 1$;
7. **end for**
8. Initialize $CC^D = \{CC_0, CC_1, \ldots CC_l\}$;
9. **for** k = 1: m **do**
10.   Category_point = CVF (h_k);
11.   idx = [10* Category_point] − 2;
12. CC_{idx} append category k;
13. **end for**
14. **function** CVF (array A)
15. A_norm = normalized(A);
16.   category_value_point = $(1 - \text{gini}(A\_norm)) + \dfrac{\text{gini}(A\_norm)}{\log_{10}(sum(A\_norm))+1}$;
17. **return** category_value_point

The proposed G-clustering is composed of three parts: initialization (line 1), construction for heat matrix (line 2 to 7), clustering for categories (line 8 to 13). First, *POI* is clustered into D_1 clusters, where D_1 is adjustable and set to 20 in our evaluation. Next, a heat matrix $H^D = \{h_{i,j}\}$ is constructed with $h_{i,j}$ representing the number of the ith category in the jth cluster according to P. Meanwhile, C_d refers to the corresponding cluster result of the jth POI in line 1. We divide each category into different groups according to its value point; the result CC^D is returned once all categories are run through and assigned to a certain cluster. Each item in CC^D indicates a set of the same level categories; meanwhile, we set $l = 6$ in the following evaluation.

Line 14 to 17 is a function that calculates the category value point. In this function, the Gini coefficient is applied to measure the distribution of each category in different clusters. The more even the distribution is, the closer this index gets to 0 (closer to the blue line in Figure 4); otherwise, it gets closer to 1 (closer to the red line in Figure 4). The function in lines 14 to 17 is designed to determine whether a category is indicative or not. The more even the distribution, the higher the value point, and the less indicative the category is. On the other hand, we also apply K-means to reallocate POI into D_2 clusters as described in line 1, where D_2 is set to 20 in our evaluation, and all other steps for G-clustering are left the same. The two types of clustering results are listed in Table 2.

Table 2. Results of G-clustering.

	POI **Categories**	
	DBSCAN	**K-Means**
Cluster 1	Reptile Per Store, Boat/Sailing, Instructor, Night Market	Archery Shop, Night Market, Squash Court
Cluster 2	Art Gallery, Local Business, Arts and Entertainment	Consulate and Embassy
Cluster 3	Elementary School, Language School, Lawyer and Law Firm	Art Gallery, Airport Lounge, Airport Terminal, Cruise Line
Cluster 4	Travel Service, Video Game, Junior High School, Public Swimming Pool	Surfing Spot, Landmark and Historical Place, College and University
Cluster 5	Music Video, Skate Shop, Football Stadium Education Company	Junior High School, Lawyer and Law Firm, Taxi Service
Cluster 6	Aquarium, Diagnostic Center, Drive-In Movie Theater	Public Swimming Pool, Bus Station, Supermarket, Pizza Place
Cluster 7	Fitness Venue, Hockey Arena, Retail Bank	Gas Station, Catholic Church

We list several representative categories in each cluster to explain the effectiveness of the G-Clustering algorithm. In the left column of Table 2 (DBSCAN), categories more evenly distributed in the area such as Fitness Venues and Retail Banks are clustered in the same class since these types of POIs have no obvious regional characteristics. In other words, there is no excessive demand from these categories in specific districts. On the contrary, the number of Night Markets and Art Galleries is obviously larger in certain areas and thus may be regarded as indicative categories in the prediction. A similar trend can also be found in the right column of Table 2 (K-means). The small difference between DBSCAB and K-means clustering results in some categories being clustered in different hierarchies. For example, Art Galleries and Junior High Schools are in different clusters, which might be due to their different local characteristics. The clustering results will then be used as important categorical features for the bike stations.

3.3. Feature Extraction

We divide all features into six categories based on their data sources. They are I.#POI and #Checkins, II. Nearby station features, III. Popular spots, IV. G-clustering, V. Bike route structure, VI. Season. In the experiment, we will evaluate the effectiveness of these six categories. In Table 3, we give an overview of features.

Table 3. All Features and their Descriptions.

Features	
Feature Name	**Description**
POIs	*#POIs* in Facebook
check-ins	# check-ins in Facebook
Nearby station features	The difference of establishing dates, the number of cumulative demands, and the Euclidean distance between the target location and their nearby stations
G-clustering (DBSCAN)	Category clustering results by applying DBSCAN
G-clustering(K-means)	Category clustering results by applying K-means
Bike route structure	Sum of total route length and the number of intersections of bike routes in the reachable region of the station
Season	Operating seasons

I. #POI and #Checkins. The number of POIs (Point-Of-Interests) and check-ins can be indicated as the level of prosperity in an area and therefore results in a higher frequency of bike demands. We extract #POI and #Checkin's based on Facebook API.

II. Nearby station features. A new station is usually highly related to the nearby stations due to spatial effect and human mobility. Three features of top-k nearby stations are considered in our work: the difference in establishing dates, the number of cumulative demands, and the Euclidean distance between the target location and their nearby stations. If a nearby station is built later than the target location, the number of cumulative demands will be set as zero. After extraction, we obtain a total of $3\,k$ features for nearby stations. Such a large number might dominate the prediction result of the classifier. Therefore, PCA (Principal Component Analysis) is applied to reduce feature dimensions.

III. Popular spots. We define popular types of POIs (e.g., over 1000 stores in New York) specifically, calculating the number of corresponding types of POIs and check-ins of each station in its reachable station region.

IV. G-clustering. We perform the G-clustering algorithm to use the clustering result as our features. We set two kinds of clustering methods in step 1 of G-clustering: one is DBSCAN, and the other is K-means.

IV-D. Category clustering results applying DBSCAN.

IV-K. Category clustering results applying K-means.

V. Bike route structure. The more bike routes near a station, the higher the probability the bikes will be rented for convenience. We then calculate the sum of total route length and the number of intersections of bike routes in the reachable region of station S_i.

VI. Season. Seasons will greatly affect people's willingness to ride a bike. For example, users tend to rent a bike in spring rather than in winter, so data in December is obviously less than in May. According to Definition 4, if station S_i starts operating in May, then the number of months in the following six months from spring to winter is 1, 3, 2, 0.

3.4. Batches Prediction

Constructing a bike-sharing system in most cities can be realized in several steps (batches). First, the government sets up a large number of bike station locations in the downtown area where lots of commercial buildings and tourist attractions are located, spreading out to nearby regions in the following months, perhaps with a short lull. However, as the frequency of shared bikes and new users increases, the government needs to distribute a wider range of bike locations to satisfy users' demand, and therefore the area expands to the suburbs and even empty districts in the city center to relieve excessive demand.

Definition 6. *Batches Prediction. Our work focuses on batch prediction; in other words, site prediction established in later stages in the suburbs or border zones, which are also defined as* *expansion areas in this paper. We propose to utilize EMA (Exponential Moving Average) to determine the periods of batches given a continuous time interval. The EMA is a type of average that applies weighting factors that decrease exponentially to the past. We define a batch that exists if the EMA values of month demands are continuously not less than a given threshold for several months. Figure 5 shows the EMA distribution that we perform using 2, 3, and 6 months as the average units. For example, if we define the threshold as 30 using the two months average of EMA for New York City, we can then identify three batches(peaks) from 2013 to 2018. The corresponding periods of the first, second and third batches of NYC are shown in Table 4. Our framework provides the government the estimation of the demands of newly established stations through given locations, and this can also be applied to the expansion of other facilities.*

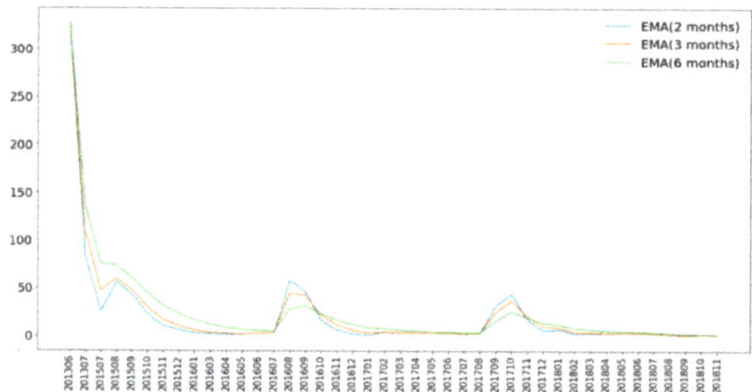

Figure 5. EMA of each month.

Table 4. Data source and detailed contents.

	New York Citi Bike System			
	Origin	Batch 1	Batch 2	Batch 3
Time Span	June 2013~June 2015	August 2015~September 2015	August 2016~September 2016	September 2017~October 2017
# Stations	329	121	127	97
	Facebook Place API			
# Check-ins	175+ billion			
# POI Categories	1279			
	New York Bike Route			
# Intersections	26,868			
Total Length	1300+ km			

In this work, we mainly use XGBoost [15] to make the prediction for each batch. Apart from XGBoost in this work, other machine learning approaches can also be applied under our framework. We will compare their effectiveness in our experiments.

4. Experiments

To evaluate the performance of our framework, we conduct experiments on a real-world dataset from New York Citi Bike. Details of multi-source open data are in Table 4. Bike station data are collected from June 2013 to November 2018, and stations operating for less than six months, or with a monthly average demand of less than 300, are removed. Batches can be realized as the time period of a relatively large number of bike stations construction. Stations with established dates from June 2013 to July 2015 are the origin. From Batch 1 to 3 prediction, we divide stations in the training set and testing set according to their established date. For instance, in Batch 2 prediction, stations established earlier than August 2016 are training data, and the other stations established during August 2016 to September 2016 are testing data. We retrieve multi-source open data from Citi Bike (the bike-sharing system in New York), bike routes data, and Facebook Place API. Detailed datasets are listed in Table 4. The settings for radius r of the reachable station region are 500 m, and we extract the top-15 nearby station features in our experiment.

4.1. Experimental Settings

We evaluate the effectiveness of different combinations of feature sets, which are listed in Table 5. A single factor is not listed due to the low performance; however, important factors such as I and II are included in each set.

Table 5. Feature set combination.

Feature Set	Features					
	I + II	III	IV-D	IV-K	V	VI
A	■		■			
B	■				■	
C	■	■	■			■
D	■			■	■	
E	■	■			■	■
F	■	■	■		■	■
G	■	■		■	■	■
CC-XGB	■	■	■	■	■	■

4.1.1. Baselines

The framework proposed in our work is denoted as **Category Clustering applying eXtremeGradient Boosting** (**CC-XGB**). XGBoost [15] is regarded as one of the most powerful techniques in the public transportation domain.

Regressors such as **RF** (Random Forest), **LR** (Linear Regression), and **SVR** (Support Vector Regression) are used in comparison; **NN** (Neural Network) is also included as a predictor. Moreover, the following compared baselines according to historical average demand are used to verify the performance of our models.

HA (**History Average**). History rental/drop-off average of stations whose established months are earlier than the predicted station S_i.

HSA (**History Similarity Average**). History rental/drop-off average of stations whose established month is earlier and is in the top-five high cosine similarity with the predicted station S_i.

HSW (**History Similarity Weight**). Let $S_{i,1} \sim S_{i,5}$ be the top-five high cosine similarity stations to the predicted station S_i.

$$\text{HSW}(S_i) = \frac{\sum_{k=1}^{5}(S_i \cdot rent) * CS(S_i, S_{i,k})}{\sum_{k=1}^{5} CS(S_i, S_{i,k})} \tag{2}$$

HSC (**History in the Same Cluster**). History rental/drop-off average of stations whose established months are earlier in the same DBSCAN cluster with station S_i.

HNN (**History Nearest Neighbors**). History rental/drop-off average of stations whose established month is earlier and distance in the top-k nearest with the predicted station S_i.

4.1.2. Evaluation Metric

Since bike demands vary dramatically due to many factors, RMSLE (**Root Mean Squared Logarithmic Error**) is a more appropriate metric to adopt.

$$\sqrt{\frac{1}{N} \sum_{i=1}^{N} \left(\log((S_i \cdot rent/drop) - (\log((S_i \cdot rent'/drop')))\right)^2} \tag{3}$$

$S_i \cdot rent/drop$ is the ground truth of demand in six months of S_i, and $S_i \cdot rent'/drop'$ is the corresponding prediction result of the ground truth.

4.2. Batch Prediction Results

4.2.1. Overall Comparison

In this part, we show the effectiveness of the proposed LDA and the comparison to the baselines.

Results of Baselines: Figure 6a,b represent the baseline results of rental and drop-off, respectively. Baselines without machine learning such as HA, HSA, and HSW are worse than regression or NN results. CC-XGB, our proposed framework, defeats the second-best with an average of 0.2 to 0.3 approximately in RMSLE, whether in a rental or drop-off situation.

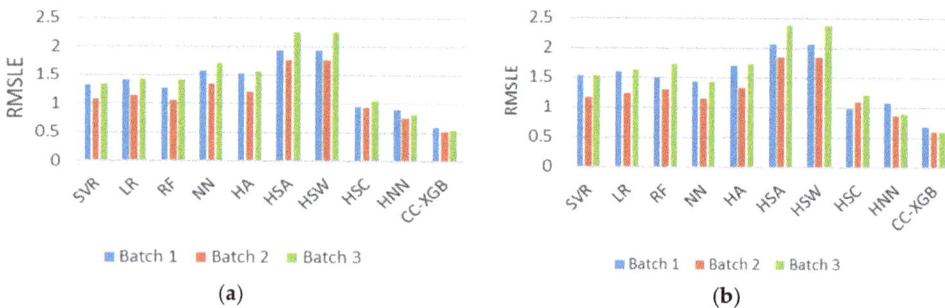

Figure 6. Performance of baselines. (**a**) represents the rental demand and (**b**) represents the drop-off one.

Results of Feature Combination: Figure 7a,b represent the results of the different combinations of features in rental and drop-off, respectively. The result of feature set E without features of category clustering in Figure 7b has poor performance evidently, confirming that G-clustering is effective. No one always performs better between IV-D and IV-K; one reason may be due to slight differences in clustering results. Though the differences in the batches are not obvious, CC-XGB performs much better than other feature sets in batch 2 and 3, confirming the applicability of our framework.

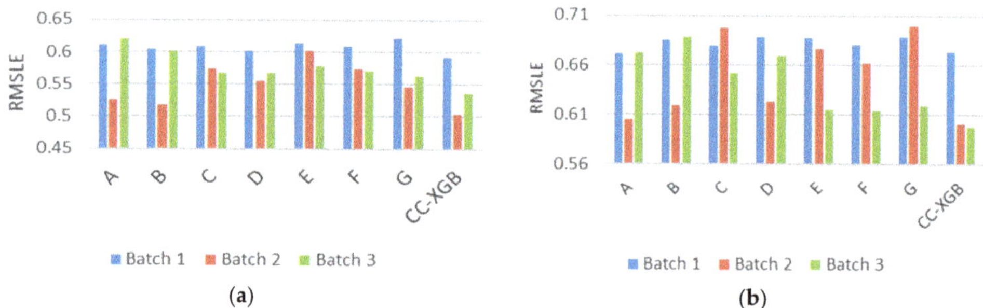

Figure 7. Performance of different feature combinations. (**a**) represents the rental demand and (**b**) represents the drop-off one.

Analyze for Batches: Under the prediction result of CC-XGB, our proposed framework, RMSLE decreases from Batch 1 to 3 in drop-off mode; yet results in Batch 3 are worse than in Batch 2 in rental mode. We infer that the demand for renting bikes downtown is more stable than in other areas; in other words, users are less willing to rent a bike from newly established stations, making the prediction difficult. On the other hand, the drop-off demand is hard to predict for the first batch stations.

4.2.2. Region Size Setting for Extracted Features

In our experiment, the reachable station region is set as 500 m (Figure 1 (left)) for the appropriate number of POIs and check-ins. In this part, we would like to compare how different radiuses affect the results. Features I, III, and V are related to the reachable station number. Experiments are conducted from 300 m to 1000 m in Figure 8. As shown in Figure 8, a larger radius does not necessarily mean a better prediction result. We can observe that in Figure 8, 500 m is a superior radius region for a target station to extract corresponding features since the RMSLE for three batches are relatively low when r = 500 m.

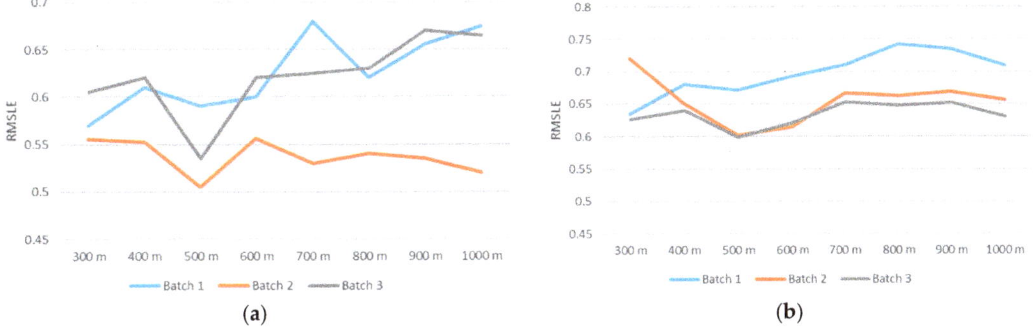

Figure 8. Results of different regions for feature extraction in (a) Rental and (b) Drop-off.

4.2.3. Feature Importance (FI)

Figures 9–11 show the feature importance for Batch 1 to Batch 3, and the detailed features whose importance is ranked in the top five are listed aside. Figure 9a, Figure 10a, and Figure 11a show rental feature importance, while Figure 9b, Figure 10b, and Figure 11b show drop-off feature importance. Overall, the nearby station features are extremely important in prediction since they have the highest scores in all situations; in particular, the score gap is more significant in Batch 3 (Figure 11a,b), explaining that nearby stations are highly correlated to newly established stations. The feature importance obtained from G-clustering is all ranked in the top five in those five figures (top-6 in Figure 11a), proving that our idea of clustering categories is reasonable and useful.

4.2.4. Prediction of Different Periods

Our work focuses on long-term prediction, e.g., six months, since the short-term prediction (e.g., one month) is too difficult to predict and not worth studying in practice due to initially unstable environments. The experiments conducted on one, three, six and nine month(s) in Figures 12 and 13 have shown that the six months' prediction has the best performance. The nine months case is worse than the six months. The reason comes from the data instead of our model. In our dataset, we observe that there are some new stations built surrounding the existing stations after six months so that the demands of some stations in a certain batch were influenced by new stations. The prediction then would become not so accurate. For batch 1, batch 2, and batch 3, the RMSLE of six months is the lowest comparing to one month, three months, and nine months. In batch 1, the gap between six-month and others for rental is from 0.02 to 0.31, and the gap for drop-off is from 0.07 to 0.36. In batch 2, the gap between six-month and others for rental is from 0.07 to 0.2, and the gap for drop-off is from 0.01 to 0.11. In batch 3, the gap between six-month and others for rental is from 0.09 to 0.2, and the gap for drop-off is from 0.03 to 0.09.

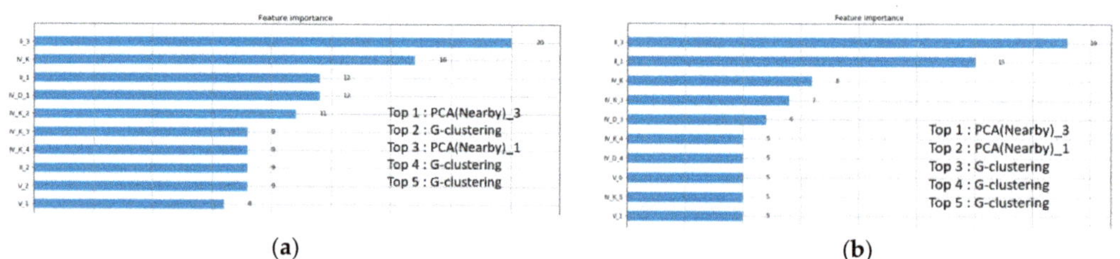

Figure 9. Feature importance of Batch 1. (**a**) Represents rental demand, and (**b**) represents the drop-off one.

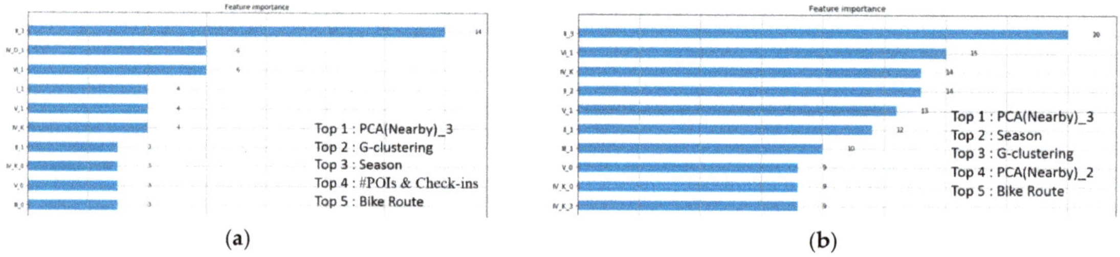

Figure 10. Feature importance of Batch 2. (**a**) Represents rental demand, and (**b**) represents the drop-off one.

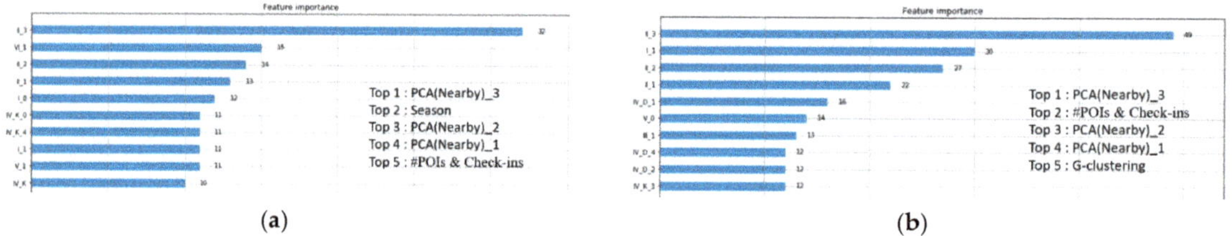

Figure 11. Feature importance of Batch 3. (**a**) Represents rental demand, and (**b**) represents the drop-off one.

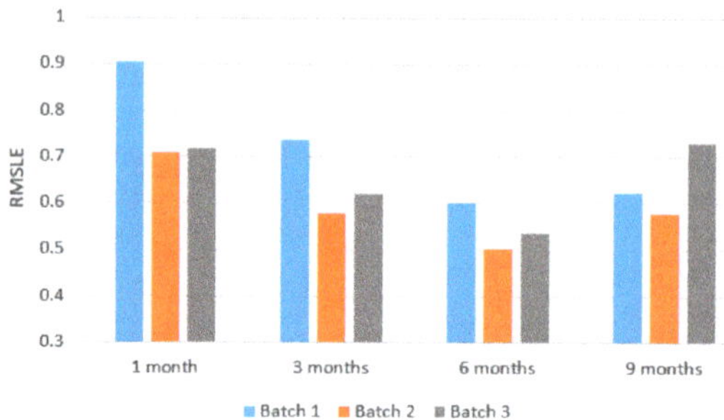

Figure 12. Different periods of prediction for Rental.

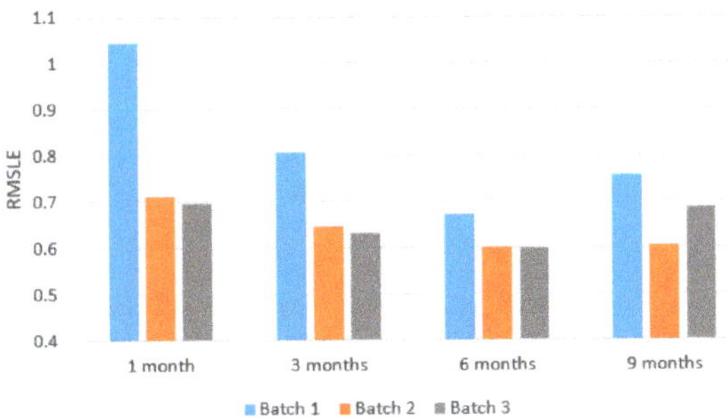

Figure 13. Different periods of prediction for Drop-off.

4.3. Random Prediction Results

Similar to works focusing on predicting demand through splitting data into the training set and testing set without considering established time, we also repeat the same steps in our experiment to verify the usefulness of our LDA framework. In other words, we conduct the prediction experiment of rental/drop-off demand 10,000 times through randomly divided stations and return the average RMSLE result (Figure 14). The result of CC-XGB still performs the best. However, our superiority is not so apparent since our proposed features are relatively suitable for batch prediction rather than random prediction.

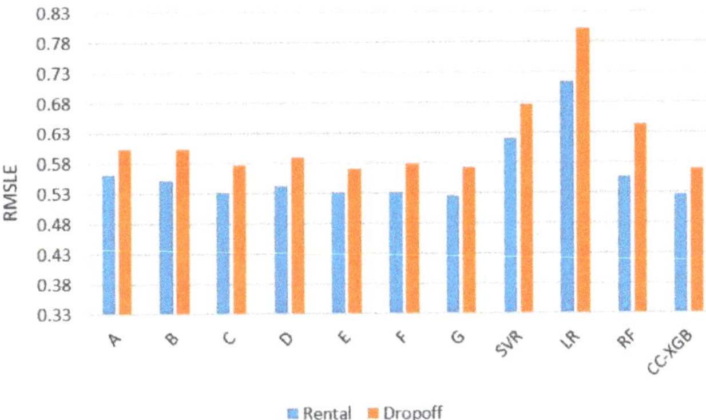

Figure 14. Random prediction for rental/drop-off.

5. Discussion of the Results

In this research, we are facing the demand prediction problem of real-world bike-sharing systems. In the previous experiments, we can observe that two important factors in LDA settings are worth discussing, considering real-world applications. One is batch deployment. Another is the prediction time period. These two factors are mutually high-correlated.

Discussion of batch deployment: In the past, existing works usually aimed to predict human flows for each individual station in a short time, such as next hour, next day, and next 1–3 days. However, in real-world applications, we claim that predicting long-term demands for station deployment is also critical for urban planning and construction. Therefore, we propose an LDA framework, which can help governments or transportation companies to make decisions for deploying bike-sharing services in a smart city. We have observed that the real-world bike stations are mainly built-in batches for expansion areas in modern cities. That is, we can use only the historical demand data from previously deployed areas for prediction. The batch consideration in the LDA framework confirms that our work is the first to address the long-term demand of new stations for future batch stations, providing the government with a tool to pre-evaluate the bike flow of new stations before deployment. LDA can avoid wasting resources such as personnel expense or budget.

Discussion of prediction periods: In Section 4.2.4, our experiment shows that the six months' prediction has the best performance. The reason is we observe that in the New York Citi bike sharing system there are some new stations built surrounding the existing stations after six months so that the demands of some stations in a certain batch were influenced by new stations. However, we believe our proposed LDA framework is also helpful for making decisions using the prediction results of periods that are more than six months since the prediction error is mainly from the crawled future data. To conclude, our LDA framework can work as a web service to assess the effectiveness of new bike stations for expansion areas in different cities.

6. Related Work

Impacts of bike-sharing systems. Many studies analyzed the impact of bike-sharing systems on different aspects of society. The work of [16] mentioned that bike-sharing programs have significantly positive externalities, including the economy, the environment, and health-related externalities. Moreover, introducing bike-sharing systems gives an opportunity to organize public transport interchanges better [17]. Shared bicycles facilitate allow getting to stops and stations for those who do not own a private bike. Additionally, bike-sharing gives more flexibility–shared bicycles users are not burdened with the threat of theft or an obligation to service the bicycle. The study of [18] developed a spatial Agent-based model to simulate the use of bike-sharing services and other transport modes in Taipei city. The simulation results indicate that free use of bike-sharing to connect the transit system can be more sustainable with 1.5 million US dollars in transportation damage cost saved per year and 22 premature deaths further prevented per year due to mode shift to cycling and walking based on the business. The work of [19] demonstrated the importance of user-interface (UI) design, social influence, and new media in affecting users' awareness of and attitude towards uncivilized behaviors, which in turn improve their intention of bike-sharing services use.

The emergence of dockless bike-sharing services has revolutionized bike-sharing markets in recent years. The work of [20] suggested that the dockless design of bike-sharing systems significantly improves users' experiences at the end of their bike trips. However, the availability and usage rates of dockless bike-sharing systems imply that they may seriously affect individuals' subjective well-being by influencing their satisfaction with their travel experiences, health, and social participation, which requires further exploration. The work of [21] mentioned that, as Chinese enterprises already invest heavily in Europe, it is crucial for policymakers to introduce rules that would counteract potentially negative consequences of the introduction of a new system of bike-sharing and support positive effects.

Behavior analysis in bike-sharing systems. The behavior patterns of users in bike-sharing systems are also worth exploring. The estimation results of [22] show that descriptive norm, conformity tendency, and past behavior are important factors that affect both e-bike riders' intention to violate traffic rules and accident proneness. The work of [23] found that perceived ease of use positively influences the attitude towards the systems and the use intention. Therefore, the bike-sharing operating companies should carefully design the usage procedures to make them as simple as possible. The work of [24] adopted machine learning to show that speed, travel distance, and the number of parks and recreational facilities seem to be critical spatial predicting factors of the travel choice in bike-sharing systems. Moreover, considering the impact of COVID-19 on bike-Sharing systems, the work of [25] indicated that usage bike-sharing is more likely to become a more preferable mobility option for people who were previously commuting with private cars as passengers and people who have already registered users in a bike-sharing system. The bike-sharing systems have proved in the study of [26] to be more resilient than the subway system, with a less significant ridership drop and an increase in its trips' average duration.

The work of [27] shows that a high availability rate, a low price, and a large difference in travel time between bike-sharing and other travel modes make potential customers more likely to use a bike-sharing program by modeling a different aspect of travel behavior: heterogeneous time-sensitive customers.

Bike station deployment. Research on bike-sharing systems is becoming more and more prevalent worldwide; topics covered range from site selection to rebalancing bike distribution. The works of [28,29] try to figure out the best locations for bike stations from candidate sites. The work of [30] proposes a mixed model to minimize fixed construction costs and variable operational costs. Research combining probability and simulation such as in [31] develops a probabilistic model to infer future demand, and the work of [32] adopts Monte Carlo to predict the over-demand probability in each bike station cluster. On the other hand, the works of [8,33–35] focus on bike imbalance and rebalancing problems, proposing methods to transfer bikes between stations.

Bike demand analysis and prediction. In all bike-related problems, the most widely studied is bike demand or traffic flow prediction. The studies of [22,36] have identified the importance of natural environmental factors such as temperature, precipitation, and humidity on cycling activities across different cities. At the feature level, studies [5,37] consider a single factor instead of multiple aspects features and thus may neglect representative elements. Other works collect historical data such as public transportation pattern records [38], crowd flow [39], meteorology data [7,8,40], and so on. Clustering methods applied to bike stations are more and more common in recent works since bike stations share partially similar regional characteristics and will reduce the variance and improve prediction accuracy. The difference between these works is what the cluster is based on. The works of [7,9,32,41] cluster stations according to bike transition pattern records, geographical locations, bike usage, etc. The study of [42] employs SimRank to calculate the similarities between stations and then adopts the density clustering algorithm OPTICS.

However, the works above are not applicable for our scenario since they rely on the historical mobility data and therefore are unavailable for batch prediction in newly established stations in expansion areas. Furthermore, they mostly aim to predict demand in a relatively short period from hourly [11,43], rush hours [9], to weekends and holidays [32], and thus cannot be applied to our long-term prediction.

7. Conclusions

In this paper, we propose a framework consisting of spatial and temporal features to predict long-term rental/drop-off demand in newly established stations, e.g., in expansion areas. Specifically, we extract features from multi-source open data, propose G-clustering, and apply regression models to predict the demand of stations in three batches according to the established periods. Experiments carried out in the New York Citi bike sharing system demonstrate that our framework for long-term prediction in expansion areas is applicable and outperforms baselines. In the future, we aim to analyze more factors, such as transfer probability from downtown to the suburbs and deal with unusual events to improve predicting accuracy.

Author Contributions: Supervision, H.-P.H.; methodology, H.-P.H., F.L. and T.-Y.K.; validation, T.-Y.K.; investigation, H.-P.H., F.L., J.J. and T.-Y.K.; writing—original draft preparation, F.L., J.J. and T.-Y.K.; writing—review and editing, H.-P.H. and Y.-E.C. All authors have read and agreed to the published version of the manuscript.

Funding: This research was funded by the Ministry of Science and Technology (MOST) of Taiwan under grants MOST 108-2221-E-006-142, MOST 108-2636-E-006-013, and MOST 109-2636-E-006-025 (MOST Young Scholar Fellowship).

Acknowledgments: This work was partially supported by the Ministry of Science and Technology (MOST) of Taiwan under grants MOST 108-2221-E-006-142, MOST 108-2636-E-006-013, and MOST 109-2636-E-006-025(MOST Young Scholar Fellowship).

Conflicts of Interest: The authors declare no actual or potential conflict of interest.

References

1. Chen, S.-Y. Using the Sustainable Modified Tam and Tpb to Analyze the Effects of Perceived Green Value on Loyalty to a Public Bike System. *Transp. Res. Part A Policy Pract.* **2016**, *88*, 58–72. [CrossRef]
2. Cohen, B.; Kietzmann, J. Ride On! Mobility Business Models for the Sharing Economy. *Organ. Environ.* **2014**, *27*, 279–296. [CrossRef]
3. Eckhardt, G.M.; Bardhi, F. The Sharing Economy Isn't About Sharing at All. *Harv. Bus. Rev.* **2015**, *28*, 881–898.
4. Schor, J.B.; Fitzmaurice, C.J. Collaborating and Connecting: The Emergence of the Sharing Economy. *Handb. Res. Sustain. Consum.* **2015**, *26*, 410–425.
5. Alvarez-Valdes, R.; Belenguer, J.M.; Benavent, E.; Bermudez, J.D.; Muñoz, F.; Vercher, E.; Verdejo, F. Optimizing the level of service quality of a bike-sharing system. *Omega* **2016**, *62*, 163–175. [CrossRef]
6. Kalvapalli, S.P.K.; Chelliah, M. Analysis and Prediction of City-Scale Transportation System Using XGBOOST Technique. In *Recent Developments in Machine Learning and Data Analytics*; Springer: Berlin/Heidelberg, Germany, 2019; pp. 341–348.
7. Li, Y.; Zheng, Y.; Zhang, H.; Chen, L. Traffic prediction in a bike-sharing system. In Proceedings of the 23rd ACM SIGSPATIAL International Conference on Advances in Geographic Information Systems, Seattle, WA, USA, 3–6 November 2015; pp. 1–10.
8. Liu, J.; Sun, L.; Chen, W.; Xiong, H. Rebalancing bike sharing systems: A multi-source data smart optimization. In Proceedings of the 22nd ACM SIGKDD International Conference on Knowledge Discovery and Data Mining, San Francisco, CA, USA, 13–17 August 2016; pp. 1005–1014.
9. Feng, Y.; Affonso, R.C.; Zolghadri, M. Analysis of bike sharing system by clustering: The Vélib'case. *IFAC-PapersOnLine* **2017**, *50*, 12422–12427. [CrossRef]
10. Lin, L.; He, Z.; Peeta, S. Predicting station-level hourly demand in a large-scale bike-sharing network: A graph convolutional neural network approach. *Transp. Res. Part C Emerg. Technol.* **2018**, *97*, 258–276. [CrossRef]
11. Liu, J.; Sun, L.; Li, Q.; Ming, J.; Liu, Y.; Xiong, H. Functional zone based hierarchical demand prediction for bike system expansion. In Proceedings of the 23rd ACM SIGKDD International Conference on Knowledge Discovery and Data Mining, Halifax, NS, Canada, 13–17 August 2017; pp. 957–966.
12. Long, Y.; Shen, Z. Discovering functional zones using bus smart card data and points of interest in Beijing. In *Geospatial Analysis to Support Urban Planning in Beijing*; Springer: Berlin/Heidelberg, Germany, 2015; pp. 193–217.
13. Yuan, N.J.; Zheng, Y.; Xie, X.; Wang, Y.; Zheng, K.; Xiong, H. Discovering urban functional zones using latent activity trajectories. *IEEE Trans. Knowl. Data Eng.* **2014**, *27*, 712–725. [CrossRef]
14. Gini, C. Measurement of Inequality of Incomes. *Econ. J.* **1921**, *31*, 124–126. [CrossRef]
15. Chen, T.; Guestrin, C. Xgboost: A scalable tree boosting system. In Proceedings of the 22nd Acmsigkdd, San Francisco, CA, USA, 13–17 August 2016; pp. 785–794.
16. Qiu, L.-Y.; He, L.-Y. Bike Sharing and the Economy, the Environment, and Health-Related Externalities. *Sustainability* **2018**, *10*, 1145. [CrossRef]
17. Brunner, H.; Hirz, M.; Hirshberg, W.; Fallast, K. Evaluation of various means of transport for urban areas. *Energy Sustain. Soc.* **2018**, *8*, 9. [CrossRef]
18. Lu, M.; Hsu, S.-C.; Chen, P.-C.; Lee, W.-Y. Improving the sustainability of integrated transportation system with bike-sharing: A spatial agent-based approach. *Sustain. Cities Soc.* **2018**, *41*, 44–51. [CrossRef]
19. Jia, L.; Liu, X.; Liu, Y. Impact of Different Stakeholders of Bike-Sharing Industry on Users' Intention of Civilized Use of Bike-Sharing. *Sustainability* **2018**, *10*, 1437. [CrossRef]
20. Chen, Z.; Lierop, D.V.; Ettema, D. Dockless bike-sharing systems: What are the implications? *Transp. Rev.* **2020**, *40*, 333–353. [CrossRef]
21. Bieliński, T.; Ważna, A. New Generation of Bike-Sharing Systems in China: Lessons for European Cities. *J. Manag. Financ. Sci.* **2018**, *11*, 25–42.
22. Tang, T.; Guo, Y.; Zhou, X.; Labi, S.; Zhu, S. Understanding electric bike riders' intention to violate traffic rules and accident proneness in China. *Travel Behav. Soc.* **2021**, *23*, 25–38. [CrossRef]
23. Yu, Y.; Yi, W.; Feng, Y.; Liu, J. Understanding the Intention to Use Commercial Bike-sharing Systems: An Integration of TAM and TPB. The Sharing Economy. In Proceedings of the 51st Hawaii International Conference on System Sciences, Waikoloa, Big Island, HI, USA, 3–6 January 2018.
24. Zhou, X.; Wang, M.; Li, D. Bike-sharing or taxi? Modeling the choices of travel mode in Chicago using machine learning. *J. Transp. Geogr.* **2019**, *79*, 102479. [CrossRef]
25. Nikiforiadis, A.; Ayfantopoulou, G.; Stamelou, A. Assessing the Impact of COVID-19 on Bike-Sharing Usage: The Case of Thessaloniki, Greece. *Sustainability* **2020**, *12*, 8215. [CrossRef]
26. Teixeria, J.F.; Lopes, M. The link between bike sharing and subway use during the COVID-19 pandemic: The case-study of New York's Citi Bike. *Transp. Res. Interdiscip. Perspect.* **2016**, *6*, 100166. [CrossRef] [PubMed]
27. Chen, Y.; Wang, D.; Chen, K.; Zha, Y.; Bi, G. Optimal pricing and availability strategy of a bike-sharing firm with time-sensitive customers. *J. Clean. Prod.* **2019**, *228*, 208–221. [CrossRef]
28. Liu, J.; Li, Q.; Qu, M.; Chen, W.; Yang, J.; Xiong, H.; Zhong, H.; Fu, Y. Station site optimization in bike sharing systems. In Proceedings of the 2015 IEEE International Conference on Data Mining, Atlantic City, NJ, USA, 15–17 November 2015; pp. 883–888.

29. Martinez, L.M.; Caetano, L.; Eiró, T.; Cruz, F. An optimisation algorithm to establish the location of stations of a mixed fleet biking system: An application to the city of Lisbon. *Procedia-Soc. Behav. Sci.* **2012**, *54*, 513–524. [CrossRef]

30. Cao, J.X.; Xue, C.C.; Jian, M.Y.; Yao, X.R. Research on the station location problem for public bicycle systems under dynamic demand. *Comput. Ind. Eng.* **2019**, *127*, 971–980. [CrossRef]

31. Gast, N.; Massonnet, G.; Reijsbergen, D.; Tribastone, M. Probabilistic forecasts of bike-sharing systems for journey planning. In Proceedings of the 24th ACM International on Conference on Information and Knowledge Management, Melbourne, Australia, 18–23 October 2015; pp. 703–712.

32. Chen, L.; Zhang, D.; Wang, L.; Yang, D.; Ma, X.; Li, S.; Wu, Z.; Pan, G.; Nguyen, T.M.; Jakubowicz, J. Dynamic cluster-based over-demand prediction in bike sharing systems. In Proceedings of the 2016 ACM International Joint Conference on Pervasive and Ubiquitous Computing, Heidelberg, Germany, 12–16 September 2016; pp. 841–852.

33. Liu, Z.; Shen, Y.; Zhu, Y. Inferring dockless shared bike distribution in new cities. In Proceedings of the Eleventh ACM International Conference on Web Search and Data Mining, Marina Del Rey, CA, USA, 5–9 February 2018; pp. 378–386.

34. Singla, A.; Santoni, M.; Bartók, G.; Mukerji, P.; Meenen, M.; Krause, A. Incentivizing users for balancing bike sharing systems. In Proceedings of the Twenty-Ninth AAAI Conference on Artificial Intelligence, Austin, TX, USA, 25–30 January 2015.

35. Vogel, P.; Greiser, T.; Mattfeld, D.C. Understanding bike-sharing systems using data mining: Exploring activity patterns. *Procedia-Soc. Behav. Sci.* **2011**, *20*, 514–523. [CrossRef]

36. Eren, E.; Uz, V.E. A review on bike-sharing: The factors affecting bike-sharing demand. *Sustain. Cities Soc.* **2020**, *54*, 101882. [CrossRef]

37. Schuijbroek, J.; Hampshire, R.C.; van Hoeve, W.-J. Inventory rebalancing and vehicle routing in bike sharing systems. *Eur. J. Oper. Res.* **2017**, *257*, 992–1004. [CrossRef]

38. Wang, D.; Wu, E.; Tan, A.-H. Analysis of Public Transportation Patterns in a Densely Populated City with Station-based Shared Bikes. In Proceedings of the 3rd International Conference on Crowd Science and Engineering, Singapore, 28–31 July 2018; pp. 1–8.

39. Hoang, M.X.; Zheng, Y.; Singh, A.K. FCCF: Forecasting citywide crowd flows based on big data. In Proceedings of the 24th ACM SIGSPATIAL International Conference on Advances in Geographic Information Systems, Burlingame, CA, USA, 31 October–3 November 2016; pp. 1–10.

40. Yang, Z.; Hu, J.; Shu, Y.; Cheng, P.; Chen, J.; Moscibroda, T. Mobility modeling and prediction in bike-sharing systems. In Proceedings of the 14th Annual International Conference on Mobile Systems, Applications, and Services, Singapore, 26–30 June 2016; pp. 165–178.

41. Etienne, C.; Latifa, O. Model-based count series clustering for bike sharing system usage mining: A case study with the Vélib'system of Paris. *ACM Trans. Intell. Syst. Technol. (TIST)* **2014**, *5*, 1–21. [CrossRef]

42. Liu, L.; Hu, Z.; Zhou, C.; Xu, G. Research on the clustering algorithm of the bicycle stations based on OPTICS. *Concurr. Comput. Pract. Exp.* **2019**, *31*, e4876. [CrossRef]

43. Hulot, P.; Aloise, D.; Jena, S.D. Towards station-level demand prediction for effective rebalancing in bike-sharing systems. In Proceedings of the 24th ACM SIGKDD International Conference on Knowledge Discovery & Data Mining, London, UK, 19–23 August 2018; pp. 378–386.

applied
sciences

Article

Cycle Logistics Projects in Europe: Intertwining Bike-Related Success Factors and Region-Specific Public Policies with Economic Results

Carlo Giglio [1,*], **Roberto Musmanno** [2] **and Roberto Palmieri** [2]

1   Department of Civil, Energy, Environmental and Material Engineering, Mediterranean University of Reggio Calabria, 89122 Reggio Calabria, Italy

2   Department of Mechanical, Energy and Management Engineering, University of Calabria, 87036 Rende, Italy; roberto.musmanno@unical.it (R.M.); roberto.palmieri@unical.it (R.P.)

*   Correspondence: carlo.giglio@unirc.it

Abstract: The aim of this paper is to investigate whether and which specific, distinctive characteristics of European cycle logistics projects and the corresponding supporting policies have an impact on their economic performances in terms of profit and profitability. First, we identify project success factors by geographic area and project-specific characteristics; then, we statistically test possible dependence relationships with supporting policies and economic results. Finally, we provide a value-based identification of those characteristics and policies which more commonly lead to better economic results. This way, our work may serve as a basis for the prioritization and contextualization of those project functionalities and public policies to be implemented in a European context. We found that cycle logistics projects in Europe achieve high profit and profitability levels, and the current policies are generally working well and supporting them. We also found that profit and profitability vary across the bike model utilized: mixing cargo bikes and tricycles generates the highest profit and profitability, whilst a trailer–tricycle–cargo bike mix paves the way for high volumes and market shares.

Keywords: cycle logistics; European projects; goods delivery; bike delivery; cargo cycle; cargo bike

Citation: Giglio, C.; Musmanno, R.; Palmieri, R. Cycle Logistics Projects in Europe: Intertwining Bike-Related Success Factors and Region-Specific Public Policies with Economic Results. *Appl. Sci.* **2021**, *11*, 1578. https://doi.org/10.3390/app11041578

Academic Editor: Michele Roccotelli

Received: 29 December 2020
Accepted: 6 February 2021
Published: 9 February 2021

Publisher's Note: MDPI stays neutral with regard to jurisdictional claims in published maps and institutional affiliations.

1. Introduction

An increasing noticeable focus on the adoption of cargo cycles for commercial deliveries and their social and economic impacts has been shown in local, national or Europe-wide projects and communication campaigns, together with a more comprehensive analysis of factors and policies characterizing non-motorized mobility programs at large [1–7]. In particular, such policies have relevant impacts on the achievement of sustainable urban mobility goals as well as on the improvement of local economy and employment— e.g., by minimizing European economy losses (ca. 1% of gross domestic product) due to the congestion and prolongation of private and commercial journeys [4,8–10].

The existing literature in this research area also identifies manifold aspects, which vary from region to region—e.g., either cities or countries [11–15]. Area- and project-specific variables include (but are not limited to) speed and size of vehicles, trip generation potential in the surroundings, driver's experience and confidence, weather conditions, number of traffic lanes and side roads, outside lane width, integration of land use and transportation planning, pavement surface quality and traffic signals [4,16–18]. Hence, analyzing differences across countries, regions and cities is crucial since it may bring significant policy implications, which vary widely throughout the European scenario [19,20].

The literature review has been partly extended to include active travel behavior since many factors and policies affecting bike delivery initiatives were found to be delved into in scientific publications concerning the field of non-motorized travel at large. In fact, factors and policies encouraging non-motorized or active travel behavior often include specific measures geared to foster cycle logistics initiatives, especially in some European regions [1–7,9,10]. This way, it is more likely to capture all possible aspects potentially impacting cycle logistics initiatives. However, the significance of such aspects has been tested with statistical analyses later on in this paper.

The same applies to the investigation of factors and policies in different regions. In fact, it has been conducted on projects run on a global basis in order to avoid excluding relevant aspects not covered by studies limited to the European scenario from the statistical analysis.

Overall, this research work proposes a cross-regional comprehensive study on cycle logistics projects in Europe. It considers both projects' features and policies in the European area, together with the corresponding economic, financial impacts. The adoption of a Europe-wide perspective is among the key contributions of this paper. In fact, the review of hitherto published works in the literature shows that existing research directs its efforts toward studies limited to urban, regional and national contexts—more than comparing Europe-wide geographical areas—as well as to themes related to public health, environment, quality of life, etc. As a result, any cross-national approach is overlooked—especially pertaining to economic, financial results—thus showing how such a Europe-wide analysis of policies and impacts is unexplored.

The aim of this paper is to statistically test whether and which specific, distinctive characteristics of European cycle logistics projects and the corresponding supporting policies have an impact on their economic performances in terms of profit and profitability.

The paper is structured as follows: after the introduction, the second section provides an extensive and complete literature review in order to identify success factors for cycle logistics projects, including brief references to some closely related aspects concerning active travel at large; in the third section, data collection and processing methods and the methodological approach are described; the fourth section includes some discussions about results; in the last section, conclusive remarks are presented.

2. Literature Review

Although the goal of this paper is intertwining success factors and local policies with economic results of cycle logistics projects in Europe, the literature review proposed in the above section embraces a wider perspective. In fact, it considers those factors and policies also affecting the adoption of bikes for the mobility of people as well as active travel behaviors at large. This was a deliberate methodological choice geared to adopt a more comprehensive perspective of analysis in order to capture all possible elements affecting the cycle logistics field. In fact, this approach avoids the siloization of cycle logistics-related factors and policies: they cannot be and are not considered separately from the "whole picture", which includes the use of bikes for personal purposes and active travel behavior at large. In fact, many factors and policies concerning cycle logistics are included in more comprehensive political programs, affecting the wider area of bike mobility and active travel. Finally, the reason behind this wider slant of analysis is justified by considering that cycle logistics solutions for freight transport always require some supporting policies from public authorities as well as the fact that they implicitly concern the adoption of bikes by individual users. Therefore, all of those factors and policies affecting the overall use of bikes for individual purposes prove to be relevant to cycle logistics solutions.

The search strategy implemented in Scopus, the Elsevier database, included the following terms: cargo bike, cargo-bike, cargobike, cyclelogistics, cycle-logistics, cycle logistics, cycle mobility, bike + mobility, active travel, bike + economy, bike + policy, cargo cycle, cargo-cycle, cargocycle. The resulting documents were then selected by analyzing their contents with a qualitative review, and all the co-authors independently reviewed each selected paper. Finally, they shared their independent evaluations with each other and identified the documents to be used for the literature review. Moreover, additional searches were conducted on the Internet by checking for additional relevant sources—e.g., from specialized websites—to integrate with the initial documents.

2.1. Experiences from the UK and Ireland

Local authorities in the UK have targeted large logistics companies in order to incentivize them to adopt cargo bikes into their supply chain. Moreover, communication campaigns of cargo cycle programs at the local level have emphasized both perception issues and lack of awareness and regulation as factors impacting negatively on the implementation of cycle logistics initiatives [7]. Ref. [21] demonstrates that promoting policies in the UK should address three pillars: incentive system, risk perception and availability and maintenance of infrastructures. Some of these key factors are common among cycle logistics and cycling at large, despite significant definitory differences and obstacles that characterize each strain. Finally, extant studies found some relationships between the impacts of both cycle logistics and cycling at large, but the definitory elements are not shared among them: it is a question of convergence of two different strains (cycle logistics and cycling at large) toward some similar results [18,21–30].

An additional factor lies in the impacts: an inverse relationship between the adoption of bike delivery solutions and the number of obese citizens, including bike messengers, has been found, thus proving the broad extent of social impacts partly attributable to cycle logistics [22–26]. In fact, cycle logistics projects in the UK have several impacts, since they contribute to reduce the pressure on the National Health Service; delivering goods by bike is positively linked to health benefits and is proven to help tackle urban mobility issues, which are directly responsible for 70% of substances threatening public health [21,27]. In Scotland, different research studies have identified a complex set of shared impacts—e.g., demographic, economic and infrastructural—related to both cycle logistics and cycling at large [18,28–30]. In Ireland, cycle logistics have been included in two ad hoc government programs in order to increase both individual and socio-economic benefits [9]. Tackling safety issues in Irish urban transportation networks is key in order to build on the reputation of cycling as safe in cities and, especially, to ensure the business viability of bike delivery projects [18]. In this context, risk perception, infrastructures and attitudinal aspects have been identified as key factors [18,31,32]. However, the current strong political will in many towns in the UK is not the only enabling factor supporting cycle logistics and cycling at large. In fact, the traditional good public transport systems and the urban infrastructures adverse to car use in city centers act as complementary factors in the British scenario since they help in decreasing traffic levels and, at the same time, enable the timely delivery of goods and, hence, the increase in customer satisfaction for cycle logistics businesses [33]. On the contrary, public transport proved to be poorly designed in Ireland since it provoked a 29.7% drop for non-motorized commuting and a 37.5% increase in car use from 1986 to 2006, thus revealing a negative context factor and discouraging the start-up of bike delivery businesses [9,34]. Many authors have investigated socio-economic and transportation- and household-specific factors in major Irish cities: supporting policies, infrastructures, car ownership and socio-demographic status at large have been identified as some of the more relevant issues impacting non-motorized transport, thus including cycle logistics [9,35–39].

2.2. Experiences from Greece and Italy

An analysis of factors impacting the adoption of bikes as a standard transportation means in Greece suggested that women's eco-cyclist inclination tends to make them ask for bike delivery services more often than men [40]. Nonetheless, the existing literature about the Greek scenario sheds light on how the gender factor is mitigated by other variables—i.e., demographic, economic and environmental—and gender may play a varying role depending on the relative significance of such factors interacting with it [40]. Demographic and cultural ones—e.g., marital and education status—often have a low impact, while age may be associated with environmental concern [41]. Finally, low income has been shown to be the most relevant economic aspect affecting the demand of bike delivery services, thus giving policy makers and managers a relevant insight into unexploited targets for new mobility solutions in Greece [40].

In Italy, factors such as the network and the topological shapes of many cities made it necessary to test several pilots which emphasize the environmental and social benefits of such initiatives [42]. These Italian projects build on previous research, which has already demonstrated that 51% of all trips for goods transportation in European cities can be realized by bike, and 19% to 48% of the overall mileage currently performed by combustion engine vehicles can be done by cargo bike [7,43]. In this context, the main factors determining the success or failure of pilots were the size and weight of goods to be delivered compared to the load capacity of cargo bikes; the relevance of time windows for the delivery; the impacts on brand image and corporate social responsibility; cost levels; availability of a supporting network and reliability of enabling technologies [42]. Finally, the cycle logistics scenario in Italy proves to be primarily affected by the social (e.g., visibility and green image, low energy consumption, service quality and coverage), physical (e.g., load capacity vs. goods size, weight and number, technological reliability and battery duration) and political (e.g., better quality of life for citizens, re-use of public facilities and incentives) factors of the socio-ecologic model proposed by [16].

2.3. Experiences from Scandinavia

An analysis of Scandinavian countries, especially Denmark, showed that accessibility and availability of both safe infrastructures and parking facilities—together with high urban density—enable the start-up and development of cycle logistics projects as well as the adoption of bikes at large [20]. Other authors recognize the key role of supporting policies [16,44–46].

2.4. Experiences from Central Europe

In the Netherlands, the use of bikes for the mobility of both goods and people already accounted for 30% of overall local trips in 1997, thus showing a strong cultural integration of bikes into Dutch society [47]. As for cycle logistics projects, commercial deliveries in Dutch cities are generally planned for short trips within the 3.5-km threshold, while in Germany, they are feasible within 2 km [47]. Differences between the average trip thresholds in the Dutch and German scenarios are due to cultural factors as well as urban density and infrastructural factors [47]. In Germany, recent research efforts on inner-city courier shipments have identified the specific vehicle choice of "messengers"—i.e., freelance couriers—as one of the main drivers for the adoption of bike delivery solutions [48]. In turn, vehicle choice is affected by several variables at the individual—e.g., demographic, attitudes and values—technological—e.g., accessibility and availability of enabling technologies—and economic—e.g., price and availability of information—levels [48]. As for technological aspects, technical innovations adopted by cycle logistics initiatives in Germany and France have been combined with new concepts and configurations of urban mobility systems. They have been successful, especially when associated with urban micro-consolidation centers as well as technologies for reduced driver fatigue and improved range and payload [48–51]. However, recent studies have shown that effective commercial transport solutions in city centers always come out of a multitude of factors harmoniously combined with each other, such as organizational structure of supplier firms, demand patterns, technical prerequisites and cultural inclination to accept a modal shift from customers, firms and messengers [48].

2.5. Experiences out of Europe: Australia and the United States of America

Many research works report that Australian bike-based businesses are worse positioned compared with their competitors in North America, China and Europe [6,52–54]. The findings suggest that adverse reactions to safety helmets being compulsory together with trip distance may affect messengers' vehicle choice and undermine the success of

Australian cycling and cycle logistics programs (see also [39,55,56]). Other studies conducted in Australian and American cities suggest that individual factors, including messengers' vehicle choice, are key in order to nurture bike delivery or bike sharing schemes as well as cycling at large [41,56–65]. As for mandatory helmet usage, it is not perceived as an advantageous factor because of the need for either purchasing or always carrying such safety accessories; studies show that it is detrimental in both Australia and the USA, with other factors being equal [17,55,66–68]. Moreover, urban density and availability of infrastructures are also recognized as relevant factors in Australia and the USA [69–71].

However, commercial deliveries in the United States are not effective because of the noticeable differences with the network and topological shapes in European inner cities [72–74]. Although there has been a sound debate about primary causes of the flop of Australian programs, the field research is still poorly grounded and calls for further empirical research efforts.

2.6. Experiences from Asian Countries

As regards the Malaysian scenario, customer needs and political factors play a key role—e.g., need for door-to-door transport, spread and availability of dedicated infrastructures and environmental aspects [4,75]. Other studies confirm that socio-economic impacts were found to be significant [4,76].

As Western societies have shown a strong commitment to cycle logistics and cycling at large (see Sections 2.1–2.5), likewise, Asian ones have proven to be strategically engaged in them as well because of their relevance to national agendas [77–79]. Policy implications also call for a dramatic change in population distribution in some Asian cities, such as in Malaysia, where a foreseeable "donut cake" distribution due to the downstreaming phenomenon in urban centers will generate a downsizing of economic activities and residential density in downtown areas, impacting, in turn, the operations and the development of bike delivery businesses [4]. Downstreaming in cities will be amplified by other local policies such as priority being afforded to motorized vehicles and lack of non-motorized promotion, resulting in a clearly disadvantageous scenario for cycle logistics projects [4]. These results call for taking the opportunity to design transportation networks suitable for cycle logistics and integrating them with existing road networks in Malaysia. Moreover, policy makers should push towards the adoption of priority policies reversing the current ones, as it happens in Europe—i.e., ensuring priority to cyclists and bike messengers, excluding them from turning or one-way direction constraints [4].

However, Malaysia and European and Northern American countries are not the only ones to cope with political, safety and socio-demographic factors, since they have proven to be relevant also to the Japanese national agenda [80,81]. In particular, the lack of effective safety regulation in Japan called for the implementation of active safety policies and countermeasures which also suggest an increase in the viability of commercial deliveries in cities [80,82–85]. This way, it has been possible to adopt more effective policies in order to tackle both bike messengers' and goods' safety in Japan as well as it is being done in other Western societies such as Ireland and in developing countries [18,20,28,35–38,47,86–89].

2.7. Experiences from Developing Countries

In the last century, policy makers in developing countries have focused on motorized transportation, thus promoting and designating urban development as a hindering context for cycle logistics projects. This approach to planning and implementation of activities has dominated the transportation arena, even though non-motorized travel is even more significant in developing countries than in Western societies. The reasons behind this have been recognized to be the poorly grounded literature and the poor dissemination of research results in this field, also at the power elite level [89]. Moreover, policy makers hold their responsibility since they did not relevantly take into account the positive impacts of non-motorized transportation in terms of environmental, energy and socio-economic benefits. Therefore, they neglected the need of promoting the start-up of bike delivery services—e.g., by incentivizing them or realizing supporting infrastructures. Their carelessness towards the multifaceted negative effects of traditional combustion engine vehicles—e.g., increase in congestion, energy consumption, pollution, costs, pressure on health system, safety and security—is even more alarming [90,91]. Policy makers and people at large have perceived, for many decades, that the dilemma between motorized versus non-motorized transportation could be associated with rural versus urban and developed versus non-developed, while the most advanced studies prove that cycle logistics may provide a significant contribution to cope with many issues affecting urban contexts worldwide [89].

A further challenge appears on the horizon: most cities in some developing countries, e.g., India, are not able to satisfy the need for investments and measures concerning infrastructures, safety, land use and incentive systems geared to serve their growing cycling population, including bike messengers, and the overall viability of their delivery services in inner cities [37,86,88,89,92–101].

In conclusion, where policy makers overlook the need for an integrated set of interventions geared to promote cycle logistics and cycling at large, this may result in limited success or even in failure of supporting policies. Sometimes, this political issue comes from the desire to maintain good relationships with relevant shares of voters, and other times from evidence of weak political capacity [102–106]. Finally, the paradigmatic shift towards bike delivery services is linked to a set of factors and interventions to tackle them. In the following, all relevant factors and policies are analyzed together with their relative significance across geographical areas and countries. Furthermore, corresponding measures are identified that prove effective in modifying the ways in which goods are delivered in cities. In this context, the comprehensive subject of this paper—i.e., Europe-wide analysis—is particularly suitable for defining policy implications in the broader field of sustainable urban transport. The focus on cities and urban areas at large is further justified by recent research results showing that even European or nationwide transport policies depend on their success at the local level [6,9].

3. Data Collection and Methodological Approach

We must point out that the profit and profitability variables help when assessing the potential of an individual project to achieve business objectives or to produce an economic result based on both an effective and efficient use of resources, within the context in which such a project is implemented. In fact, in general, profit in itself is not sufficient to prove the economic appeal of an investment in a project and whether it is worth pursuing, except when dealing with a business company. On the contrary, the concept of profitability applies, more generally, to all kinds of economic organizations, including non-profit. Profitability has been calculated as a dimensionless value according to the standard definition—i.e., sum of present value of cash flows over 5 years divided by initial investment. On the contrary, profit is not dimensionless—the currency is euros—and it has been calculated by subtracting the normalized costs in the Cyclelogistics project (e.g., for bike purchasing, maintenance, insurance and messenger pay) from the overall income [107,108].

Such hypotheses depending on profit and profitability as economic, financial results are tested by utilizing normalized data that are calculated at the European level [108]. Therefore, such data do not reflect any region-specific features, as they are generated by definition and by construction with a normalization process throughout the Europe-wide scenario, which was the context of the analysis considered by the source study [108]. To the best of the authors' knowledge, the data included in this research document are the most comprehensive and reliable, so far, and are endowed with an official element, being the result of activities supported also by the European Commission.

Both profit and profitability are calculated with reference to the specific organization implementing the cycle logistics service. Therefore, profit and profitability refer to the organization's business/project level, whereas those values of profit and profitability are registered as economic, financial results.

The overall methodological approach and the corresponding steps are summarized below (Figure 1) and discussed in detail in the rest of this section.

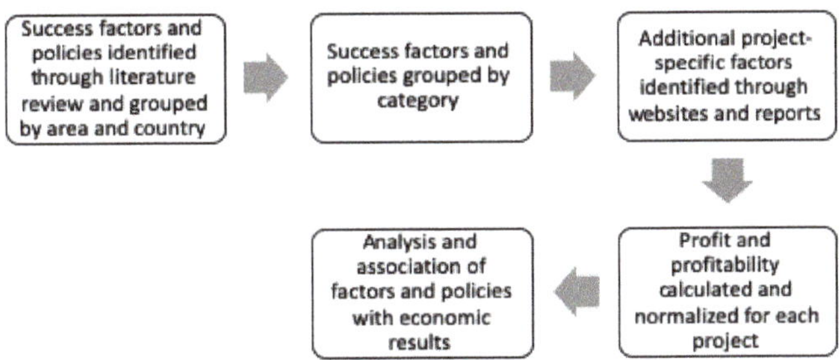

Figure 1. Methodological approach and stages.

Following this approach, in the first stage, critical factors and policies mentioned in Section 2 have been clustered by area and country. The results are shown in Appendix A (Table A1) and Appendix B (Table A2).

In the second stage, both critical factors and policies relevant to those projects run in Europe have been selected and grouped by category (Table 1a,b).

Table 1. (**a**) Success factors by category. (**b**) Supporting policies by category.

(a) Success Factors Category
Good information, communication
Health benefits
Economic appeal and ownership
Behavioral propensity as a consequence of individual values, needs and socio-demographic variables at large
Perceived safety and security
Favorable weather
Favorable urban density, traffic and distance
Accessibility, availability of infrastructures, services and enabling technologies
Environmental and energy consumption benefits

(b) Supporting Policies Category
Communication policies
Regulatory policies
Public transport policies
Architectural and infrastructural policies
Safety policies
Economic policies
Environmental policies

In the third stage, beyond factors and policies identified with the above analysis, a complementary study has been conducted in order to take into account also project-specific factors characterizing 50 cycle logistics solutions implemented in Europe (Table 2a). The main type of bike is considered as the key feature since other project-specific factors—such as size range of delivery, ease of driving and parking, price per delivery, cost per bike, etc.—result to be affected by it. In particular, the bike models considered are traditional, trailer and cargo bikes as well as tricycles. Traditional bikes are those standard models used also by urban citizens for their own private purposes. If used for commercial deliveries, they only allow transporting small-sized goods and in a limited quantity, since they do not have large cargo accessories, neither in front nor behind. They are generally cheaper than other bike models in terms of purchase, maintenance and insurance costs and ensure a higher speed and ease of driving and parking in cities. Cargo bikes have a large box, generally in front, which allows transporting even big-sized goods as well as a large quantity of small- to medium-sized items. They are much more expensive than traditional bikes and more difficult to park and drive. Trailer bikes lie in the middle between traditional and cargo ones and are endowed with a small cargo behind. As for tricycles, they are the largest and more capacious bikes as well as the ones with the highest costs. Moreover, they have the lowest average speed as well as the lowest ease of driving and parking in cities.

Table 2. (**a**) Project-specific factors (sources: [107,109] and projects' websites). (**b**) Average economic arguments from [108]. (**c**) Average data on the European cities scenario from [108].

(a)								
Country	Main Type of Bike	Max. Load (km)	Cost of Bike (EUR)	Average Speed in Cities (km/h)	Ease of Driving	Ease of Driving in Adverse Conditions	Ease of Parking	Distinctive Size Range of Delivery
UK	Trailer bike	80	250–500	20	High	High	High	Small/Medium
UK	Cargo bike; Tricycle	250	2000–12,000	15	High	Medium/Low	Medium	Small/Medium
UK	Cargo bike; Tricycle	250	2000–12,000	15	High	Medium/Low	Medium	Small/Medium
Ireland	Trailer bike	80	250–500	20	High	High	High	Small/Medium
Ireland	Cargo bike	80	2000–5000	20	High	High	High	Small/Medium/Big
Ireland	Cargo bike	80	2000–5000	20	High	High	High	Small/Medium/Big
Greece	Traditional	40	80–300	20	High	High	High	Small
Italy	Traditional	40	80–300	20	High	High	High	Small/Medium
Italy	Traditional	40	80–300	20	High	High	High	Small
Italy	Trailer bike	80	250–500	20	High	High	High	Small
Italy	Tricycle	250	3000–12,000	15	High	Medium/Low	Medium	Big
Italy	Cargo bike	80	2000–5000	20	High	High	High	Small/Medium/Big
Italy	Cargo bike	80	2000–5000	20	High	High	High	Small/Medium/Big
Greece	Traditional	40	80–300	20	High	High	High	Small
Greece	Traditional	40	80–300	20	High	High	High	Small/Medium
Greece	Traditional	40	80–300	20	High	High	High	Small
Italy	Trailer bike	80	250–500	20	High	High	High	Small
Italy	Tricycle	250	3000–12,000	15	High	Medium/Low	Medium	Big
Italy	Cargo bike	80	2000–5000	20	High	High	High	Small/Medium/Big
Italy	Cargo bike	80	2000–5000	20	High	High	High	Small/Medium/Big
Cross-country (Denmark, UK)	Traditional	40	80–300	20	High	High	High	Small
Cross-country (Denmark, UK)	Traditional	40	80–300	20	High	High	High	Small
Cross-country (Denmark, UK)	Traditional	40	80–300	20	High	High	High	Small
Cross-country (Denmark, UK)	Traditional	40	80–300	20	High	High	High	Small
The Netherlands	Cargo bike	80	2000–5000	20	High	High	High	Small/Medium/Big
The Netherlands	Cargo bike	80	2000–5000	20	High	High	High	Small/Medium/Big
The Netherlands	Cargo bike	80	2000–5000	20	High	High	High	Small/Medium/Big
The Netherlands	Cargo bike	80	2000–5000	20	High	High	High	Small/Medium/Big
Germany	Cargo bike	80	2000–5000	20	High	High	High	Small/Medium/Big
Austria	Cargo bike	80	2000–5000	20	High	High	High	Small
Germany	Cargo bike	80	2000–5000	20	High	High	High	Small/Medium/Big
Germany	Cargo bike	80	2000–5000	20	High	High	High	Small/Medium/Big
Germany	Cargo bike	80	2000–5000	20	High	High	High	Small/Medium/Big
Germany	Cargo bike	80	2000–5000	20	High	High	High	Small/Medium/Big
Germany	Cargo bike	80	2000–5000	20	High	High	High	Small/Medium/Big
Germany	Cargo bike	80	2000–5000	20	High	High	High	Small/Medium/Big

Table 2. *Cont.*

(a)

Country	Main Type of Bike	Max. Load (km)	Cost of Bike (EUR)	Average Speed in Cities (km/h)	Ease of Driving	Ease of Driving in Adverse Conditions	Ease of Parking	Distinctive Size Range of Delivery
Austria	Cargo bike	80	2000–5000	20	High	High	High	Small
France	Cargo bike; Tricycle	250	2000–12,000	15	High	Medium/Low	Medium	Small/Medium/Big
France, Denmark	Traditional	40	80–300	20	High	High	High	Small
Switzerland	Trailer bike	80	250–500	20	High	High	High	Small
France	Cargo bike; Tricycle	250	2000–12,000	15	High	Medium/Low	Medium	Small/Medium/Big
France	Cargo bike; Tricycle	250	2000–12,000	15	High	Medium/Low	Medium	Small/Medium/Big
France, Denmark	Traditional	40	80–300	20	High	High	High	Small
France	Cargo bike; Tricycle	250	2000–12,000	15	High	Medium/Low	Medium	Small/Medium/Big
France	Cargo bike; Tricycle	250	2000–12,000	15	High	Medium/Low	Medium	Small/Medium/Big
France	Cargo bike; Tricycle	250	2000–12,000	15	High	Medium/Low	Medium	Small/Medium/Big
Spain	Tricycle	250	3000–12,000	15	High	Medium/Low	Medium	Small/Medium/Big
Spain	Tricycle	250	3000–12,000	15	High	Medium/Low	Medium	Small/Medium/Big
Spain	Tricycle	250	3000–12,000	15	High	Medium/Low	Medium	Small/Medium/Big
Hungary	Traditional	40	80–300	20	High	High	High	Small/Medium

(b)

Tangible Costs	Cargo Bike	Van
Set up costs		
Purchase cost	EUR 2483.00	EUR 3310 per annum
Running costs		
Annual maintenance	EUR 237.00	Included in the hire cost
Fuel	Zero	EUR 1334.00 per annum
Excise duty	Zero	EUR 201.00 per annum
Insurance	EUR 154.00 per annum	EUR 591.00 per annum
Rider/Driver costs		
Hourly pay rate	EUR 9.60	Usually self-employed paid by delivery (EUR 1.59 per delivery)
Intangible costs		
Emissions Contribution	Zero	152 g/km CO_2
Congestion Contribution	Minimal impact	Another vehicle on the road contributing to congestion
Noise	None	Diesel Clatter
Average speed in city	10–12 mph	5–15 mph
Parking	Not a problem	Restricted (risk of parking ticket)
Flexibility	Access to restricted areas and cycle paths	Restricted to the road network
Range	50 miles per day	Unlimited
Contribution to rider/ driver health	Rigorous daily workout	Sedentary

(c)

Average European city with 240,000 inhabitants and 1,000,000 trips per day			
All Trips	**Bicycle, Pedestrian, Public Transport**	**Motor Vehicle Trips**	**Motorized Trips Related to Goods Transport**
1,000,000	400,000	600,000	490,000
Among motorized trips related to good transport:	**Number of trips per day**	**Number of trips to shift to bicycle and cargo bike**	**Relative % of shift within motorized trips related to goods**
Motorized trips related to goods transport	490,000	250,000	51%
Delivery	100,000	25,000	25%
Service and business	110,000	55,000	50%
Shopping	130,000	100,000	77%
Leisure	90,000	40,000	44%
Commuter	60,000	30,000	50%

In the fourth stage, economic results have been calculated and normalized with reference to the data referring to the average European city as provided by the Cyclelogistics project [108], as reported in Table 2b,c. In detail, profit-and-loss data have been calculated and normalized coherently with the consolidated average economic arguments on page 19 of the aforementioned data from the Cyclelogistics project [108], whilst the average data on the European cities scenario have been extracted from page 9 of the same reference [108]. In this context, normalization means using the same average economic arguments in the aforementioned report in order to calculate profit and profitability of cycle logistics projects. This way, the methodological approach adopted allowed us to compare project data homogeneously. Profit and profitability measures for each project have been calculated over 5 years depending on available data from the Cyclelogistics project [107,108] and by rounding yearly profits down with a 0.1 correction coefficient. This way, we have reduced the overall profit (and also the profitability inferred from it). This helps us in further challenging our research hypotheses by assuming a more pessimistic scenario in terms of profit and profitability levels as well as further mitigating the risk of their overestimation. In addition, profits have been prudently calculated by considering only the share of bike-based trips explicitly devoted to goods delivery in European cities, thus assuming a scenario with an even more pessimistic underestimation.

At first, these assumptions could sound like an attempt to simplify calculations in order to be able to perform a rough estimate and comparison among projects. Anyhow, as a matter of fact, that is currently the only way to perform such an analysis because of the lack of sufficient and homogeneous data. Such a shortage of data concerns both the non-profit projects funded by the European Commission and those projects run by private start-ups. Profit and profitability have been calculated in order to obtain an insight into the potential of each individual project to achieve business objectives or to produce an economic result based on both an effective and efficient use of resources. In fact, in general, profit in itself is not sufficient to prove the economic appeal of an investment in a project and whether it is worth pursuing, except when dealing with a business company. On the contrary, the concept of profitability applies, more generally, to all kinds of economic organizations, including non-profit. Profitability has been calculated as a dimensionless value according to the standard definition—i.e., sum of present value of cash flows over 5 years divided by initial investment. On the contrary, profit is not dimensionless—the currency is euros—and it has been calculated by subtracting the normalized costs in the Cyclelogistics project (e.g., for bike purchasing, maintenance, insurance and messenger pay) from the overall income [108]. Economic results and their descriptive statistics calculated in Microsoft Excel® are reported below (Tables 3–5).

Table 3. Area-/country-specific policy and factor categories, average profit and profitability levels.

Area	Profitability over 5 Years	Average Estimated Profit/Year (EUR)
	2.43	76,225.00
	4.08	476,945.00
UK and Ireland	2.43	76,225.00
	5.29	619,060.00
	5.29	619,060.00
	4.08	476,945.00

Table 3. *Cont.*

Area	Profitability over 5 Years	Average Estimated Profit/Year (EUR)
Greece, Italy and Mediterranean islands	5.09	80,850.00
	5.09	80,850.00
	5.09	80,850.00
	2.43	76,225.00
	4.92	616,560.00
	4.08	476,945.00
	4.08	476,945.00
	4.92	616,560.00
	5.09	80,850.00
	5.09	80,850.00
	2.43	76,225.00
	4.08	476,945.00
	5.09	80,850.00
	4.08	476,945.00
Central Europe °	4.08	476,945.00
	4.08	476,945.00
	4.08	476,945.00
	4.08	476,945.00
	4.08	476,945.00
	5.29	619,060.00
	5.29	619,060.00
	2.43	76,225.00
	4.92	616,560.00
	5.09	80,850.00
	2.43	76,225.00
	4.92	616,560.00
	4.92	616,560.00
	5.09	80,850.00
	5.09	80,850.00
	5.29	619,060.00
	4.08	476,945.00
	4.08	476,945.00
	4.08	476,945.00
	5.09	80,850.00
	4.08	476,945.00
	5.09	80,850.00
	2.43	76,225.00
	5.09	80,850.00

Table 3. *Cont.*

Area	Profitability over 5 Years	Average Estimated Profit/Year (EUR)
	5.09	80,850.00
	5.09	80,850.00
Cross-area	5.09	80,850.00
	5.09	80,850.00
	5.09	80,850.00
	5.09	80,850.00

° In this paper, "Central Europe" includes only continental European countries and excludes the Italian peninsula, Greece, the Balkans and the Scandinavian peninsula.

Table 4. Profit and profitability statistics by area-/country-specific policy and factor category.

Area-/Country-Specific Policy Category	Area-/Country-Specific Factor Category	Profit Statistics		Profitability Statistics	
		Mean	Std. Deviation	Mean	Std. Deviation
UK and Ireland	Communication campaigns; Well-designed regulation system; Supporting an integrated public transport system; Accessibility and availability of infrastructures; Educating, designing and implementing safety; Supporting economic measures	390,743.33	251,778.4817	3.93	1.28
Greece, Italy and Mediterranean islands	Communication campaigns; Supporting economic measures	269,889.29	232,259.18	4.40	0.94
Scandinavia and Central Europe °	Communication campaigns; Supporting economic measures; Accessibility and availability of infrastructures Supporting environmental protection measures	363,047.50	231,328.93	4.52	0.85
Overall	-	306,423.00	234,417.89	4.42	0.93

° In this paper, "Central Europe" includes only continental European countries and excludes the Italian peninsula, Greece, the Balkans and the Scandinavian peninsula.

Table 5. Project-specific factors, average profit and profitability levels.

Main Type of Bike	Profit		Profitability	
	Mean	Std. Deviation	Mean	Std. Deviation
Trailer bike	76,225.00	0.00	2.43	0.00
Cargo bike	476,945.00	0.00	4.08	0.00
Tricycle	616,560.00	0.00	4.92	0.00
Traditional bike	80,850.00	0.00	5.09	0.00
Cargo bike; Tricycle	619,060.00	0.00	5.29	0.00
Overall	373,928.00	275,273.72	4.37	1.05

In the fifth and last stage, possible associations of factors and policies with economic results have been analyzed with IBM® SPSS® Statistics 24. Some analyses have been made in order to make inferences about data and to understand whether the observed pattern is real or due to chance. Before using IBM® SPSS® Statistics 24, the dataset was cleaned up by deleting overlapping data concerning cross-country-specific factors and policies.

Therefore, in the fourth section, mean and standard deviation were calculated again in IBM® SPSS® Statistics 24, together with other advanced statistics.

As an additional note, despite the many variables that could be identified in the literature review, the availability of data was limited to some of them. Moreover, data for some variables were only partly available. Finally, the only variables with full data available for a quantitative analysis were related to profit, profitability, geographical area and type of bike.

Finally, we state our four research hypotheses, based on the above explication of the variables:

Hypothesis 1 (H1). *The profit distribution varies across categories of geographic area, that is, across the different geographic areas, not across each country pertaining to a specific geographic area.*

Hypothesis 2 (H2). *The profitability distribution varies across categories of geographic area, that is, across the different geographic areas, not across each country pertaining to a specific geographic area.*

Hypothesis 3 (H3). *The profit distribution varies across categories of bike model.*

Hypothesis 4 (H4). *The profitability distribution varies across categories of bike model.*

4. Results and Discussion

The preliminary step of the statistical analysis was conducted in order to verify whether data distributions of profit and profitability are normal or not. Checking the normality of distributions is relevant since this methodological step impacts the choice of the statistical tests to adopt (e.g., parametric vs non-parametric tests) in order to ensure the reliability of results.

In the following, Table 6a,b show normality tests on area-specific profit and profitability. Table 7a,b show normality tests on project-specific profit and profitability.

Table 6. (a) Preliminary analysis of normal distribution hypothesis of area-specific profit and profitability: skewness and kurtosis. (b) Preliminary analysis of normal distribution hypothesis of area-specific profit and profitability: Kolmogorov–Smirnov and Shapiro–Wilk tests.

(a)				
	Descriptives			
			Statistic	**Std. Error**
	Mean		750,672.7778	407,828.85784
	95% Confidence Interval for Mean	Lower Bound	−109,770.8996	
		Upper Bound	1,611,116.4551	
	5% Trimmed Mean		406,373.9198	
	Median		476,945.0000	
	Variance		2,993,838,791,115.359	
Profit	Std. Deviation		1,730,271.30564	
	Minimum		76,225.00	
	Maximum		7,622,500.00	
	Range		7,546,275.00	
	Interquartile Range		535,710.00	
	Skewness		4.117	0.536
	Kurtosis		17.256	1.038

Table 6. *Cont.*

(a)

Descriptives			Statistic	Std. Error
Profitability	Mean		4.3133	0.23310
	95% Confidence Interval for Mean	Lower Bound	3.8215	
		Upper Bound	4.8051	
	5% Trimmed Mean		4.3637	
	Median		4.5000	
	Variance		0.978	
	Std. Deviation		0.98898	
	Minimum		2.43	
	Maximum		5.29	
	Range		2.86	
	Interquartile Range		1.01	
	Skewness		−1.058	0.536
	Kurtosis		0.039	1.038

(b)

Tests of Normality	Kolmogorov–Smirnov [a]			Shapiro–Wilk		
	Statistic	df	Sig.	Statistic	df	Sig.
Profit	0.475	50	0.000	0.359	40	0.000
Profitability	0.240	50	0.007	0.794	40	0.001

[a] Lilliefors Significance Correction.

Table 7. (a) Preliminary analysis of normal distribution hypothesis of project-specific profit and profitability: skewness and kurtosis. (b) Preliminary analysis of normal distribution hypothesis of project-specific profit and profitability: Kolmogorov–Smirnov and Shapiro–Wilk tests.

(a)

Descriptives			Statistic	Std. Error
Profit	Mean		306,376.7500	53,249.06184
	95% Confidence Interval for Mean	Lower Bound	194,925.1827	
		Upper Bound	417,828.3173	
	5% Trimmed Mean		301,791.6667	
	Median		278,897.5000	
	Variance		56,709,251,742.829	
	Std. Deviation		238,137.04404	
	Minimum		76,225.00	
	Maximum		619,060.00	
	Range		542,835.00	
	Interquartile Range		396,095.00	
	Skewness		0.146	0.512
	Kurtosis		−1.984	0.992

Table 7. *Cont.*

(a)

Descriptives			Statistic	Std. Error
Profitability	Mean		4.3910	0.21590
	95% Confidence Interval for Mean	Lower Bound	3.9391	
		Upper Bound	4.8429	
	5% Trimmed Mean		4.4500	
	Median		4.9200	
	Variance		0.932	
	Std. Deviation		0.96554	
	Minimum		2.43	
	Maximum		5.29	
	Range		2.86	
	Interquartile Range		1.01	
	Skewness		−1.213	0.512
	Kurtosis		0.413	0.992

(b)

Tests of Normality

	Kolmogorov–Smirnov [a]			Shapiro–Wilk		
	Statistic	df	Sig.	Statistic	df	Sig.
Profit	0.328	50	0.000	0.738	40	0.000
Profitability	0.258	50	0.001	0.767	40	0.000

[a] Lilliefors Significance Correction.

The results show that all data distributions are not normal since the prevailing tests of normality—i.e., Kolmogorov–Smirnov and Shapiro–Wilk—lead to reject the normal distribution hypothesis.

Therefore, parametric tests—i.e., one-way ANOVA—were not conducted, whilst nonparametric ones were conducted: the Kruskal–Wallis H test was applied [110] since it is more appropriate than the Mann–Whitney U one. In fact, in our analysis, all independent categorical variables—i.e., both area- and project-specific factors—have more than two levels. In Table 8a,b, the results of the Kruskal–Wallis H tests are reported.

The results of the first statistical tests conducted–from the Kolmogorov–Smirnov to the Shapiro–Wilk one—are aimed at proving the reliability and suitability of the second step of the statistical analysis—i.e., Kruskal–Wallis H test [110]—which challenges the research hypotheses with corresponding null hypotheses. The final results prove that the only null hypothesis rejected is the one related to the bike model, thus confirming that the hypothesized dependence is true and significant. On the other hand, the across-the-region hypothesis is not supported, thus showing that there are no specific regional features affecting profit and profitability results more than others.

4.1. Discussion on the Statistical Tests of H1 and H2

In the following, statistical results concerning our research hypotheses on categories of geographic area are discussed. In particular, such hypotheses—which are negatives of the null ones reported in Table 8a—state that:

Hypothesis 1 (H1). *The profit distribution varies across categories of geographic area, that is, across the different geographic areas, not across each country pertaining to a specific geographic area.*

Hypothesis 2 (H2). *The profitability distribution varies across categories of geographic area, that is, across the different geographic areas, not across each country pertaining to a specific geographic area.*

Table 8. (a) Kruskal–Wallis H test of area-specific factors and policies. (b) Kruskal–Wallis H test of project-specific factors.

(a) Hypothesis Test Summary				
	Null Hypothesis (H0$_i$) Challenging the Corresponding i-th Hypothesis (Hi)	**Test**	**Sig.**	**Decision**
H0$_1$ vs. H1	The distribution of profit is the same across categories of area	Independent Samples Kruskal–Wallis Test	0.151	Retain the null hypothesis
H0$_2$ vs. H2	The distribution of profitability is the same across categories of area	Independent Samples Kruskal–Wallis Test	0.828	Retain the null hypothesis
(b) Hypothesis Test Summary				
	Null Hypothesis (H0$_i$) Challenging the Corresponding i-th Hypothesis (Hi)	**Test**	**Sig.**	**Decision**
H0$_3$ vs. H3	The distribution of profit is the same across categories of main type of bike	Independent Samples Kruskal–Wallis Test	0.001	Reject the null hypothesis
H0$_4$ vs. H4	The distribution of profitability is the same across categories of main type of bike	Independent Samples Kruskal–Wallis Test	0.001	Reject the null hypothesis

Asymptotic significances are displayed. The significance level is 0.05.

As a first remark, such hypotheses are rejected. In fact, both profit and profitability distributions are the same across categories of geographic area (Table 8a). In this case, multiple comparisons were not performed because the overall test does not show significant differences across such categories.

Considering profit and profitability performances shown in Tables 3 and 4, the statistical analysis of area-specific variables (Table 8a) shows an overall significance of success factors and policies in the European context. It also proves that there are no single factors or policies having a relevantly higher impact than others on the likelihood of success of cycle logistics projects. Although differences between distributions depending on the area are not significant, we can still analyze data in Table 4 in order to obtain a deeper understanding of the determinants of such a phenomenon. Data concerning profit and profitability by area highlight somewhat high values in terms of mean and standard deviation, except for Scandinavia. Mean values associated with projects in "Central Europe" prove to be higher than the overall average, whilst those in the "UK and Ireland" and "Greece, Italy and the Mediterranean islands" are just below it. "Scandinavia" has the lowest mean values. One of the main reasons behind that may be found by considering that cycling is an activity deeply rooted in Scandinavian cultures, especially in Denmark, which is the application context of the two projects considered for this area. Therefore, Danish people are used to transporting both small- and big-sized goods by themselves, thus not calling for bike delivery services. For instance, in 2008, IKEA invested in bikes—and trailers, if needed—at selected stores in Denmark (and also in Sweden) so that customers can ride home for free with their new purchases [111,112]. Although Danish projects show the lowest profit level, their profitability level goes high, much more than in any other area in Europe. A possible interpretation is that the high degree of accessibility and availability of suitable infrastructures in those countries helps with lowering the cost of some items—e.g., vehicle maintenance, insurance and, hence, purchase of new bikes. On the other hand, it helps with increasing service levels—e.g., deliveries are more likely to be made on time and customers are more satisfied. Moreover, supporting economic measures prove to nurture bike delivery organizations as well. Finally, such delivery services tend to be more effective, efficient and, hence, with a higher profitability in those countries than in others.

4.2. Discussion on the Statistical Tests of H3 and H4

In the following, statistical results concerning our research hypotheses on categories of main type of bike are discussed. In particular, such hypotheses—which are negatives of the null ones reported in Table 8b—state that:

Hypothesis 3 (H3). *The profit distribution varies across categories of bike model.*

Hypothesis 4 (H4). *The profitability distribution varies across categories of bike model.*

As a first remark, both profit and profitability distributions vary across categories of main type of bike (Table 8b). In this case, multiple comparisons were performed because the overall test does show significant differences across such categories.

In detail, the statistical analysis proves how profit and profitability distributions change according to the main type of bike utilized in cycle logistics projects and, thus, recognizes that some bike models have a relevantly higher impact than others on the likelihood of success of bike delivery businesses. Multiple comparisons concerning profit and profitability distributions across categories of the project-specific variable have been performed in order to understand which types of bike have the most significant impact on profit and profitability. The multiple comparisons have been analyzed by means of adjustment using the Bonferroni correction for multiple tests and they are reported in Figure 2 (concerning profit) and Figure 3 (concerning profitability).

In Figure 2, the pairwise comparison for profit shows the highest value for the node representing projects using contemporarily cargo bikes and tricycles. Note that this result is coherent with the data reported in Table 5, where the mean values associated with the two bike models used together are the highest ones for profit and also for profitability. Hence, that appears to be the best modal configuration. Furthermore, tricycles and cargo bikes taken separately prove to have a relatively significant impact on the profit variable and to generate relatively high values for both profit and profitability. By contrast, traditional or trailer bikes seem to not to provide such a relevant impact on profit, even though traditional bikes have a profitability mean value higher than the overall average in Table 5. On the contrary, if we analyze the paired use of different modal configurations, trailer bikes combined with big-sized bike models—i.e., either cargo bikes or tricycles or both of them—appear to be the best matched choice. In particular, trailer bikes combined with cargo bikes and/or tricycles seem to broadly cover the whole demand span. In fact, such configurations of paired bike models can satisfy altogether the delivery needs of different goods— i.e., small- to big-sized—and cope flexibly with different topological features and network shapes at the same time. As further proof, the trailer–tricycle–cargo bike configuration including all those types of bikes is the one having the highest impact on profit distribution among the three significant configurations in Figure 2. In fact, it covers a more extended demand span than either the trailer–tricycle or trailer–cargo bike configurations.

In Figure 3, the pairwise comparison for profitability shows the two highest values for those nodes representing projects exploiting cargo bikes and tricycles together as well as those adopting traditional bikes. Again, note that this result is coherent with the data reported in Table 5, where the mean values associated with the two bike models used together are the highest ones for profitability and also for profit. Analogously, traditional bikes have one of the highest values in terms of average profitability in Table 5. Hence, those appear to be the best modal configurations. The node representing tricycles, taken separately, has a relatively high value as well, but the possible matches with other modal solutions do not show statistical significance. Similarly, in Table 5, tricycles alone generate the highest mean value in terms of profit, as well as a profitability level above the overall average. On the contrary, the paired use of different modal configurations shows how trailer bikes are the most flexible bike model since they can be fruitfully combined with either traditional bikes or cargo bikes and tricycles. The significant impact on profitability shown by trailer bikes together with cargo bikes and tricycles could be explained by the broader coverage of the whole demand span, since those matched bike models satisfy the delivery needs of heterogeneously sized goods, similarly to data concerning the profit variable (Figure 2) already discussed. On the other hand, trailer and traditional bikes together also significantly affect profitability. In this case, it is worth mentioning that those bike models imply quite a low cost (Table 2a) in terms of purchase, maintenance and insurance. Moreover, they require quite a short time for goods delivery due to the high ease of driving and parking, low load capacity and high average speed in cities. Those features characterizing both trailer and traditional bikes prove how they allow efficient use of resources, thus justifying the significant performance in terms of profitability in Figure 3.

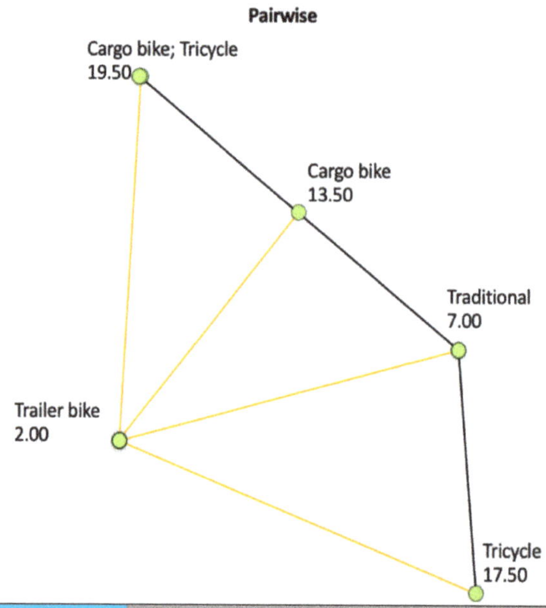

Sample1-Sample2	Test Statistic	Std. Error	Std. Test Statistic	Sig.	Adj. Sig.
Trailer bike-Traditional	5.000	3.931	1.272	0.203	1.000
Trailer bike-Cargo bike	11.500	4.028	2.855	0.004	0.043
Trailer bike-Tricycle	-15.500	5.200	-2.981	0.003	0.029
Trailer bike-Cargo bike; Tricycle	17.500	5.200	3.365	0.001	0.008
Traditional-Cargo bike	6.500	3.169	2.051	0.040	0.403
Traditional-Tricycle	-10.500	4.567	-2.299	0.022	0.215
Traditional-Cargo bike; Tricycle	12.500	4.567	2.737	0.006	0.062
Cargo bike-Tricycle	-4.000	4.651	-0.860	0.390	1.000
Cargo bike-Cargo bike; Tricycle	-6.000	4.651	-1.290	0.197	1.000
Tricycle-Cargo bike; Tricycle	2.000	5.696	0.351	0.726	1.000

Figure 2. Multiple comparisons of differences concerning profit reported in Table 8b. Each row tests the null hypothesis that the Sample 1 and Sample 2 distributions are the same. Asymptotic significances (2-sided tests) are displayed. The significance level is 0.05. Significance values have been adjusted by the Bonferroni correction for multiple tests.

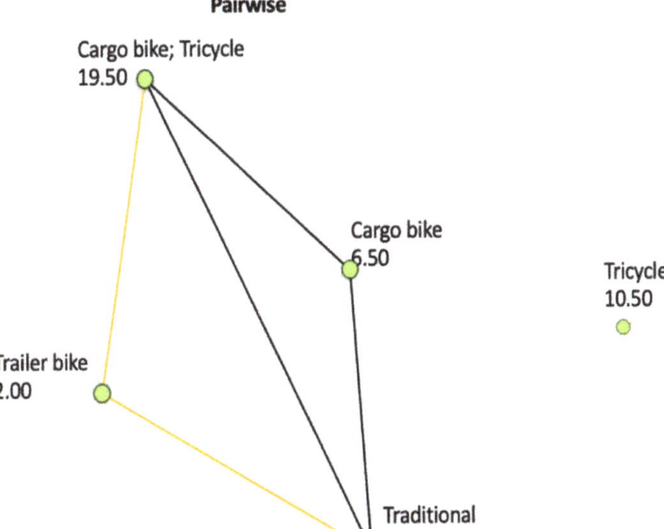

Sample1-Sample2	Test Statistic	Std. Error	Std. Test Statistic	Sig.	Adj. Sig.
Trailer bike-Cargo bike	4.500	4.028	1.117	0.264	1.000
Trailer bike-Tricycle	-8.500	5.200	-1.635	0.102	1.000
Trailer bike-Traditional	13.000	3.931	3.307	0.001	0.009
Trailer bike-Cargo bike; Tricycle	17.500	5.200	3.365	0.001	0.008
Cargo bike-Tricycle	-4.000	4.651	-0.860	0.390	1.000
Cargo bike-Traditional	-8.500	3.169	-2.682	0.007	0.073
Cargo bike-Cargo bike; Tricycle	-13.000	4.651	-2.795	0.005	0.052
Tricycle-Traditional	-4.500	4.567	0.0985	0.324	1.000
Tricycle-Cargi bike; Tricycle	9.000	5.696	1.580	0.114	1.000
Traditional-Cargo bike; Triciyle	4.500	4.567	0.985	0.324	1.000

Figure 3. Multiple comparisons of differences concerning profitability reported in Table 8b. Each row tests the null hypothesis that the Sample 1 and Sample 2 distributions are the same. Asymptotic significances (2-sided tests) are displayed. The significance level is 0.05. Significance values have been adjusted by the Bonferroni correction for multiple tests.

5. Conclusions

This paper proposes a comprehensive study on cycle logistics projects in Europe, focusing on their characteristics and policies and subsequent economic impacts. Such a Europe-wide perspective is one of the key contributions of this work. In fact, to the best of our knowledge, the existing literature includes only studies limited to urban, regional and national contexts. Hence, the cross-national dimension is missing, and as a consequence, the possible comparison of policies and their impact on cycle logistics solutions is poorly explored so far.

As a first remark, all of the 50 initiatives analyzed in our work are successful. This means that the overall European scenario is favorable to the start-up and development of cycle logistics projects. In fact, the data analysis performed clearly suggests that cycle logistics can generate both high profit and profitability levels. This result is coherent and further reinforced by other studies proving analogous advantages, not only from an economic perspective but also in terms of public health, environment, quality of life, etc. [4,6,8–10,21–27,33,40,42,76,80,82–85,90,91,113–123].

As for the area- and country-specific policies and factors, none of them have an impact more significant than the others on the likelihood of success of cycle logistics projects. A deeper analysis of such a finding leads to a threefold understanding. Firstly, a sound interpretation of this statistical result is that cycle logistics projects achieve comparably relevant profit and profitability levels in Europe, regardless of the specific factors and policies implemented in each country or area. Hence, there are no clearly superior policies compared to each other. Secondly, all of the different area- and country-specific factors and policies throughout Europe foster the economic performance of those projects in terms of profit and profitability. Thirdly, for each area, it is proven that the corresponding factors and policies are beneficial to cycle logistics projects, and hence, they should be kept in order to generate the same benefits also for future initiatives in the same geographical area. Finally, for each area in Tables A1 and A2 in the Appendices A and B, we summarize those success factors and supporting policies which we recommend to maintain.

On the contrary, profit and profitability results vary significantly depending on the main type of bike utilized in the European projects at hand. This means that the project-specific bike model affects the economic performance of different cycle logistics solutions. In particular, profit is strongly, positively affected by the adoption of cargo bikes or tricycles taken separately and even more by their combined use. Furthermore, trailer bikes are positively associated with profit performance when paired with cargo bikes and/or tricycles. A possible interpretation concerns the higher demand coverage of a trailer–tricycle–cargo bike configuration compared with other bike models combined together, especially in terms of different sizes of the delivered goods. In fact, the transport of both small- and big-sized goods contributes to the achievement of high volumes and market shares and, in turn, of positive economic performances at large. As for profitability, the two solutions that showed the highest impacts on it are (1) the combination of cargo bikes and tricycles and (2) traditional bikes. Trailer bikes also show a significant impact on profitability when they are associated with either traditional bikes or cargo bikes and tricycles. Again, a possible explanation for the significance of trailer–tricycle–cargo bike configurations lies in the possibility of delivering heterogeneously sized goods. On the contrary, the paired use of trailer and traditional bikes has a significant impact on profitability because of the low costs and low delivery time concerning such bike models (Table 2a).

Hence, the findings clearly prove that cycle logistics projects in Europe achieve high profit and profitability levels, and the current policies are generally working well and supporting them. Moreover, profit and profitability vary across the bike models utilized. In fact, mixing cargo bikes and tricycles generates the highest profit and profitability, whilst a trailer–tricycle–cargo bike mix paves the way for high volumes and market shares.

From the methodological perspective, a twofold original contribution is claimed, since the overall approach provides a well-structured research method geared to identify relevant policies and success factors. Firstly, our literature review has covered studies focusing on cycle logistics from a global perspective, thus broadening the traditional methodological approach based on research conducted in a local or national context. Furthermore, studies about some relevant factors concerning active travel behavior and private use of bikes have been considered in order to reduce the likelihood of erroneously overlooking potentially relevant elements. Hence, the overall set of area- and country-specific aspects captured through this enlarged view is more complete than in previous studies. Secondly, the widely different and partly overlapping nature of the resulting factors and policies called for grouping them into categories. Then, such categories were tested against profit and profitability distributions. As a result, each category in Table 1a,b includes at least one factor or policy from project experiences run in Europe. Hence, it is worth mentioning that the statistical analysis embracing also non-European contexts enriched the overall set of factors and policies captured but did not significantly affect results and conclusions on area- and country-specific aspects concerning European projects.

This paper also has some limitations and gives room to future scenarios at the same time. First, it has been developed by using available data on 50 European projects and by normalizing the dataset in order to analyze and compare metrics homogeneously. A larger set of projects and corresponding data at both European and global levels are not available to date, but it would be useful to create and exploit some datasets in future research efforts. This way, researchers would be able to conduct more detailed analyses and to obtain a deeper understanding on cycle logistics at large. Second, statistical tests were conducted on European projects. A global analysis of cycle logistics experiences would also be beneficial in order to capture additional aspects and data to be further introduced into the research area. Third, this study does not compare bikes for goods delivery with other transportation means. We suggest that such a comparison should be included in the research agenda of future studies.

In conclusion, this work has implications for policy makers, managers and researchers. In fact, policy makers may use the results of this research in order to design and implement specific policies adopted in analogous areas or countries where context-dependent factors apply. This way, they could support cycle logistics projects within a consolidated framework of working policies, especially at the European Commission level. Moreover, managers of public projects as well as private firms may exploit the analysis conducted in order to design and implement successful projects. In particular, they may take into account those results related to the best bike model configuration and make decisions accordingly. Finally, researchers may exploit a new and consolidated approach and statistical results in order to conduct even more comprehensive and advanced studies on cycle logistics projects, thus overcoming the narrow local or national dimension. From this perspective, this work may give a relevant contribution in order to pave the way for future research efforts on cycle logistics.

Author Contributions: Conceptualization, C.G.; methodology C.G.; software, C.G., R.M., and R.P.; validation, C.G., R.M., and R.P.; formal analysis, C.G.; investigation, C.G., R.M., and R.P.; resources, C.G., R.M., and R.P.; data curation, C.G., R.M., and R.P.; writing—original draft preparation, C.G.; writing—review and editing, C.G., R.M., and R.P.; visualization, C.G.; supervision, C.G.; project administration, C.G.; funding acquisition, R.M. All authors have read and agreed to the published version of the manuscript.

Funding: This research received no external funding.

Institutional Review Board Statement: Not applicable.

Informed Consent Statement: Not applicable.

Data Availability Statement: Data are contained within the article.

Conflicts of Interest: The authors declare no conflict of interest.

Appendix A

Table A1. Success factors by area and country.

Area °	Country *	Success Factor
UK and Ireland	UK	• Good communication (perception, awareness) • Health benefit perception • Ownership model • Favorable socio-demographic status at large • Favorable safety perception • Favorable weather • Favorable urban density
	Ireland	• Accessibility • Health benefit perception • Favorable urban density • Travel cost reduction • Travel infrastructure cost reduction • Traffic level reduction • Commitment towards energy savings • Commitment towards pollution reduction • Perception of safety and security level increased • Favorable weather • Behaviors and attitudes of local road users • Gender factor (fostering female commuting) • Favorable socio-demographic status at large • Journey distance • Ownership model
	UK—Scotland	• Suitable age of targeted users, attitude and ability • Favorable socio-demographic status at large • Favorable urban density • Favorable weather • Low risk/high safety perception • Health benefit perception • Compatibility with family and household responsibilities • Significant income • Ownership model • Perception of active travel as the normal means of travel in cities • Short distance to customers for commercial deliveries
Greece, Italy and Mediterranean islands	Greece	• Age (young users are more likely to be eco-cyclists than others) • Gender (women are more likely to be eco-cyclists than men) • Income (low income encourages active travel)
	Italy	• Suitable size and weight of goods compared to load capacity of cargo bikes • Relevance of time windows for the delivery • Positive impacts on Brand Image and Corporate Social Responsibility • Cost reduction • Availability of a supporting network • Reliability of enabling technologies • Reduced time for private and commercial journeys • Lower energy consumption • Lower congestion and emissions • High service quality and coverage • Better quality of life for citizens (not only users) • Short charging time for goods • Sufficient battery duration of e-vehicles
Scandinavia	Denmark	• Sufficient safety perception • Availability of water, woods and overwhelming natural landscapes on the road • Closeness to home

Table A1. *Cont.*

Area °	Country *	Success Factor
Central Europe	Germany	• Short, local trips (below 2-km threshold) • Closeness to relevant centers • Closeness to relevant services, economic boroughs, schools, etc. • Compatible demographic status, cultural inclination, professional practice and attitudes towards a modal shift from customers, firms and messengers • Favorable price and sufficient market conditions (demand patterns, etc.) • Availability of information • Sufficient technological conditions (battery duration, electric range) • Combination of technical innovations and new concepts/configurations of urban mobility systems associated with urban micro-consolidation centers • Reduced driver fatigue through e-cargo bike adoption • Improved range through e-cargo bike adoption • Improved payload through e-cargo bike adoption
	The Netherlands	• Short, local trips (below 3.5-km threshold) • Attitude culturally entrenched
	France	• Combination of technical innovations and new concepts/configurations of ur-ban mobility systems associated with urban micro-consolidation centers • Reduced driver fatigue through e-cargo bike adoption • Improved range through e-cargo bike adoption • Improved payload through e-cargo bike adoption • Compatible demographic status, cultural inclination, professional practice and attitudes towards a modal shift from customers, firms and messengers • Favorable price and sufficient market conditions (demand patterns, etc.) • Availability of information • Sufficient technological conditions (battery duration, electric range)
Australia and the USA	Australia	• Increasing acceptance vs. adverse reaction to helmet compulsoriness • Distance from work or school • Income levels • Variety-seeking • Buffer solutions • Independence, refusal of compulsory measures, policies, etc. • Cost of mandatory helmet (purchasing or always carrying it) • Weather
	USA	• Educational status of individuals and families • Health benefit perception • Low income is positively linked • Features of suitable e-cargo bike for American cities • Variety-seeking • Buffer solutions • Independence, refusal of compulsory measures, policies, etc. • Cost of mandatory helmet (purchasing or always carrying it) • Weather

Table A1. *Cont.*

Area °	Country *	Success Factor
Asia	Malaysia	• Individual needs and values (recreation, door-to-door, minimized waiting times and stops, health and environmental benefits) • Aesthetics • "Donut-cake" population distribution • Demographic factors (age, residential density) • Topographic and spatial factors (waterways, ocean, lakes, hills, flats, industrial and residential areas, schools, hospitals, economic districts) • Environmental factors (climate, temperature, tropical rainforests, rain, humidity)
	Japan	• Age—i.e., elderly drivers, cyclists and pedestrians' behaviors, skills and (for cyclists and pedestrians) possession of driving license • Hazardous inclination of elderly people • Availability and reliability of safety devices
Developing countries	India	• Lowering costs for low income citizens • Giving citizens free access • High benefit for low income citizens • Safety perception • Use of telecommunication technology as a substitute for physical transport • "Time pollution" • Accident, crimes, arsons • Drivers' behavior, training and skills • Gender issues • Social and full cost of transportation systems

° In this paper, "Central Europe" includes only continental European countries and excludes the Italian peninsula, Greece, the Balkans and the Scandinavian peninsula. * The list of countries for each area includes only those with dedicated relevant data, but is not limited to them as some data concern the whole area and are not attributable to a single country in it.

Appendix B

Table A2. Supporting policies by geographical area and country.

Area °	Country * (Policy Program)	Supporting Policy
UK and Ireland	UK	• Campaigns on public health benefits for individuals and NHS • Suitable regulation • Traditional good public transport system in cities centers • Urban infrastructures adverse to car use/for non-motorized travel in cities centers and land use at large • Incentives to logistics companies using cargo bikes in urban contexts • Incentives to new adopters (with social and economic support programs) • Disincentives to car use • Educating motorists to pay more attention to other road users • Making cycling and walking risk-free

Table A2. *Cont.*

Area °	Country * (Policy Program)	Supporting Policy
	Ireland (Agreed Program for Government between Fianna Fàil and the Progressive Democrats, etc.)	• Campaigns on individual and societal benefits • Urban infrastructures adverse to car use/for non-motorized travel in cities centers and land use at large • Disincentives to motorized travel behavior • Favoring safety experiences of cyclists in urban contexts • Availability of infrastructures, facilities and operating systems • Women-specific supporting policies
	UK—Scotland	• Backing of intermodal systems • Urban infrastructures adverse to car use/for non-motorized travel in cities centers and land use at large • Suitable road network
Greece, Italy and Mediterranean islands	Greece; Malta (Master Plan for the restructuring of the road network, etc.)	• Reducing the use of motorized vehicles • Traffic reduction
	Italy	• Backing pilots on e-vehicles in inner cities • Emphasizing environmental and social effects of e-vehicles in inner cities • Incentives to adopt e-vehicles and re-use of public facilities entrusted to logistics companies
Scandinavia	Denmark; Norway (National Transport Plan, etc.) Finland (Cycling and Walking Policy Programs, etc.)	• Availability of safe infrastructures • Availability of parking facilities • Availability of walking and cycling routes • Availability of active travel-specific features in urban green spaces more than just creating such areas per se • Increase safety • Increase appeal • Pollution reduction • Distribution of urban green spaces (suitable distance between and size of them) • Accessibility of public open spaces
Central Europe	Germany (National Cycling Plan: "Ride Your Bike", etc.)	• Accessibility to (non-motorized) e-vehicle choice in inner-city courier shipments • Increased safety
	The Netherlands (Dutch Bicycle Master Plan, etc.) Slovakia (National Action Plan of Environment and Health, etc.) Slovenia (National Cycling Network Development Strategy, etc.) Switzerland (Mission statement for human powered mobility, etc.) Hungary (Position of cycle traffic and main directions of its development in Hungary, etc.) Latvia (Cycle Transport Development State Program for 1999–2015, etc.)	• High accessibility to (non-motorized) vehicle choice • Highly subsidized accessibility • Accidents and deaths reduction • Pollution reduction • Availability of infrastructures • Increased health benefits and physical activity of citizens • Increased cycle tourism • Cooperation with EU countries to create a EuroVelo (European Cycle Route network) • Increased safety
	France	• Incentives to commute by bike

Table A2. *Cont.*

Area °	Country * (Policy Program)	Supporting Policy
Australia and the USA	Australia	• Availability of open spaces • Well-distributed parks and green spaces in cities • Closeness to home and to other parks • Size and number of public areas • Incentives to bike sharing solutions
	USA	• Availability of open spaces • Well-distributed parks and green spaces in cities • Closeness to home and to other parks • Size and number of public areas
Asia	Malaysia (Vision 2020)	• Inclusion in the national agenda and Vision 2020 • Spread and availability of dedicated infrastructures • Backing of multimodal transport systems • Removal of physical and socio-economic barriers • Integration of cycling in the design of transportation networks • Reversal of the current priority policies (incentivizing, ensuring priority to cyclists, excluding them from turning or one-way direction constraints, improving quality of life, etc.)
	Japan	• Inclusion in the national agenda • Active safety policies implemented ("crash severity mitigation systems", camera- and sensor-based "collision damage mitigation braking systems")
Developing countries	India	• Focus of policy makers in the political agenda • Ensuring environmental justice • Ensuring equity and citizen involvement in decision-making • Suitable topological features and network shape (accessibility, growth potential, flexibility, density, zoning strategies, comfort, directness of routes, speed potential, lane width, vehicle occupancy rate, etc.) • Road conditions, infrastructures and investments

° In this paper, "Central Europe" includes only continental European countries and excludes the Italian peninsula, Greece, the Balkans and the Scandinavian peninsula. * The list of countries for each area includes only those with dedicated relevant data, but is not limited to them as some data concern the whole area and are not attributable to a single country in it.

References

1. Grimm, N.B.; Faeth, S.H.; Golubiewski, N.E.; Redman, C.L.; Wu, J.; Bai, X.; Briggs, J.M. Global change and the ecology of cities. *Science* **2008**, *319*, 756–760. [CrossRef]
2. Russo, F.; Comi, A. A classification of city logistics measures and connected impacts. *Procedia Soc. Behav. Sci.* **2010**, *2*, 6355–6365. [CrossRef]
3. Shaheen, S.; Guzman, S.; Zhang, H. Bikesharing in Europe, the Americas, and Asia. *Transp. Res. Rec.* **2010**, *2143*, 159–167. [CrossRef]
4. Mat Yazid, M.R.; Ismail, R.; Atiq, R. The use of non-motorized for sustainable transportation in Malaysia. *Procedia Eng.* **2011**, *20*, 125–134. [CrossRef]
5. European Commission. Living Well, Within the Limits of Our Planet. 7th EAP—The New General Union Environment Action Programme to 2020. 2014. Available online: http://ec.europa.eu/environment/pubs/pdf/factsheets/7eap/en.pdf (accessed on 29 December 2020).

6. Fishman, E.; Washington, S.; Haworth, N.; Watson, A. Factors influencing bike share membership: An analysis of Melbourne and Brisbane. *Transp. Res. Part A* **2015**, *71*, 17–30. [CrossRef]
7. Schliwa, G.; Armitage, R.; Aziz, S.; Evans, J.; Rhoades, J. Sustainable city logistics–Making cargo cycles viable for urban freight transport. *Res. Transp. Bus. Manag.* **2015**, *15*, 50–57. [CrossRef]
8. European Commission. Impact Assessment Accompanying Document to the White Paper: Roadmap to a Single European Transport Area—Towards a Competitive and Resource Efficient Transport System. Available online: https://eur-lex.europa.eu/LexUriServ/LexUriServ.do?uri=SEC:2011:0358:FIN:EN:PDF (accessed on 29 December 2020).
9. Lawson, A.R.; McMorrow, K.; Ghosh, B. Analysis of the non-motorized commuter journeys in major Irish cities. *Transp. Policy* **2013**, *27*, 179–188. [CrossRef]
10. European Commission. Urban Mobility. 2015. Available online: http://ec.europa.eu/transport/themes/urban/urban_mobility/index_en.htm (accessed on 29 December 2020).
11. Harkey, D.L.; Stewart, J.R. Bicycle and motor vehicle operations on wide curb lanes, bicycle lanes, and paved shoulders. In Proceedings of the 1997 Conference on Traffic Congestion and Traffic Safety in the 21st Century, Chicago, IL, USA, 8–11 June 1997; pp. 139–145.
12. Landis, B.W.; Vattikuti, V.R.; Ottenberg, R.M.; Petritsch, T.A.; Guttenplan, M.; Crider, L.B. Intersection level of service for the bicycle through movement. *Transp. Res. Rec.* **2003**, *1828*, 101–106. [CrossRef]
13. Klobucar, M.S.; Fricker, J.D. Network evaluation tool to improve real and perceived bicycle safety. *Transp. Res. Rec.* **2007**, *2031*, 25–33. [CrossRef]
14. Parkin, J.; Wardman, M.; Page, M. Models of perceived cycling risk and route acceptability. *Accid. Anal. Prev.* **2007**, *39*, 364–371. [CrossRef]
15. Møller, M.; Hels, T. Cyclists' perception of risk in roundabouts. *Accid. Anal. Prev.* **2008**, *40*, 1055–1062. [CrossRef]
16. Sallis, J.F.; Cervero, R.B.; Ascher, W.; Henderson, K.A.; Kraft, M.K.; Kerr, J. An ecological approach to creating active living communities. *Annu. Rev. Public Health* **2006**, *27*, 297–322. [CrossRef]
17. Bopp, M.J.; Kaczynski, A.T.; Besenyi, G.M. Active commuting influences among adults. *Prev. Med.* **2012**, *54*, 237–241. [CrossRef] [PubMed]
18. Lawson, A.R.; Pakrashi, V.; Ghosh, B.; Szeto, W.Y. Perception of safety of cyclists in Dublin City. *Accid. Anal. Prev.* **2013**, *50*, 499–511. [CrossRef] [PubMed]
19. Lachowycz, K.; Jones, A.P. Greenspace and obesity: A systematic review of the evidence. *Obes. Rev.* **2011**, *12*, e183–e189. [CrossRef] [PubMed]
20. Schipperijn, J.; Bentsen, P.; Troelsen, J.; Toftager, M.; Stigsdotter, U.K. Associations between Physical activity and characteristics of urban green space. *Urban For. Urban Green.* **2013**, *12*, 109–116. [CrossRef]
21. Pooley, C.G.; Horton, D.; Scheldeman, G.; Mullen, C.; Jones, T.; Tight, M.; Jopson, A.; Chisholm, A. Policies for promoting walking and cycling in England: A view from the street. *Transp. Policy* **2013**, *27*, 66–72. [CrossRef]
22. NICE-National Institute for Health and Clinical Excellence. *Transport Interventions Promoting Safe Cycling and Walking: Evidence Briefing*; National Institute for Health and Clinical Excellence: London, UK, 2006.
23. Ogilvie, D.; Foster, C.; Rothnie, H.; Cavill, N.; Hamilton, V.; Fitzsimons, C.; Mutrie, N. Interventions to promote walking: Systematic review. *Br. Med. J.* **2007**, *334*, 1204–1207. [CrossRef]
24. DfT—Department for Transport (DfT). *Sustainable Travel Guide*; DfT: London, UK, 2010.
25. DfT—Department for Transport (DfT); DoH—Department of Health. *Active Travel Strategy*; DfT: London, UK, 2010.
26. Travel Actively: Walking and Cycling for Health and Wellbeing. Available online: https://www.travelactivelydata.org.uk/ (accessed on 29 December 2020).
27. Silva, A.; Ribero, A. An integrated planning for cities to promote sustainable mobility. In Proceedings of the European Transport Conference, Leiden Leeuwenhorst Conference Centre, Noordwijkerhout, The Netherlands, 5–7 October 2009.
28. Ryley, T.J. Use of non-motorized modes and life stage in Edinburgh. *J. Transp. Geogr.* **2006**, *14*, 367–375. [CrossRef]
29. Shove, E. Beyond the ABC: Climate change policies and theories of social change. *Environ. Plan. A* **2010**, *42*, 1273–1285. [CrossRef]
30. HoL—House of Lords Science and Technology Select Committee. *Behaviour Change Report*; TSO: London, UK, 2011. Available online: http://www.publications.parliament.uk/pa/ld201012/ldselect/ldsctech/179/179.pdf (accessed on 29 December 2020).
31. Zegeer, C. *FHWA Study Tour for Pedestrian and Bicyclist Safety in England, Germany, and The Netherlands*; US Department of Transportation: Washington, DC, USA, 1994.
32. Aultman-Hall, L.; Kaltenecker, M.G. Toronto bicycle commuter safety rates. *Accid. Anal. Prev.* **1999**, *31*, 675–686. [CrossRef]
33. TfL—Transport for London. *Central London Congestion Charging: Sixth Annual Report*; TfL: London, UK, 2008. Available online: http://content.tfl.gov.uk/central-london-congestion-charging-impacts-monitoring-sixth-annual-report.pdf (accessed on 29 December 2020).
34. DoT—Department of Transport. *Statement of Strategy: 2003–2005*; Department of Transport: Dublin, Ireland, 2003.
35. Tolley, R. *The Greening of Urban Transport: Planning for Walking and Cycling in Western Cities*; Belhaven Press: London, UK, 1990.
36. Cervero, R. Mixed land uses and commuting: Evidence from the American Housing Survey. *Transp. Res. Part A Policy Pract.* **1996**, *30*, 361–377. [CrossRef]
37. Pucher, J.; Komanoff, C.; Shimek, P. Bicycling renaissance in North America? Recent trends and alternative policies to promote bicycling. *Transp. Res. Part A Policy Pract.* **1999**, *33*, 625–654. [CrossRef]

38. Plaut, P.O. Non-motorized commuting in the US. *Transp. Res. Part Transp. Environ.* **2005**, *10*, 347–356. [CrossRef]
39. Oliver, M.; Badland, H.; Mavoa, S.; Witten, K.; Kearns, R.; Ellaway, A.; Hinckson, E.; Mackay, L.; Schluter, P.J. Environmental and socio-demographic associates of children's active transport to school: A cross-sectional investigation from the URBAN Study. *Int. J. Behav. Nutr. Phys. Act.* **2014**, *11*, 1–12. [CrossRef] [PubMed]
40. Sardianou, E.; Nioza, E. Who are the eco-bicyclists? *Transp. Res. Part D* **2015**, *34*, 161–167. [CrossRef]
41. Sigurdardottir, S.B.; Kaplan, S.; Moller, M.; Teasdale, T.W. Understanding adolescents' intentions to commute by car or bicycle as adults. *Transp. Res. Part D* **2013**, *24*, 1–9. [CrossRef]
42. Nocerino, R.; Colorni, A.; Lia, F.; Luè, A. E-bikes and E-scooters for smart logistics: Environmental and economic sustainability in pro-E-bike Italian pilots. *Transp. Res. Procedia* **2016**, *14*, 2362–2371. [CrossRef]
43. Gruber, J.; Ehrler, V.; Lenz, B. Technical potential and user requirements for the implementation of electric cargo bikes in courier logistics services. In Proceedings of the 13th World Conference on Transport Research, Berlin, Germany, 15–18 July 2013.
44. Bedimo Rung, A.L.; Mowen, A.J.; Cohen, D.A. The Significance of Parks to Physical Activity and Public Health: A Conceptual Model. *Am. J. Prev. Med.* **2005**, *28* (Suppl. 2), 159–168. [CrossRef]
45. Public Health Office Copenhagen. Sunde Københavnere i alle Alder–Københavns Kommunes Sundhedspolitik 2006–2010 (Healthy Copenhageners in All Ages–Health Policy of the Municipality of Copenhagen for 2006–2010). Municipality of Copenhagen. 2006. Available online: https://www.kk.dk/sites/default/files/edoc_old_format/Borgerrepraesentationen/12-10-20 06%2017.30.00/Dagsorden/24-10-2006%2010.44.52/Kbhs_Kom_Sundhspol_250906.PDF (accessed on 29 December 2020).
46. Toftager, M.; Ekholm, O.; Schipperijn, J.; Stigsdotter, U.K.; Bentsen, P.; Grønbæk, M.; Randrup, T.B.; Kamper-Jørgensen, F. Distance to green space and physical activity: A Danish national representative survey. *J. Phys. Act. Health* **2011**, *8*, 741–749. [CrossRef]
47. Pucher, J. Bicycling boom in Germany: A revival engineered by public policy. *Transp. Q.* **1997**, *51*, 31–46.
48. Gruber, J.; Kihm, A.; Lenz, B. A new vehicle for urban freight? An ex-ante evaluation of electric cargo bikes in courier services. *Res. Transp. Bus. Manag.* **2014**, *11*, 53–62. [CrossRef]
49. Tfl—Transport for London. *Cycle Freight in London: A Scoping Study*; Transport for London, Mayor of London: London, UK, 2009. Available online: http://content.tfl.gov.uk/cycle-as-freight-may-2009.pdf (accessed on 29 December 2020).
50. Dablanc, L. Transferability of Urban Logistics Concepts and Practices from a Worldwide Perspective—Deliverable 3.1—Urban Logistics Practices—Paris Case Study. TURBLOG_WW Project. 2011. Available online: https://trimis.ec.europa.eu/project/transferability-urban-logistics-concepts-and-practices-world-wide-perspective#tab-outline (accessed on 29 December 2020).
51. Leonardi, J.; Browne, M.; Allen, J. Before-after assessment of a logistics trial with clean urban freight vehicles: A case study in London. *Procedia Soc. Behav. Sci.* **2012**, *39*, 146–157. [CrossRef]
52. Fishman, E. The impacts of public bicycle share schemes on transport choice. In Proceedings of the Asia-Pacific Cycle Congress. Brisbane Convention and Exhibition Centre, Brisbane, Australia, 18–21 September 2011.
53. Meddin, R. The Bike-Sharing World: First Days of Summer 2011. 2011. Available online: http://bike-sharing.blogspot.com/search?q=Brisbane (accessed on 29 December 2020).
54. Rojas-Rueda, D.; De Nazelle, A.; Tainio, M.; Nieuwenhuijsen, M.J. The health risks and benefits of cycling in urban environments compared with car use: Health impact assessment study. *Br. Med. J.* **2011**, *343*, d4521. [CrossRef]
55. Alta Bike Share. *Melbourne Bike Share Survey*; Prepared for VicRoads: Melbourne, Australia, 2011.
56. Fan, J.X.; Wen, M.; Kowaleski-Jones, L. Sociodemographic and environmental correlates ofactive commuting in rural America. *J. Rural Health* **2014**, *31*, 176–185. [CrossRef] [PubMed]
57. Mokhtarian, P.L. Travel as a desired end, not just a means. *Transp. Res. A* **2005**, *39*, 93–96. [CrossRef]
58. Bassett, D.R., Jr.; Pucher, J.; Buehler, R.; Thompson, D.L.; Crouter, S.E. Walking, cycling, and obesity rates in Europe, North America, and Australia. *J. Phys. Act. Health* **2008**, *5*, 795–814. [CrossRef] [PubMed]
59. Shephard, R.J. Is active commuting the answer to population health? *Sports Med.* **2008**, *38*, 751–758. [CrossRef]
60. TfL—Transport for London. *Travel in London Report 3*; Transport for London: London, UK, 2010.
61. Bopp, M.J.; Kaczynski, A.T.; Wittman, P. The relationship of eco-friendly attitudes with walking and biking to work. *J. Public Health Manag. Pract.* **2011**, *17*, 9–17. [CrossRef]
62. Midgley, P. *Bicycle-Sharing Schemes: Enhancing Sustainable Mobility in Urban Areas*; United Nations: New York, NY, USA, 2011.
63. Kaczynski, A.T.; Bopp, M.J.; Wittman, P. To drive or not to drive: Factors differentiating active versus non-active commuters. *Health Behav. Public Health* **2012**, *2*, 14–19.
64. LDA Consulting. *Capital Bikeshare 2011 Member Survey Report*; LDA Consulting: Washington, DC, USA, 2012.
65. Fishman, E.; Washington, S.; Haworth, N. Bike share's impact on car use: Evidence from the United States, Great Britain, and Australia. *Transp. Res. D* **2014**, *31*, 13–20. [CrossRef]
66. Fischer, C.M.; Sanchez, C.E.; Pittman, M.; Milzman, D.; Volz, K.A.; Huang, H.; Sanchez, L.D. Prevalence of bicycle helmet use by users of public bikeshare programs. *Ann. Emerg. Med.* **2012**, *60*, 228–231. [CrossRef] [PubMed]
67. Fishman, E.; Washington, S.; Haworth, N. Barriers and facilitators to public bicycle scheme use: A qualitative approach. *Transp. Res. F* **2012**, *15*, 686–698. [CrossRef]
68. Fishman, E.; Washington, S.; Haworth, N. Bike share: A synthesis of the literature. *Transp. Rev.* **2013**, *33*, 148–165. [CrossRef]
69. Giles-Corti, B.; Broomhall, M.H.; Knuiman, M.; Collins, C.; Douglas, K.; Ng, K.; Lange, A.; Donovan, R.J. Increasing walking. How important is distance to, attractiveness, and size of public open space? *Am. J. Prev. Med.* **2005**, *28*, 169–176. [CrossRef] [PubMed]

70. Kaczynski, A.T.; Potwarka, L.R.; Saelens, B.E. Association of park size, distance and features with physical activity in neighborhood parks. *Am. J. Public Health* **2008**, *98*, 1451–1456. [CrossRef] [PubMed]

71. Kaczynski, A.T.; Potwarka, L.R.; Smale, B.J.A.; Havitz, M.E. Association of parkland proximity with neighborhood and park-based physical activity: Variations by gender and age. *Leis. Sci.* **2009**, *31*, 174–191. [CrossRef]

72. Le Galès, P.; Zagrodzki, M. *Cities Are Back in Town: The US/Europe Comparison. Cahier Européen Numéro 05/06 du Pôle Ville/métropolis/Cosmopolis*; Centre d'Etudes Européennes de Sciences Po: Paris, France, 2006.

73. Conway, A.; Fatisson, P.; Eickemeyer, P.; Cheng, J.; Peters, D. Urban micro-consolidation and last mile goods delivery by freight-tricycle in Manhattan: Opportunities and challenges. In Proceedings of the Transportation Research Board 91st Annual Meeting, Washington, DC, USA, 22–26 January 2012.

74. Giuliano, G.; O'Brien, T.; Dablanc, L.; Holliday, K. *NCFRP Project 36 Synthesis of Freight Research in Urban Transportation Planning*; National Cooperative Freight Research Program: Washington, DC, USA, 2013.

75. Rietveld, P. *Biking and Walking: The Position of Non-Motorized Transport Modes in Transport Systems, Tinbergen Institute Discussion Papers 01-111/3*; Tinbergen Institute: Amsterdam, The Netherlands, 2001.

76. Litman, T.; Burwell, D. Issues in sustainable transportation. *Int. J. Glob. Environ. Issues* **2006**, *6*, 331–347. [CrossRef]

77. EPU-Economic Planning Unit under Malaysia's Prime Minister's Department. Vision 2020. 1991. Available online: https://www.epu.gov.my/en (accessed on 29 December 2020).

78. Richardson, B.C. Sustainable transport: Analysis frameworks. *J. Transp. Geogr.* **2005**, *13*, 29–39. [CrossRef]

79. Chapman, L. Transport and climate change: A review. *J. Transp. Geogr.* **2007**, *15*, 354–367. [CrossRef]

80. Matsui, Y.; Oikawa, S.; Hitosugi, M. Analysis of carto-bicycle approach patterns for developing active safety devices. *Traffic Inj. Prev.* **2015**, *17*, 434–439. [CrossRef]

81. Tada, M.; Okada, M.; Renge, K. An analysis of elderly drivers', cyclists' and pedestrians' behaviors using wearable sensors. *Trans. Jpn. Soc. Med. Biol. Eng.* **2016**, *54*, 129–134.

82. Kuzumaki, S. Our approach to a safe sustainable society. *J. Soc. Automot. Eng. Jpn.* **2009**, *63*, 11–19. (In Japanese)

83. Shibata, E. Development of driving assist system "EyeSight" by new stereo camera. *J. Soc. Automot. Eng. Jpn.* **2009**, *63*, 93–98. (In Japanese)

84. Saneyoshi, K. Recent trends of real-time stereo vision for automobile and application to pedestrian detection. *J. Soc. Automot. Eng. Jpn.* **2013**, *67*, 84–89. (In Japanese)

85. Behera, R.; Gangadharan, J.; Kutty, K.; Nair, S.; Vaidya, V. A novel method for day time pedestrian detection. *SAE Int. J. Passeng. Cars-Electron. Electr. Syst.* **2015**, *8*, 406–412. [CrossRef]

86. Fazio, J. Nonmotorized-Motorized Traffic Accidents and Conflicts on Delhi Streets. In *Transportation Research Record 1487, TRB*; National Research Council: Washington, DC, USA, 1995; pp. 68–74.

87. Graham, S.; Marvin, S. *Telecommunication and the City*; Routledge: London, UK, 1996.

88. Tolley, R. *The Greening of Urban Transport*, 2nd ed.; John Wiley and Sons: Chichester, UK, 1997.

89. Khisty, C.J. A systemic overview of non-motorized transportation for developing countries: An agenda for action. *J. Adv. Transp.* **2003**, *37*, 273–293. [CrossRef]

90. Khisty, C.J.; Zeitler, U. Is Hypermobility a Challenge for Transport Ethics and Systemicity? *Syst. Pract. Action Res.* **2001**, 597–613. [CrossRef]

91. Khisty, C.J.; Ayvalik, C. Automobile Dominance and the Tragedy of the Land-Use/Transport System: Some Critical Issues. *Syst. Pract. Action Res.* **2003**, 53–73. [CrossRef]

92. Maunder, D. *Household and Travel Characteristics in Two Residential Areas in New Delhi, India*; TRRL. Supplement Report, 673; Department of the Environment: Crowthome, UK, 1981.

93. Fruin, J.J.; Benz, G.P. Pedestrian Time-Space Concept for Analyzing Comers and Crosswalks. In *Transportation Research Record 959, TRB*; National Research Council: Washington, DC, USA, 1984; pp. 18–24.

94. Replogle, M.A. Sustainable Transportation Strategies for Third-World Development. In *Transportation Research Record 1294, TRB*; National Research Council: Washington, DC, USA, 1991.

95. Fruin, J.J. *Designing for Pedestrians, Chapter 8. Public Transportation*; Gray, G., Hoel, L., Eds.; Prentice Hall: Englewood Cliffs, NJ, USA, 1992.

96. Khisty, C.J. Transportation in Developing Countries: Obvious problems, Possible Solutions. In *Transportation Research Record 1396, TRB*; National Research Council: Washington, DC, USA, 1993.

97. Whitelegg, J. *Transport for a Sustainable Future*; Belhaven Press: London, UK, 1993.

98. Taaffe, E.G.; Gauthier, H.L.; O'Kelly, M.E. *Geography of Transportation*; Prentice Hall: Upper Saddle River, NJ, USA, 1996.

99. Forkenbrock, D.J.; Schweitzer, L.A. *Environmental Justice and Transportation Investment Policy*; Public Policy Center, University of Iowa: Iowa City, IA, USA, 1997.

100. Tellis, R.; Khisty, C.J. Social Cost Component of an Efficient Toll. In *Transportation Research Record 1576, TRB*; National Research Council: Washington, DC, USA, 1997.

101. Peters, D. Gender Issues in Transportation: A Short Introduction. In Proceedings of the UNEP Regional Workshop-Deals on Wheels, San Salvador, Salvador, 28–30 July 1999.

102. Mackett, R. Policies to attract drivers out of their cars for short trips. *Transp. Policy* **2001**, *8*, 295–306. [CrossRef]

103. Mackett, R. Why do people use their cars for short trips? *Transportation* **2003**, *30*, 329–349. [CrossRef]

104. Jarvis, H. Dispelling the myth that preference makes practice in residential location and transport behaviour. *Hous. Stud.* **2003**, *18*, 587–606. [CrossRef]

105. Alfonzo, M. To walk or not to walk? The hierarchy of walking needs. *Environ. Behav.* **2005**, *37*, 808–836. [CrossRef]

106. Anable, J. 'Complacent car addicts' or 'Aspiring environmentalists'? Identifying travel behaviour segments using attitude theory. *Transp. Policy* **2005**, *12*, 65–78. [CrossRef]

107. Cyclelogistics. Cargo Bikes—The Economic Argument. 2011. Available online: http://one.cyclelogistics.eu/docs/113/the_economic_argument_printingmarks_Euro_Pound.pdf (accessed on 29 December 2020).

108. Cyclelogistics–Moving Europe Ahead–Final Public Report. Available online: http://one.cyclelogistics.eu/docs/117/D6_9_FPR_Cyclelogistics_print_single_pages_final.pdf (accessed on 29 December 2020).

109. Poliedra-Politecnico di Milano. A New Move for Business. Available online: http://www.poliedra.polimi.it/wp-content/uploads/PROEBIKE_Final-Report_it.pdf (accessed on 29 December 2020). (In Italian)

110. Siegel, S.; Castellan, N.J. *Nonparametric Statistics for the Behavioral Sciences*, 2nd ed.; McGraw-Hill: New York, NY, USA, 1988.

111. Copenhagenize. IKEA Idea with Velorbis Bikes. 2008. Available online: http://www.copenhagenize.com/2008/06/ikea-idea-with-velorbis-bikes.html (accessed on 29 December 2020).

112. MyAmsterdamBike. 2016. Available online: https://myamsterdambike.com/5-businesses-embracing-cargo-bike/ (accessed on 29 December 2020).

113. Ljung Aust, M.; Jakobsson, L.; Lindman, M.; Coelingth, E. Collision avoidance systems—Advancements and efficiency. In Proceedings of the SAE 2015 World Congress & Exposition, Detroit, MI, USA, 21–23 April 2015.

114. Hess, A.-K.; Schubert, I. Functional perceptions, barriers, and demographics concerning e-cargo bike sharing in Switzerland. *Transp. Res. Part D* **2019**, *71*, 153–168. [CrossRef]

115. Fikar, C.; Hirsch, P.; Gronalt, M. A decision support system to investigate dynamic last-mile distribution facilitating cargo-bikes. *Int. J. Logist. Res. Appl.* **2018**, *21*, 300–317. [CrossRef]

116. Binetti, M.; Caggiani, L.; Camporeale, R.; Ottomanelli, M. A Sustainable Crowdsourced Delivery System to Foster Free-Floating Bike-Sharing. *Sustainability* **2019**, *11*, 2772. [CrossRef]

117. Krause, K.; Assman, T.; Schmidt, S.; Matthies, E. Autonomous driving cargo bikes–Introducing an acceptability-focused approach towards a new mobility offer. *Transp. Res. Interdiscip. Perspect.* **2020**, *6*, 100135. [CrossRef]

118. Bielinski, T.; Wazna, A. Electric Scooter Sharing and Bike Sharing User Behaviour and Characteristics. *Sustainability* **2020**, *12*, 9640. [CrossRef]

119. Assman, T.; Lang, S.; Müller, F.; Schenk, M. Impact Assessment Model for the Implementation of Cargo Bike Transshipment Points in Urban Districts. *Sustainability* **2020**, *12*, 4082. [CrossRef]

120. Sheth, M.; Butrina, P.; Goodchild, A. Measuring delivery route cost trade-offs between electric-assist cargo bicycles and delivery trucks in dense urban areas. *Eur. Transp. Res. Rev.* **2019**, *11*, 11. [CrossRef]

121. Becker, S.; Rudolf, C. Exploring the Potential of Free Cargo-Bikesharing for Sustainable Mobility. *GAIA* **2018**, *27*, 156–164. [CrossRef]

122. Baum, L.; Assman, T.; Strubelt, H. State of the art—Automated micro-vehicles for urban logistics. *IFAC PapersOnLine* **2019**, *52*, 2455–2462. [CrossRef]

123. Rajesh, P.B.; Rajan, A.J. Sustainable Performance of Cargo Bikes to Improve the Delivery Time Using Traffic Simulation Model. *FME Trans.* **2020**, *48*, 411–418. [CrossRef]

applied sciences

Article

Campus City Project: Challenge Living Lab for Smart Cities

José I. Huertas [1,*], Jürgen Mahlknecht [1], Jorge de J. Lozoya-Santos [1], Sergio Uribe [1], Enrique A. López-Guajardo [1,*] and Ricardo A. Ramirez-Mendoza [1,2]

[1] Escuela de Ingeniería y Ciencias, Tecnologico de Monterrey, Ave. Eugenio Garza Sada 2501, Monterrey 64849, Nuevo León, Mexico; jurgen@tec.mx (J.M.); jorge.lozoya@tec.mx (J.d.J.L.-S.); sergio.uribe@tec.mx (S.U.); ricardo.ramirez@tec.mx (R.A.R.-M.)

[2] Laboratory of Machine Vision and Intelligent System, Jiangxi University of Science and Technology, No. 86, Hongqi Avenue, Zhanggong District, Ganzhou City 341000, China

* Correspondence: jhuertas@tec.mx (J.I.H.); enrique.alopezg@tec.mx (E.A.L.-G.)

Abstract: This work presents the Campus City initiative followed by the Challenge Living Lab platform to promote research, innovation, and entrepreneurship with the intention to create urban infrastructure and creative talent (human resources) that solves different community, industrial and government Pain Points within a Smart City ecosystem. The main contribution of this work is to present a working model and the open innovation ecosystem used in Tecnologico de Monterrey that could be used as both, a learning mechanism as well as a base model for scaling it up into a Smart Campus and Smart City. Moreover, this work presents the Smart Energy challenge as an example of a pedagogic opportunity for the development of competencies. This included the pedagogic design of the challenge, the methodology followed by the students and the results. Finally, a discussion on the findings and learnings of the model and challenge implementation. Results showed that Campus City initiative and the Challenge Living Lab allows the identification of highly relevant and meaningful challenges while providing a pedagogic framework in which students are highly motivated, engaged, and prepared to tackle different problems that involve government, community, industry, and academia.

Keywords: smart water/energy/mobility; open innovation; challenge living lab; smart city; challenge-based learning

Citation: Huertas, J.I.; Mahlknecht, J.; Lozoya-Santos, J.d.J.; Uribe, S.; López-Guajardo, E.A.; Ramirez-Mendoza, R.A. Campus City Project: Challenge Living Lab for Smart Cities. *Appl. Sci.* **2021**, *11*, 11085. https://doi.org/10.3390/app112311085

Academic Editors: Edris Pouresmaeil and Andreas Sumper

Received: 31 August 2021
Accepted: 28 October 2021
Published: 23 November 2021

Publisher's Note: MDPI stays neutral with regard to jurisdictional claims in published maps and institutional affiliations.

1. Introduction

Uncontrolled and rapid urban growth has given rise to different issues affecting Quality of Life (QoL) [1–5]. QoL is defined by the World Health Organization as "individuals' perception of their position in life in the context of the culture and value systems in which they live, and in relation to their goals, expectations, standards and concerns" [6] and relates to their happiness, security, well-being, ecology, resilience, and global awareness. Some of these problems affecting QoL include energy generation and distribution, traffic (mobility), unequal housing, health, education, environment (air, water, soil), etc. [1,7,8]. A collaborative framework involving the citizens, government, academia, and private sector is crucial to minimize the impact of these problems. The use of integrated and interconnected technological developments, supported by information and data, enables this framework to propose different solutions to improve the overall QoL of the citizens [2,9–11]. This concept is referred to as a Smart City and could be described as a living laboratory or hubs driven by innovation to meet global standards [12,13], in which political, social and environmental decisions are made based on data [14].

Figure 1 depicts the Smart City logical framework, including its various components and stakeholders or key actors. This framework begins with the needs and challenges the city is facing, called Pain Points [15]. A crucial step is an initial selective process or screening, during which the technical, economical, and social feasibility is considered by all stakeholders participating in the decision making process: citizens, government, academia, private sector, investors, and entrepreneurs (upper right part of Figure 1) [1,16]. The creation or modernization of public policies is a crucial step within this framework, especially in regards to the material, economic and human resources available to contribute to these projects and reach the goal of transforming a community or a city into a smart environment [16]. Moreover, feedback cycles are important to ensure an effective and efficient implementation of the project solutions. Feedback is based on data, digital technologies and interconnected visualization dashboards (Internet of Things, Information Systems, Artificial Intelligence etc. [2,11]) that allow the dissemination and socialization of the finished and ongoing projects. Finally, the framework's main goal is to affect the different dimensions related to the Smart City's habitability and QoL.

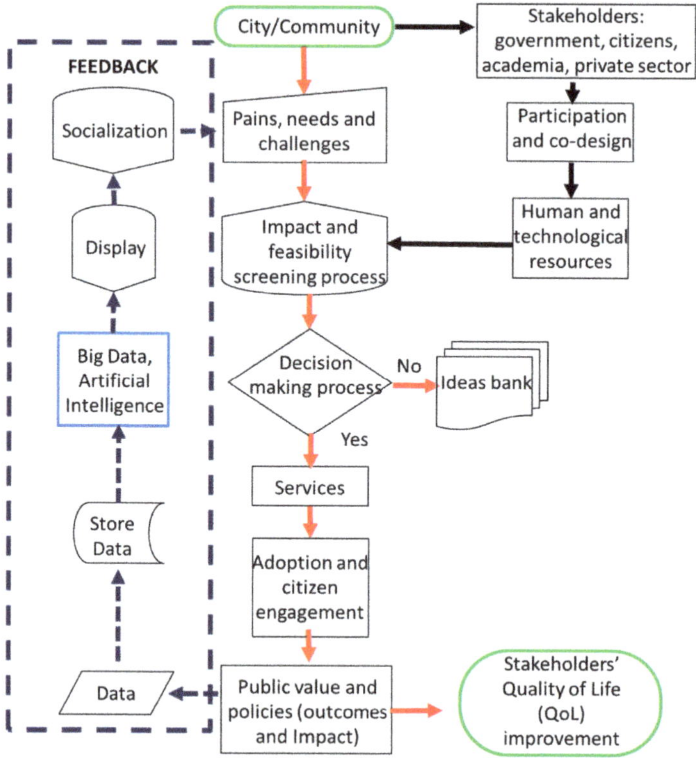

Figure 1. Logical framework of a Smart City.

At this point, it is important to highlight the role of the academic institution. The university becomes a key stakeholder for: (1) the identification of solutions for the Pain Points based on innovation, interconnectivity, and research and, (2) the development of human resources with the necessary skills to provide solutions, management, and technology at different levels in a Smart City. Additionally, the active involvement of research institutes and universities is required to sustain an open innovation ecosystem that could drive technological development [17,18].

In 2015, the United States launched an initiative to promote research, innovation, and entrepreneurship (RIE) in university campuses called the "Smart City Challenge" project [19]. With the participation of 78 cities, the project concluded that citizen wellbeing is highly influenced by connectivity, information about the city's resources and the need for better mobility. Moreover, various initiatives have studied and developed the transformation of university campuses into smart living laboratories, through which different projects could potentially be scaled up to solve the Pain Points of a city [20–23]. Even though these works provide insight into the operation, organization, technological infrastructure, and methods used, only few studies provide some information on how they could use the Smart Campus concept as a learning mechanism for students [22,24,25]. Further work is required to develop Smart Campuses with educational models that can meet the needs of a Smart City while enhancing the learning experience of the students and providing them with the knowledge and skills required to solve different real-life challenges or Pain Points.

Inspired by the RIE results and the above-mentioned needs, the transformation of the Tecnologico de Monterrey university's main campus in the city of Monterrey, Mexico, and of its neighboring communities was established as a key part of its strategic plan for 2020: to generate sustainable spaces and sustainable conditions for RIE. This project, called Distrito Tec, focuses on enabling the creation of a dynamic, safe and inspiring community, one that attracts and retains talent while promoting the development and positioning of the city and the country in general. Distrito Tec´s objective is to improve the urban area and the quality of life of nearby communities (with over 26,333 residents). This includes offering open and renewed spaces, as well as access to different campus facilities and social programs.

The inherent challenges of the Monterrey campus´ transformation into a smart community represents the perfect opportunity for its students to develop the competencies needed to assess various real-life industrial and environmental problems and to understand key concepts of a Smart City [26]. This particular plan to use the Smart Campus concept as a learning mechanism for students within the Distrito Tec project is called the Campus City initiative. The main objectives of this initiative are: (1) to establish the idea among the student community that the campus is their home and that, as its citizens, it is their responsibility to take care of it; (2) to establish a relationship between the university's researchers and professional students to jointly develop solutions through the application of science, technology, engineering and mathematics (STEM); (3) to propose solutions that improve the citizens´ experience under the premise "society, planet, profit" (triple bottom line) through research, applied research, innovation and collaboration; and (4) to create a living laboratory (Challenge Living Lab) where teachers systematically identify, define and implement learning challenges based on the main problems that cities face. The Campus City initiative involves collaboration between the university's academic community, industries and the government, using Tecnologico de Monterrey's infrastructure of innovation laboratories to answer the main research question: how to promote a scalable Smart City framework that also provides a learning environment to engage and motivate students while helping them develop the necessary competencies to solve the smart community´s Pain Points through innovation.

Additionally, Tecnologico de Monterrey launched its new educational model "Tec21" in August 2019, with "Challenge-Based Learning" (CBL) as its central axis, where the definition and development of real-world challenges are used to guide and accelerate the learning process. Tec21 [27,28] is a unique and customizable model that promotes the development of competitive, competent individuals that can tackle any real-world challenge through research and innovation. This is catalyzed by inspiring professors who employ significant real-life challenges that motivate and engage students to create a memorable experience and trigger the learning process that is vital for their formation. Fundamentally, the Tec21 educational model could be described as a student-centered model characterized by four main components/pillars: (a) challenge-based learning, (b) flexibility, (c) highly trained and inspiring professors and, (d) memorable educational experiences [28–30]. All undergraduate programs at Tecnologico de Monterrey follow this disruptive model, which has been implemented in all 26 Tecnologico de Monterrey campuses with promising results regarding its implementation and the students´ learning experience [31–33].

Therefore, this work answers the need for a Smart Campus City framework which could be used as a base model to be scaled up and applied in a Smart City, while developing competent professionals prepared to face these challenges. Specifically, the main objectives of this work are: (1) to present an overall framework and methodology based on an innovation ecosystem that could be used to select the community's Pain Points; (2) to provide a dynamic platform in which Pain Points from the different verticals axes could be used as pedagogic opportunities to favor the development of competencies in an engineering syllabus; and (3) to present an example of the pedagogic design and implementation of a Campus City Challenge, while discussing the involvement of different stakeholders and the pedagogic learnings obtained from the experience.

This work is organized as follows:

- Section 2 introduces the Smart City´s main components, starting with the definition of the Smart City verticals, followed by the open innovation ecosystem on which the Campus City working model is based (including main stakeholders and step-by-step project selection process). Finally, the Challenge Living Lab´s objective and components are described; this is the methodology used to select real-world challenges with high pedagogic value.
- Section 3 presents an overall description of a Smart Energy challenge as an example of the Campus City Challenge Living Lab platform. This section starts with the pedagogic objective of the challenge followed by the pedagogic design. This includes the context of the challenge and how it was presented to the engineering students. Moreover, the challenge solution methodology and results/discussion are presented.
- Section 4 presents the overall findings and learnings regarding the implementation of the Campus City Challenge Living Lab from the perspective of the external stakeholders, the professors (pedagogic perspective) and the students.
- Finally, Section 5 presents the overall conclusions of this work, limitations and future work.

2. Campus City Initiative Main Components

2.1. Smart City Verticals: Smart Mobility, Water and Energy Definitions

Mobility, Energy and Water are the main vertical axes of a smart city, united under a common premise: reducing economic and environmental costs, and saving time through the use of data, information and telecommunication. The citizens' quality of life and their perception of the city they live in will improve through intelligent systems that can optimize the administration of resources and inform them about the status and availability of mobility, water and energy resources.

Smart Mobility—a series of initiatives, policies and actions whose main objective is to promote cleaner, safer and more efficient forms of transportation and to facilitate mobility via public or private transportation throughout the city.

Smart Water—the use of data acquisition systems, prediction and cognition models, as well as information systems to allow better decision-making by the users and the water infrastructure agency in terms of its accumulation, monitoring, distribution and traceability.

Smart Energy—to achieve a transition towards greater balance in the distribution and use of energy from renewable sources (sun and wind among others) and fossil fuels, with the purpose of polluting less and improving energy consumption through the use of environmentally friendly and safer technologies.

2.2. Open Innovation Ecosystem

An open innovation ecosystem is fundamental for delivering high-quality service. This is facilitated by the interconnectedness of different technological platforms, services and providers [4]. The Campus City initiative is based on this model of open innovation (shown in Figure 2). The value proposition of this initiative is to offer challenges that are relevant to the Tec21 model and to create high-impact innovation projects for the industry and the community. This could be achieved through the implementation of the academic innovation platform and using Distrito Tec as a Challenge Living Lab (described in Section 2.3).

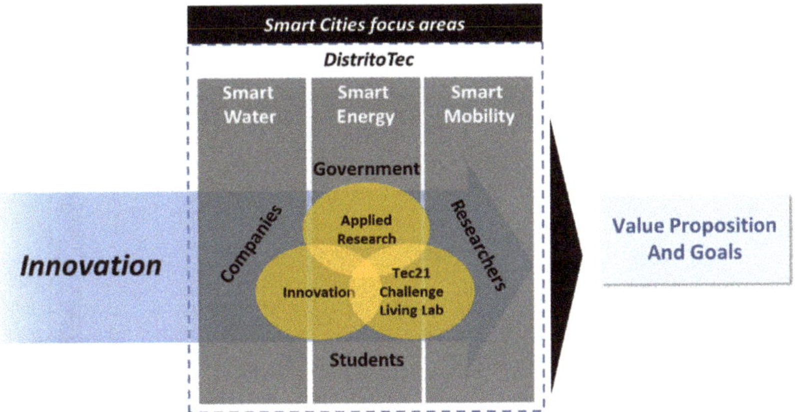

Figure 2. Campus City initiative general innovation model.

The Campus City initiative was designed to achieve the following goals:

1. Tec21 challenges
   - Identification and design of meaningful challenges that support our learning model.
   - Strong involvement of the Academic Community (lecturers, students, researchers, collaborators).

2. Industry-community innovation projects
   - Increased competitiveness through technology development to solve the requested Pain Points.
   - Creation of new businesses through the implementation of disruptive technologies.
   - Creation of high-social impact technologies, which reduce or eliminate major community challenges.

3. Applied Research Projects
   - Development of applied research to close the science-technology gap and solve complex challenges for industry and the community.

These goals are focused on the three Campus City verticals (Smart Water, Smart Energy and Smart Mobility) affecting the entirety of the Distrito Tec infrastructure, as shown in Figure 2. These three verticals will dictate the focus of the project and if a project does not comply with the objectives of one of the three main verticals, then the project is rejected. This strategy allows a better allocation of efforts and resources, increasing the chance of success.

The Campus City initiative working model is shown in Figure 3. This working model comprises three major stages: value discovery, execution and tech transfer. Throughout these stages, two main groups of actors have been defined to guide the initiatives. The first main actor corresponds to the stakeholders, in this case the industry, academia (campus) and government (top horizontal axis in Figure 3). The stakeholders provide problems and challenges from their specific sector. In this sense, the stakeholders can be considered as the Market Pull. The second main actor is the Campus City core team from Tecnologico the Monterrey, which includes the innovation area, steering committee, post-docs, partner professors and students (bottom horizontal axis in Figure 3). The Campus City core team offers the Technology Push. Information will flow between these two main groups of actors throughout the different steps or processes, allowing all parties to reach a consensus on the challenges to be solved.

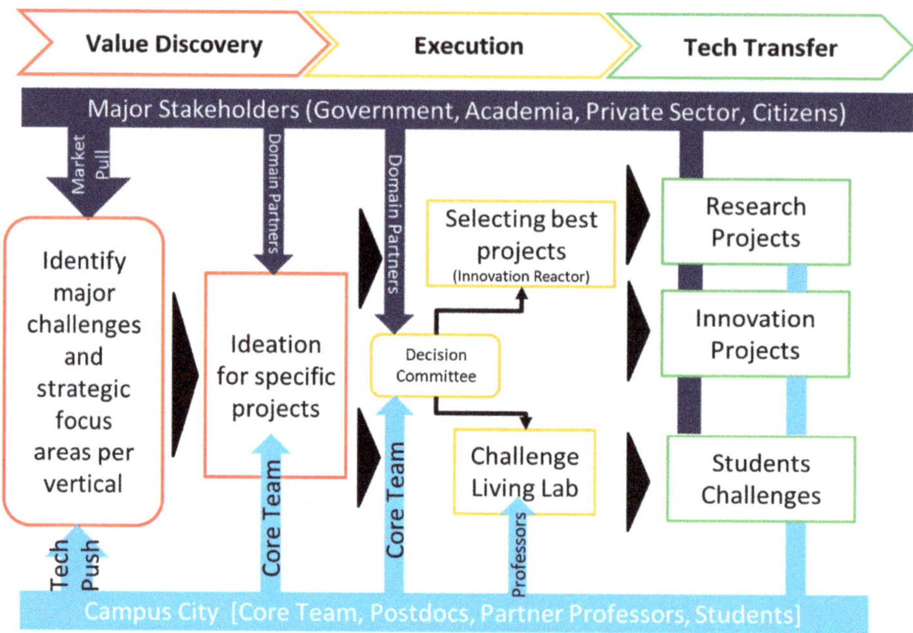

Figure 3. Campus City working model.

Regarding the stakeholders in Figure 3, the industry refers to companies that are related to the production, consumption and distribution of water and energy resources, and to how people move around urban settings. These companies have already determined their strategies and technological roadmaps. They have their own research groups, but they are always on the lookout for highly disruptive external partners. The opportunity for universities lies in solving their current challenges from a disruptive perspective and identifying new challenges that they had not even imagined or thought of, complementing the blind spot that all companies develop. The government has designed and implemented public policy directives relating to water, energy and mobility. For this reason, the authorities must also contribute and make recommendations regarding the analysis and development of these technologies and solutions.

Moreover, government entities have been supporting the development of these solutions by:

1. Proposing solutions and providing feedback on the feasibility of the potential solutions, especially with regard to social aspects and political context.
2. Funding, not only through direct sponsoring but also through partnership in the search for financial support with third parties at national and international levels.
3. First-hand knowledge of the results obtained, favoring the transfer of solutions within government institutions for their future implementation on a larger scale or replication in other regions.
4. Facilitating the implementation of the proposed solutions by authorizing actions on territories of public use. For example, authorizing the installation of video cameras to monitor the flow of vehicles on public roads.

The accompaniment of government entities, through the previous concrete actions, has encouraged students and researchers in the search for solutions to the country's problems under the Campus City initiative.

As mentioned above, the stakeholders and the core team will manage and implement different processes to reach an agreement on which challenges could meet the Campus City goals. This four-step process is described below.

Step #1: Identifying the main challenges, priorities and strategic focuses in each area.

Step #2: Breaking down the large problems into strategic problem areas that require the development of specific technological solutions and identifying a list of projects that are of common interest. Step 1 and 2 should be revisited and discussed once a year by the stakeholders and the core team.

Step #3: Obtaining and evaluating input from both internal and external stakeholders to select the best projects, those with a higher priority and that deserve the allocation of resources.

Step #4: Separating the specific problems into innovation, applied research, and Tec21 challenges. The Decision Committee, formed by the Campus City core team, analyzes the ideas that qualify to become projects, classifies them into innovation projects or applied research projects, and determines which problems could be introduced as Tec21 challenges.

Once those projects are identified and classified, they are assigned resources, researchers/professors, students, etc. in order to be executed and verified. The intention is for these projects is to end up as a functional prototype proven on Campus City (in the field). Moreover, the goal is to innovate, and innovation is achieved when the technological development is adopted by a user or a market, thus transforming science and technology into a profitable solution (recognition). In innovation it is important to stay focused and to be able to act fast. Therefore, the above-mentioned steps are distributed into three main categories to manage this innovation (Figure 3): (1) value discovery (or the discovery of the opportunity), (2) the execution, and (3) the technological transfer towards the final users or clients.

The ideas selected for Tec21 from the project list are sent to the Challenge Living Lab (Section 2.3) where these challenges will be transformed into suitable projects for classes. This transformation requires a methodology that complies with certain pedagogical aspects. After going through the Challenge Living Lab, the projects can be released as challenges that are executable by the students.

2.3. Challenge Living Lab

The Challenge Living Lab is a platform through which a problem is proposed and continuously developed as part of a project selected by the stakeholders and the core team. These projects must have a high social and pedagogic impact because the students' motivation increases when they are involved in projects where the knowledge learned has a meaningful purpose [34–36]. Moreover, this platform is key for the development of a Smart Campus, since it introduces or develops different technologies in an educational environment [25].

The main objectives of the Challenge Living Lab are systematic innovation, promotion of research efforts, formation of leaders, interconnectivity/data use and fostering multidisciplinary cooperation between public and private institutions with society. This includes a multidisciplinary communication effort between different institutions (university-companies) that provides information, data, technological, economical and human resources and different methodologies needed to promote research ventures that deal with different Pain Points. Moreover, the Challenge Living Lab strengthens and synergistically combines with Tecnologico de Monterrey's academic programs; that is, the students are actively involved in different learning environments that lead to the development of solutions for the problems being experienced by the community, companies or campus users. This involves close cooperation between these stakeholders and collaboration with colleagues to develop socially responsible ventures [34]. To achieve this, Campus City can be integrated into the Tecnologico de Monterrey's educational programs. Figure 4 illustrates the stages of the Challenge Living Lab, as well as the Challenge Based Learning pedagogic model.

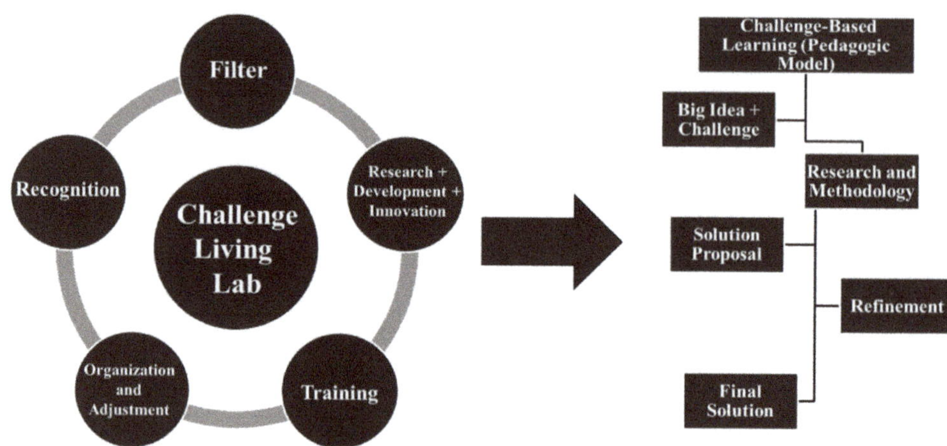

Figure 4. Campus City Challenge Living Lab virtuous loop and Challenge-Based Learning pedagogic methodology.

The Challenge Living Lab's virtuous loop consists of different components (Figure 4) that synergistically work towards the development of a pedagogic challenge based on the sustainable problems of the community, in this case the university district. The first component, Filter, refers to the selection of a project or challenge from both a technical-economic feasibility and cost-effective perspective, while considering the project's main impact on the different Campus City verticals. The second component, Research-Development-Innovation, denotes the availability of the technological and human resources to address the selected challenges. Within this component, different research groups and students are selected to address the challenge. The third component, Training, refers to the formative aspect of the challenge; that is, the challenge becomes a means for the development of competencies and skills in students. This component is naturally aligned with the objectives of Tecnologico de Monterrey's Tec21 educational model. Organization and Adjustment denotes the continuous improvement of the model based on feedback from the stakeholders and students. This component allows the documentation of the challenges, their successes and their areas of opportunity. Finally, Recognition refers to acknowledging the participation of all the parties involved.

As mentioned above, Challenge-Based Learning becomes the fundamental pedagogic approach for Campus City project. CBL focuses on the development on competencies as well as hard and soft-skills required to solve the Pain Points of a community or stakeholder [34,37–39] within an innovative and flexible learning environment [34,40,41]. Combined with Flipped Classroom techniques, this pedagogic approach promotes high levels of engagement and motivation within the students since they develop a sense of meaningful purpose through the learning process [34].

The following section describes in detail the challenge obtained from the above-mentioned process for the Smart Energy vertical axis as an example of the structure, pedagogic intentions and objectives of a typical challenge under the Campus City project. This description includes the overall objectives of the challenge, the pedagogic intentions and methodology followed by the students. For Smart Mobility and Smart Water, a brief description of the challenge and their components (stakeholders, methodology and objectives) could be found in the Supplementary Material Document S1 (Figures S1 and S2) [41–46]. Finally, Section 4 presents an overall discussion on the learnings and results.

3. Campus City Challenge—Smart Energy

3.1. Smart Energy Challenge—Smart Classrooms for the Post-COVID Era

The pedagogic objective of this challenge is to engage students in the topic of energy consumption in buildings and the operation of heating and ventilation air conditioning (HVAC) systems. Since the operation of HVAC systems is the main source of energy consumption in buildings, this challenge looks for alternatives to minimize this energy consumption and possibly transform new and existing buildings into buildings with net zero energy consumption. External partners involved in the challenge were Distrito Tec and the Managing divisions related to physical infrastructure of Tecnologico de Monterrey (maintenance/physical plant department).

3.2. Pedagogic Design

The objective of the fall 2019 and spring 2020 Smart Energy Challenge, was to verify that randomly selected classrooms at Tecnologico de Monterrey satisfy these requirements for air quality and thermal comfort, and to propose strategies to achieve this at the lowest energy cost in preparation for the return to in-person activities following the COVID-19 pandemic. The challenge required that the students search for the regulations of the specific case, the required instrumentation and, if necessary, its manufacturer. Moreover, the students were asked to implement a detailed plan to generate a baseline, and to compile a database of the rooms´ conditions in terms of temperature, humidity and CO_2 concentrations. Finally, with the obtained data and its analysis, the students were asked to propose solutions, considering energy efficiency aspects in conjunction with other aspects, such as thermal comfort and air quality.

An overview of the Smart Energy challenge is shown on Figure 5. The big idea behind the challenge corresponds to the need for decreasing energy consumption in buildings while simultaneously meeting with the comfort needs of their occupants. Students were asked to propose strategies to accomplish this purpose.

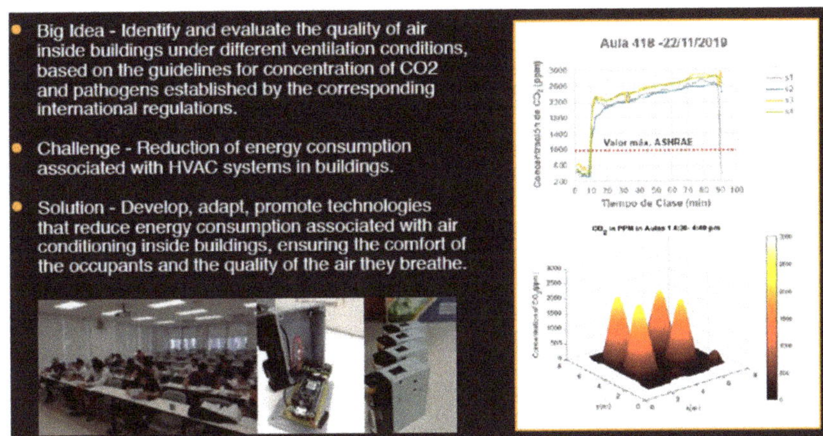

Figure 5. Use of low-cost sensors to monitor CO_2 concentration in classrooms.

An example of the challenge introduction or motivation for the students is presented below.

"In 1982, the WHO defined the sick building syndrome as a set of annoyances and diseases caused, among others, by poor ventilation and temperature decompensation, where at least 20% of the people inside the building feel unwell. More than 30% of the buildings that surround us could cause permanent discomfort to their occupants. If one of its occupants suffers from asthma, allergies, or has respiratory problems, these could be aggravated. The number of bacteria and viruses could increase, even increasing this increases the probability of becoming infected with COVID-19. Reducing the CO_2 concentration from 2000 ppm to 1000 ppm increases human efficiency by

12% and reduces the possibility of making mistakes by 3%. It is urgent to increase awareness of the importance of indoor air quality and generate real, high-impact alternatives to improve it in Tec's classrooms.

People currently spend 90% of their time indoors (homes, workplaces, offices, schools, hospitals, factories, or even shopping centers) [47,48]. Therefore, it is necessary to provide satisfactory indoor air quality while guaranteeing the energy-efficiency of the buildings. A healthy environment could be favored by having an adequate Indoor Air Quality (IAQ) level and ventilation system. Airborne virus and bacteria transmission is favored as a result of having poor IAQ, which could generate different health problems [48].

The main aspects that influence air quality are vehicle traffic, fuel burning, industries [47,49,50], and the low performance of air conditioning systems (HVAC) [51]. Carbon dioxide (CO_2) from indoor air is one of the critical factors in determining IAQ [47,52–55]. In the outside ambient air, CO_2 varies typically between 250 ppm and 350 ppm. The CO_2 in the indoor air must be below 1000 ppm to avoiding negative impacts on the occupant's health [48,56,57].

Several works have focused on assessing air ventilation on office buildings. However, not enough attention has been paid to school buildings [58]. A classroom is a tight space with several people inside, therefore, the air quality can deteriorate over time [50]. Students comfort, health and productivity (learning efficiency and attention during classes) could be dependent of the IAQ. This includes a decrease in students' performance, spread of viruses such as SARS-CoV-2 [52] and different social and economic repercussions [49,59]. The American Society of Heating, Refrigerating and Air-Conditioning Engineers (ASHRAE) recommends to increase natural ventilation, improve central air filtration or other HVAC systems that can be operated for extended hours (24/7 if possible), and the use portable HEPA purifiers. Moreover, having acceptable concentration values of CO_2 (at levels established by regulations) ensure the comfort level of the occupants and efficient energy consumption [60].

The objective of the smart energy challenge presented to the student is the following:

"Evaluate the indoor air quality under different ventilation conditions according to international regulatory standards and assessing CO_2 concentrations. Offer proposals to improve the air quality in classrooms as well as energy savings."

3.3. Challenge Methodology

Students were provided with a general strategy as a guide or methodology. The strategy involved four steps described in the following subsection, including a brief example of the information that would be expected from the students (text in italic format).

3.3.1. Step 1—Study Case 3

The first stage is the definition of the boundary conditions required to solve the challenge. Moreover, the students are required to define the type and number of study rooms, the location of the study, the scenario in which they will carry out measurements and analysis, as well as the specific identification of the energy system to be studied.

Example of obtained result: "In this work, we analyze two classrooms classified as non-residential buildings located at the Tecnologico de Monterrey, Campus Monterrey, Mexico. These classrooms correspond to the buildings A and B, both with northwest orientation, the number of occupants per class is 15 and 25, respectively. The installed HVAC systems have a constant evaporator and enthalpy, are kept continuously in operation to maintain the comfort of the occupants when modulating the equipment injection."

3.3.2. Step 2—Instrumentation

In this step, the students must list how they intend to monitor current conditions to determine the baseline to be compared with ideal situations.

Example of obtained result: "Particle devices, Xenon, and Boron modules were used. Through a communication protocol I2C and the SCD30 sensor, real-time monitoring of the CO_2 concentration (ppm), temperature, and relative humidity of the classrooms will be performed."

3.3.3. Step 3—Monitoring Campaign

The students are advised to carry out a monitoring campaign, using the proposed instrumentation while considering the moment, duration, location and number of the monitoring instruments and the frequency of data acquisition to construct the baseline. It is worth noting that the students must justify each decision and assumption made based on the literature available.

Example of obtained result: "Case A was monitored on November 15 and 22, 2019, for 100 min from 08:30 to 10:10 a.m. The monitoring time for Case B was 90 min, on November 12, 22, and 26, from 4:05 to 5:35 p.m. The placement of sensors varies, depending on the number of occupants and the size of the classrooms. An SD card incorporated into the Boron module was used to collect the data in real-time. Data were captured every 10 s."

3.3.4. Step 4—Evaluation

Finally, from the data collected, the students need to consider the aspects necessary for its evaluation, such as calculations and regulations, and to investigate the necessary basic theory and numerical methods to obtain a baseline while comparing it with the specifications of the regulations.

Example of obtained result: "In the last stage, the analysis of the CO_2 concentration data was made, the results of both cases were verified with the international standards, ASHRAE 62 and 62.1, ASTM D6245-12 [56,57,61]. Besides, some actions were designed that could be implemented to improve IAQ. CO_2 concentration values, occupancy of 25 people, and the dimensions of the classroom were used to calculate the ventilation requirements per person. This number was compared with the ASHRAE 62 standard."

From the development of these four steps, it is expected that the students will develop and apply the concepts, knowledge, and skills that correspond to their academic program. For example:

- To know the operation of adequate instrumentation for the current evaluation of the classrooms. For example, temperature, humidity, lighting, and CO_2 sensors.
- To design experiments, sampling, and statistics to obtain valid measurements for obtaining results.
- To make decisions about the placement of the sensors for representative data collection.
- The use of software such as Matlab®, to simulate the current and ideal conditions of the classrooms, using their different measurements.
- To access databases and information to obtain and perform the necessary data processing calculations. For example, calculating the volume of air per second that the room must have, ventilation level, comparison values for air quality, and comfort levels.

However, some assumptions and simplifications may be necessary to conduct the challenge. Specifically, the degrees of freedom must be clear, the type of sensors that are available (provide them with the sensors and guide them in obtaining the measurements), provide information regarding the study system, for example, current HVAC models and features in classrooms. Finally, the student must be provided with information on the available hours and the policies for the use of the rooms and the scope of the project.

No physical risks are expected from this challenge. Nevertheless, there is a risk that students will not adequately define the situation or the scope of the project. For example, to be aware of external factors that may affect their results or their proposals, such as the physical conditions of the city where they are conducting the study, meteorological conditions (temperature, humidity profiles), classroom characteristics (number of windows, dimensions, materials), description of the air conditioning system, georeferencing, etc. Thus, the professor must play an essential role in guiding the students with questions that may motivate them to make more in-depth insights into the solutions that they are proposing, in order to help them to develop the desired competencies.

3.4. Challenge Implementation Results

The challenge was implemented during August 2019–June 2020. A team consisting of three 9th-semester Mechanical Engineering students was assigned to solve the Smart Energy challenge. Some results of the challenge and the student experience from the teacher's perspective are discussed below.

Figure 6 shows that the students were able to obtain data through low-cost sensors. However, they did not demonstrate the same ability in the use of Matlab® to interpolate the conditions at points where there was no sampling, nor the use of scales to demonstrate a real behavior of CO_2 concentration. During the challenge, it was noted that the students struggled to use the mathematical equation to calculate the number of room air changes necessary to maintain the CO_2 concentration at acceptable levels, according to regulations. However, they demonstrated competency in database consulting.

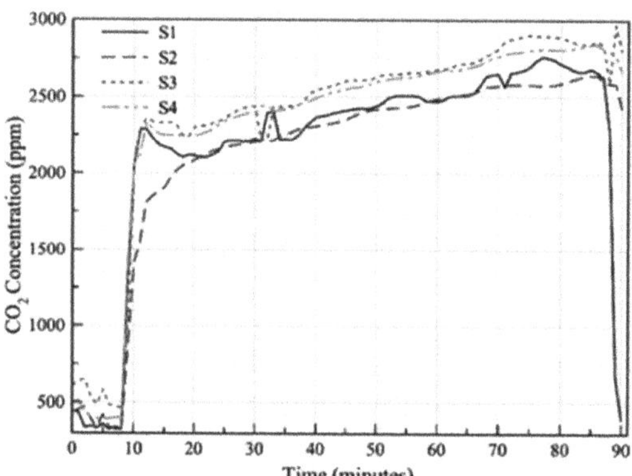
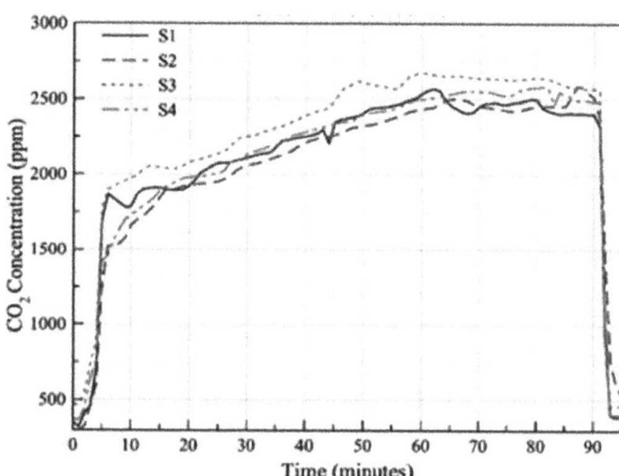

Figure 6. CO_2 concentration profile in two classrooms at Tecnologico de Monterrey.

An excerpt of the students' conclusions is presented below:

"Inexpensive sensors were used to perform classroom measurements that measure CO_2 concentration, temperature, and percentage of humidity within the classroom. According to the data obtained, color maps were constructed in Matlab® to visualize the CO_2 concentration during a class session. The concentration of CO_2 remains over 2000 ppm, reaching a peak value of up to 2883 ppm in the afternoon, which is well above the limit recomended by the regulations that tell us the standard of 1000 ppm."

"Likewise, calculations of the liters of air per second that each person should have in the room were made. Taking into account an average concentration from outside such as 400 ppm, the highest amount of CO_2 concentration, and the variable found with the average weight and height of a Mexican, we obtain 2.17 L per second for each person, being well below what gives us the regulation of 5 L per second per person."

"The air quality inside the classrooms of the Tecnologico de Monterrey is not adequate to ensure the well-being of its occupants. For this reason, it is proposed that installed air conditioners achieve adequate ventilation so that students and teachers have the security that they will stay healthy and that academic performance increases. Likewise, a filter system in the air conditioners capable of eliminating viruses is necessary to avoid contingencies such as COVID. In this way, students and teachers will feel protected by the Institution since the necessary hygiene measures will be taken to eliminate any virus."

The COVID-19 Pandemic affected the normal development of the Smart Energy Challenge since by the middle of spring 2020 students could not get access to the campus and therefore, they could not obtain new data. This fact affected their motivation, and their achievements were much lower compared to the students that had participated in the previous semester (fall 2019).

3.5. Discussion on the Challenge Implementation and Results

The students made fair use of the instrumentation that was available to them. They also designed and manufactured an interconnected system of low-cost sensors to obtain CO_2 concentration, temperature, and humidity data in the cloud. In addition, they compared this data with the recommended values defined by international organizations.

It was observed that students focused on CO_2 concentrations, while their analysis does not take into account temperatures or humidity concentrations, and gave more value to the health aspect than to the energy aspect. This situation was expected, since the challenge had double goals, which in this case was not only to propose alternatives to improve indoor air quality but also to validate the energy efficiency of these solutions. One possible explanation for the student' performance could be that the time available for the challenge was not enough, nor was the emphasis given to specific topics such as sensors and not energy aspects. This is an area of opportunity for the present challenge.

In terms of motivation, the students appeared to be encouraged and willing to show their results to share the importance of making improvements in the ventilation of the classrooms. Proof of that was submitting their work to the *Conexión Tec* contest. It is carried out every semester organized by Tecnologico de Monterrey to connect students and industry through their classroom projects. In the next version of this challenge, designers will include tools to evaluate the effect of the challenge on the students' engagement on the proposed topics. Finally, from the technology side, students demonstrated that the classrooms that they monitored had problems in their HVAC system and that those problems affect students' performance and will limit the use of these classrooms in the post-COVID era. Future challenges will focus on providing alternative solutions for Tecnologico de Monterrey administrators.

The following section presents the overall findings and learnings of the implementation of the Campus City and Challenge Living Lab framework. This includes the stakeholders perspective and involvement, as well as the perspectives from the professors and students that carried out the challenges.

4. Findings and Learning

4.1. Stakeholder's Perspective

A city's Pain Points, defined from the citizens' point of view, lead to new ideas for solutions. The goal of Campus City is to activate the Smart City ecosystem, where the private sector, the government and academia interact to solve contemporary needs. The Challenge Living Lab enables the students to better comprehend and participate in this interaction by playing an active role in developing innovative solutions to those Pain Points. In the Smart Mobility Challenge, the stakeholder shared a first approximation of energy requirement computation as well as a driving cycle from a local urban zone. The main contributions of the industrial stakeholder were: (a) their experience and coaching on vehicle dynamics, (b) their approach to projects in an industrial context, where the focus is on economically sustainable results, and (c) their openness to create new trusted relationships with our university based on these challenges, and to explore new projects based on these results.

In each challenge, the participation of an external mentor (their point of view, experience, and innovation focus), the students (their need to learn how to apply theoretical knowledge to solve real-life challenges) and the professor (their guidance, tutoring and accompaniment) creates an ecosystem of interaction, leadership, application of knowledge and follow up.

The different stakeholders had three main roles during the challenges presented: (1) to provide problems and aid in the selection process; (2) to co-design the challenge; (3) to mentor the students while providing information/data/experience to enrich the students' learning experience. The stakeholder plays an important educational role, acting as a link between the students' theoretical and academic environment and the arena of practical and real-world experience. Thus, the stakeholder acts as both mentor and client.

For instance, the maintenance/physical plant department, specifically the HVAC team, was the main stakeholder and co-designer in the Smart Energy Challenge. Initially, the challenge was developed by the academic advisor (professor) and then presented to the HVAC team for their inputs and suggestions. This was a crucial step since the HVAC team provided the students with critical information on the current ventilation and air conditioning system. The students constantly reported their advances to their academic mentor and to the HVAC team to receive feedback, adjusting their calculations and models to meet necessary international regulations and standards (ASHRAE 62 and 62.1, ASTM D6245-12). The results obtained, reinforced by the students-stakeholder interaction, led to the conclusion that the currently operation HVAC system does not meet the values required by the international standards. Consequently, the maintenance/physical plant department designed a strategy that avoids internal air circulation in the classroom while installing specialized air filters in the HVAC system.

For the Smart Water Challenge, an international academic stakeholder contributed to the co-design of the challenge. Their main contribution during the screening of the Pain Points and challenge design was to focus the study on emerging water pollutants, such as the per- and polyfluoroalkyl substances (PFAS). Moreover, the following challenge objectives were developed between the academic advisor and stakeholder:

1. To understand the environmental threats of chemicals derived from PFAs.
2. To understand the sources of PFAs and why they are found in drinking water.
3. To explore guidelines, government regulations, and action plans for significant concentrations of PFAs.
4. To survey data and trends to recognize concentrations that pose potential health hazards.
5. To develop corrective solutions in cities based on existing data, considering the implementation of various technologies and their respective costs.

Similar to the Smart Energy Challenge, the Smart Water Challenge stakeholder was actively involved with the students, providing guidance and mentorship throughout the challenge; this included an introductory lecture on the history of civil and environmental engineering. The Smart Mobility Challenge stakeholder's contribution to the academic environment was the context of actual practice in automotive engineering and the executive focus on generating technically sound solutions in record time.

The Challenge Living Lab framework contributed to building trust between stakeholders, the creation of academia-industry relationship, and the exploration of future projects (sponsored research). After the end of the course, the external partners (government, private sector and/or citizens) mentioned new ideas for future challenges in the next academic periods, thus beginning the virtuous-loop of the Challenge Living Lab and opening the possibility for further applied research projects.

4.2. Pedagogic Perspective

Students reported feeling more engaged with the syllabus content after the conclusion of the challenge. Additionally, students strengthened their understanding of different concepts, competencies and skills by systematically presenting, reporting and discussing their results to the different stakeholders. This interaction prepares the students for a real-world situation in which different points of view (citizens, academia, government, industries) contribute to the development of solutions. In a similar manner, students mentioned they enjoyed seeing real-life situations from different perspectives, learning new skills and tools, reviewing the concepts learned in class in a more practical hands-on manner, and acquiring experience in real-life problem solving.

However, it was also observed that despite the students' motivation and creativity, they initially require step-by-step directions on the activities that they should perform to solve the challenge, especially in challenges that involve experimental work.

The learnings, advantages and disadvantages of the Challenge Living Lab and challenge implementation are presented in Table 1.

Table 1. Campus City Challenge Living Lab learnings, advantages and disadvantages.

Learnings	Advantages	Disadvantages
1. The students are sensitized to actual real-world challenges and respond quickly to the smart city ecosystem's need for solutions.	1. The participation of government and private sector provides "real-world experience" for students.	1. Real world challenges sometimes are very focused on specific components of the full syllabus, so it is complex to evaluate course knowledge with one big challenge.
2. The students feel more confident in an industrial environment as they develop their competences.	2. The challenge structure in combination with nano/micro-challenges reinforces the students' critical thinking competency (systematic methodology).	2. The duration of the big challenge mostly is not enough to achieve and apply deep knowledge (this varies according to the students).
3. The syllabus is more advanced than required by regional industries.	3. The government and private sector take advantage of their involvement by recruiting talented students.	3. The students require intensive guidance (step-by-step instructions) at the beginning of the challenge.
4. The framework motivates the private sector to try new ideas.	4. The challenge fosters research and innovation.	4. The projects tend to be ambitious for the course duration.
5. The dimension and scope of the challenges must include at least 80% of the syllabus in order to engage students.	5. The challenge provides the students with a high-responsibility active role.	
6. The field validation of solutions is required in order to achieve the goal of the challenge.	6. The students are engaged and motivated throughout the challenge.	
7. The micro-challenge contents must lead up to the big challenges, rather than being independent from them.		
8. The multidisciplinary mentors enrich the framework, the challenge and the students' experience.		

5. Conclusions

The main contribution of this work is to present the overall framework of Distrito Tec's Campus City project that includes an open innovation ecosystem and the Challenge Living Lab as a platform for the identification of high-socially meaningful projects that could be used as a pedagogic opportunity for the development of competencies needed to solve different community-industrial-governmental Pain Points.

It is concluded that Distrito Tec's Campus City is an ambitious initiative that aims to provide the industry, government, academia and especially the community with innovative and smart solutions to the current high impact and complex Water, Energy and Mobility problems. The Campus City open innovation ecosystem allows the identification of these highly relevant challenges that could be solved through the use or design of disruptive technology. By solving these challenges, the wellness of the community and the efficient use of resources could be achieved.

Consequently, the Challenge Living Lab is highly relevant as the selected challenge provides the ideal pedagogic framework for the students to develop different competencies and skills that they will require throughout their professional life while fostering the identification and solution of highly relevant problems for the community, industry, government and academia.

Moreover, Challenge Living Lab platform opens the possibility for international and multidisciplinary projects and collaboration with different research institutions and governments. For instance, different challenges from the Challenge Living Lab, Smart Mobility vertical were tested in a pilot program as a "4.0 Energy Harvest Challenge" with the Indian Institute of Technology, Kharagpur. Students engage in a design challenge for different interconnected technologies for energy harvesting. More research is needed to scale-up this program to other institutions and include different common problems for the institutions/communities involved, which could generate logistical problems among faculty and students.

The scope of this work is defined within the Distrito Tec transformation initiative, in which the Tecnologico de Monterrey university, the surrounding Community, the Municipal

Authorities and Private Companies are committed to generate an urban transformation that inspires other districts and cities to undertake continuous transformation to improve their Quality of Life. However, the Distrito Tec projects and their implementation depend on the willingness, commitments and agreements between the community and the municipal authorities. This represents a challenge, because the Distrito Tec project has a horizon of 15 years during which the renewal of municipal authorities will take place several times, disrupting the continuity of the project. Therefore, the active involvement of the academia through Campus City projects could play an important role as a catalyst that constantly provides solutions, technology, and human resources to keep track of those agreements and aid in the continuity of the Distrito Tec project despite government transitions.

Future work should address: (1) the scaling up of the proposed framework into a model that could integrate different sectors, stakeholders, and communities, (2) follow-up and implementation, on a city-level or industrial scale, of the proposed challenges' solutions based on interconnected technology and (3) the evolution of the socio-cultural impact and Quality of Life by the implementation of the Campus City project (triple bottom line, people-planet-profit).

Supplementary Materials: The following are available online at https://www.mdpi.com/article/10.3390/app112311085/s1, Figure S1. Flow diagram of the overall methodology for the smart; Figure S2. Objectives and tasks to be completed for the Sustainable Water II course challenge mobility challenge. nCh and mCH refers to the nano-challenge and micro-challenge, respectively.

Author Contributions: Data curation, E.A.L.-G.; Funding acquisition, R.A.R.-M.; Investigation, J.I.H., J.M., J.d.J.L.-S. and S.U.; Project administration, R.A.R.-M., E.A.L.-G.; Supervision, R.A.R.-M.; Visualization, E.A.L.-G.; Writing—original draft, E.A.L.-G.; Writing—review & editing, J.I.H., J.M., J.d.J.L.-S., S.U. and R.A.R.-M. All authors have read and agreed to the published version of the manuscript.

Funding: The current project was funded by Tecnologico de Monterrey and Fundación FEMSA (Grant No. 0020206BB3, CAMPUSCITY Project).

Institutional Review Board Statement: Not applicable.

Informed Consent Statement: Not applicable.

Acknowledgments: The authors would like to acknowledge the financial and the technical support of Vicerrectoria Academica y de Innovacion Educativa of the Tecnologico de Monterrey and Faculty Development and Innovative Education Center for the pedagogic support during this work. The authors would like to acknowledge the Edrick Ramos, Diego Padilla, O. P. Vazquez, Mauricio Ramírez for their support and helping the authors during the challenge design and implementation.

Conflicts of Interest: The authors declare no conflict of interest.

References

1. Kumar, H.; Singh, M.K.; Gupta, M.P.; Madaan, J. Moving towards smart cities: Solutions that lead to the Smart City Transformation Framework. *Technol. Forecast. Soc. Change* **2020**, *153*, 119281. [CrossRef]
2. Ismagilova, E.; Hughes, L.; Dwivedi, Y.K.; Raman, K.R. Smart cities: Advances in research—An information systems perspective. *Int. J. Inf. Manage.* **2019**, *47*, 88–100. [CrossRef]
3. Breetzke, T.; Flowerday, S.V. The Usability of IVRs for Smart City Crowdsourcing in Developing Cities. *Electron. J. Inf. Syst. Dev. Ctries.* **2016**, *73*, 1–14. [CrossRef]
4. Embarak, O. Smart Cities New Paradigm Applications and Challenges. In *Immersive Technology in Smart Cities; EAI/Springer Innovations in Communication and Computing*; Aurelia, S., Paiva, S., Eds.; Springer International Publishing: Cham, Switzerland, 2022; pp. 147–177. ISBN 978-3-030-66607-1.
5. El Mdari, Y.; Daoud, M.A.; Namir, A.; Hakdaoui, M. Casablanca Smart City Project: Urbanization, Urban Growth, and Sprawl Challenges Using Remote Sensing and Spatial Analysis. In Proceedings of the Sixth International Congress on Information and Communication Technology, London, UK, 25–26 February 2021; Yang, X.-S., Sherratt, S., Dey, N., Joshi, A., Eds.; Springer: Singapore, 2022; pp. 209–217, ISBN 978-981-16-1781-2.
6. World Health Organization. WHOQOL: Measuring Quality of Life. Available online: https://www.who.int/tools/whoqol (accessed on 10 October 2021).

7.	Neirotti, P.; De Marco, A.; Cagliano, A.C.; Mangano, G.; Scorrano, F. Current trends in Smart City initiatives: Some stylised facts. *Cities* **2014**, *38*, 25–36. [CrossRef]

8.	Lee, J.H.; Phaal, R.; Lee, S.-H. An integrated service-device-technology roadmap for smart city development. *Technol. Forecast. Soc. Change* **2013**, *80*, 286–306. [CrossRef]

9.	Carvalho, L.; Santos, I.P.; van Winden, W. Knowledge spaces and places: From the perspective of a "born-global" start-up in the field of urban technology. *Expert Syst. Appl.* **2014**, *41*, 5647–5655. [CrossRef]

10.	Schaffers, H.; Ratti, C.; Komninos, N. Special Issue on Smart Applications for Smart Cities—New Approaches to Innovation: Guest Editors' Introduction. *J. Theor. Appl. Electron. Commer. Res.* **2012**, *7*, 9–10. [CrossRef]

11.	Samih, H. Smart cities and internet of things. *J. Inf. Technol. Case Appl. Res.* **2019**, *21*, 3–12. [CrossRef]

12.	Ojo, A.; Curry, E. Designing next generation Smart City initiatives-harnessing findings and lessons from a study of ten Smart City programs. In Proceedings of the Second European Conference on Information Systems, Tel Aviv, Israel, 9–11 June 2014.

13.	Yigitcanlar, T.; Kamruzzaman, M.; Buys, L.; Ioppolo, G.; Sabatini-Marques, J.; da Costa, E.M.; Yun, J.J. Understanding 'smart cities': Intertwining development drivers with desired outcomes in a multidimensional framework. *Cities* **2018**, *81*, 145–160. [CrossRef]

14.	Allam, Z.; Dhunny, Z.A. On big data, artificial intelligence and smart cities. *Cities* **2019**, *89*, 80–91. [CrossRef]

15.	Niaros, V.; Kostakis, V.; Drechsler, W. Making (in) the smart city: The emergence of makerspaces. *Telemat. Inform.* **2017**, *34*, 1143–1152. [CrossRef]

16.	De Guimarães, J.C.F.; Severo, E.A.; Felix Júnior, L.A.; Da Costa, W.P.L.B.; Salmoria, F.T. Governance and quality of life in smart cities: Towards sustainable development goals. *J. Clean. Prod.* **2020**, *253*, 119926. [CrossRef]

17.	Yun, J.J.; Jeong, E.; Yang, J. Open innovation of knowledge cities. *J. Open Innov. Technol. Mark. Complex.* **2015**, *1*, 16. [CrossRef]

18.	Yun, J.J.; Leem, Y.T. Editorial: The smart city as an open innovation knowledge city. *Int. J. Knowl. Based Dev.* **2016**, *7*, 103–106.

19.	US Department of Transportation. Smart City Challenge. Available online: Available online: https://www.transportation.gov/smartcity (accessed on 12 May 2021).

20.	Benltoufa, A.N.H.S.; Jaafar, F.; Maraoui, M.; Said, L.; Zili, M.; Hedfi, H.; Labidi, M.; Bouzidi, A.; Jrad, B.B.; Belhadj Salah, H. From smart campus to smart city: Monastir living lab. In Proceedings of the 2017 International Conference on Engineering and Technology (ICET), Antalya, Turkey, 21–23 August 2017; IEEE: New York, NY, USA, 2017; pp. 1–6.

21.	Flammini, A.; Pasetti, M.; Rinaldi, S.; Bellagente, P.; Ciribini, A.C.; Tagliabue, L.C.; Zavanella, L.E.; Zanoni, S.; Oggioni, G.; Pedrazzi, G. A Living Lab and Testing Infrastructure for the Development of Innovative Smart Energy Solutions: The eLUX Laboratory of the University of Brescia. In Proceedings of the 2018 AEIT International Annual Conference, Bari, Italy, 3–5 October 2018; IEEE: New York, NY, USA, 2018; pp. 1–6.

22.	Fortes, S.; Santoyo-Ramón, J.; Palacios, D.; Baena, E.; Mora-García, R.; Medina, M.; Mora, P.; Barco, R. The Campus as a Smart City: University of Málaga Environmental, Learning, and Research Approaches. *Sensors* **2019**, *19*, 1349. [CrossRef]

23.	Min-Allah, N.; Alrashed, S. Smart campus—A sketch. *Sustain. Cities Soc.* **2020**, *59*, 102231. [CrossRef]

24.	Mazutti, J.; Londero Brandli, L.; Lange Salvia, A.; Fritzen Gomes, B.M.; Damke, L.I.; Tibola da Rocha, V.; Santos Rabello, R. dos Smart and learning campus as living lab to foster education for sustainable development: An experience with air quality monitoring. *Int. J. Sustain. High. Educ.* **2020**, *21*, 1311–1330. [CrossRef]

25.	Pupiales-Chuquin, S.-A.; Tenesaca-Luna, G.-A.; Mora-Arciniegas, M.-B. Proposal of a Methodology for the Implementation of a Smart Campus. In Proceedings of the Sixth International Congress on Information and Communication Technology, London, UK, 25–26 February 2021; Yang, X.-S., Sherratt, S., Dey, N., Joshi, A., Eds.; Springer: Singapore, 2022; pp. 589–602, ISBN 978-981-16-2380-6.

26.	Pellegrino, M.A.; Roumelioti, E.; Gennari, R.; D'Angelo, M. Smart City Design as a 21st Century Skill. In *Methodologies and Intelligent Systems for Technology Enhanced Learning. In Proceedings of the 11th International Conference, MIS4TEL 2021, Salamanca, Spain, 6–8 October 2021*; De la Prieta, F., Gennari, R., Temperini, M., Di Mascio, T., Vittorini, P., Kubincova, Z., Popescu, E., Rua Carneiro, D., Lancia, L., Addone, A., Eds.; Springer International Publishing: Cham, Switzerland, 2022; pp. 271–280. ISBN 978-3-030-86618-1.

27.	Tecnológico de Monterrey Modelo Tec21. Available online: https://tec.mx/en/model-tec21 (accessed on 5 September 2021).

28.	Membrillo-Hernández, J.; de Jesús Ramírez-Cadena, M.; Ramírez-Medrano, A.; García-Castelán, R.M.G.; García-García, R. Implementation of the challenge-based learning approach in Academic Engineering Programs. *Int. J. Interact. Des. Manuf.* **2021**, *15*, 287–298. [CrossRef]

29.	López, H.A.; Ponce, P.; Molina, A.; Ramírez-Montoya, M.S.; Lopez-Caudana, E. Design Framework Based on TEC21 Educational Model and Education 4.0 Implemented in a Capstone Project: A Case Study of an Electric Vehicle Suspension System. *Sustainability* **2021**, *13*, 5768. [CrossRef]

30.	Tec de Monterrey: How a Top University in Mexico Radically Overhauled its Educational Model. Available online: https://www.ifc.org/wps/wcm/connect/industry_ext_content/ifc_external_corporate_site/education/publications/tec+de+monterrey (accessed on 10 October 2021).

31.	Pérez, J.A.; Campos, J.-M. Tec 21: First outcomes of a new integral university framework for long-life education through challenge-based learning. In Proceedings of the 2021 IEEE Global Engineering Education Conference (EDUCON), Vienna, Austria, 21–23 April 2021; pp. 316–321.

32. Medina, R.B.; Ordoñez, S.J.; Packza, M.E.U. The Tec21 educational model and its perception. An educational innovation for student-based learning. In Proceedings of the 2021 IEEE Global Engineering Education Conference (EDUCON), Vienna, Austria, 21–23 April 2021; pp. 861–866.

33. Reyna-González, J.M.; Ramírez-Medrano, A.; Membrillo-Hernández, J. Challenge Based Learning in the 4IR: Results on the Application of the Tec21 Educational Model in an Energetic Efficiency Improvement to a Rustic Industry. In *Advances in Intelligent Systems and Computing*; Auer, M.E., Hortsch, H., Sethakul, P., Eds.; Springer International Publishing: Cham, Switzerland, 2020; pp. 760–769. ISBN 978-3-030-40274-7.

34. Ramirez-Mendoza, R.A.; López-Guajardo, E.A.; Morales-Menéndez, R.; Díaz de León, E.; Lopez-Cruz, C.S.; Jianhong, W.; Lozoya-Santos, J.; de, J.; Enriquez de la O., J.F. Teaching Strategies for Massive Digital Courses: A Flexible Learning Model Applied During the COVID-19. *Int. J. Interact. Des. Manuf.* **2021**. (Accepted in Press).

35. Savery, J.R.; Duffy, T.M. Problem based learning: An instructional model and its constructivist framework. *Educ. Technol.* **1995**, *35*, 31–38. [CrossRef]

36. Savery, J.R. Overview of Problem-based Learning: Definitions and Distinctions. *Interdiscip. J. Probl. Learn.* **2006**, *1*, 1–59. [CrossRef]

37. Membrillo-Hernández, J.; Ramírez-Cadena, M.J.; Martínez-Acosta, M.; Cruz-Gómez, E.; Muñoz-Díaz, E.; Elizalde, H. Challenge based learning: The importance of world-leading companies as training partners. *Int. J. Interact. Des. Manuf.* **2019**, *13*, 1103–1113. [CrossRef]

38. Nichols, M.; Cator, K.; Torres, M. *Challenge Based Learning Guide*; Digital Promise: Redwood City, CA, USA, 2016; pp. 1–59.

39. Morales-Avalos, J.R.; Heredia-Escorza, Y. The academia–industry relationship: Igniting innovation in engineering schools. *Int. J. Interact. Des. Manuf.* **2019**, *13*, 1297–1312. [CrossRef]

40. Al-Maroof, R.A.; Al-Emran, M. Research Trends in Flipped Classroom: A Systematic Review. In *Studies in Systems, Decision and Control*; Al-Emran, M., Shaalan, K., Hassanien, A., Eds.; Springer: Cham, Switzerland, 2021; pp. 253–275.

41. Gao, D.; Jin, Z.; Zhang, J.; Li, J.; Ouyang, M. Development and performance analysis of a hybrid fuel cell/battery bus with an axle integrated electric motor drive system. *Int. J. Hydrogen Energy* **2016**, *41*, 1161–1169. [CrossRef]

42. Lobelles Sardiñas, G.O.; López Bastida, E.J.; Pedraza Gárciga, J.; Debora Mira, L. Aplicación de la tecnología Water Pinch para minimizar aguas residuales sulfurosas en una refinería de petróleo. *Cent. Azúcar* **2017**, *44*, 1–10.

43. Oliver, P.; Rodríguez, R.; Castro, M.R.; Echegaray, M.; Palacios, C.; Héctor, K.; Stella, M.U.; Asociación Interamericana de Ingeniería Sanitaria y Ambiental Uruguay. Análisis water-pinch en la industria del vino. In Proceedings of the XXX Congreso Interamericano de Ingeniería Sanitaria y Ambiental, Montevideo, Uruguay, 5 March 2006; p. 44.

44. Nemati-Amirkolaii, K.; Romdhana, H.; Lameloise, M.-L. Pinch Methods for Efficient Use of Water in Food Industry: A Survey Review. *Sustainability* **2019**, *11*, 4492. [CrossRef]

45. Abdeveis, S.; Sedghi, H.; Hassonizadeh, H.; Babazadeh, H.; Samaneh Abdeveis. Application of Water Quality Index and Water Quality Model QUAL2K for Evaluation of Pollutants in Dez River, Iran. *Water Resour.* **2020**, *47*, 892–903. [CrossRef]

46. Ahmad Kamal, N.; Muhammad, N.S.; Abdullah, J. Scenario-based pollution discharge simulations and mapping using integrated QUAL2K-GIS. *Environ. Pollut.* **2020**, *259*, 113909. [CrossRef] [PubMed]

47. Usmeldi, U.; Amini, R.; Trisna, S. The Development of Research-Based Learning Model with Science, Environment, Technology, and Society Approaches to Improve Critical Thinking of Students. *J. Pendidik. IPA Indones.* **2017**, *6*, 318. [CrossRef]

48. Park, J.-H.; Lee, T.J.; Park, M.J.; Oh, H.; Jo, Y.M. Effects of air cleaners and school characteristics on classroom concentrations of particulate matter in 34 elementary schools in Korea. *Build. Environ.* **2020**, *167*, 106437. [CrossRef]

49. Soomro, M.A.; Memon, S.A.; Shaikh, M.M.; Channa, A. Indoor air CO2 assessment of classrooms of educational institutes of hyderabad city and its comparison with other countries. In Proceedings of the AIP Conference Proceedings 2119, Jamshoro, Pakistan, 11 July 2019; p. 020014.

50. Yildiz, Y.; Koçyiğit, M. Evaluation of indoor environmental conditions in university classrooms. *Proc. Inst. Civ. Eng. Energy* **2019**, *172*, 148–161. [CrossRef]

51. Board on Population Health and Public Health Practice; Health and Medicine Division; National Academies of Sciences, Engineering, and Medicine. *Health Risks of Indoor Exposure to Particulate Matter*; Butler, D.A., Madhavan, G., Alper, J., Eds.; National Academies Press: Washington, DC, USA, 2016; ISBN 978-0-309-44362-3.

52. Querol, X.; Minguillón, M.C.; Moreno, T.; Alastuey, A.; López, D.; Bañares, M.A.; Capdevila, C.; Lagarón Cabello, J.M.; Alcamí, A.; Rubio Alonso, F.; et al. *Informe Sobre Filtros de Aire en Diferentes Sectores Industriales y Posibilidad de Eliminación del Virus SARS-CoV-2*; Consejo Superior de Investigaciones Científicas: Madrid, Spain, 2020.

53. Saini, M.K.; Goel, N. How Smart Are Smart Classrooms? A Review of Smart Classroom Technologies. *ACM Comput. Surv.* **2020**, *52*, 1–28. [CrossRef]

54. Mendell, M.J.; Heath, G.A. Do indoor pollutants and thermal conditions in schools influence student performance? A critical review of the literature. *Indoor Air* **2005**, *15*, 27–52. [CrossRef]

55. Uzelac, A.; Gligoric, N.; Krco, S. A comprehensive study of parameters in physical environment that impact students' focus during lecture using Internet of Things. *Comput. Human Behav.* **2015**, *53*, 427–434. [CrossRef]

56. Hernandez Luna, M.; Robledo Fava, R.; Fernandez De Cordoba Castella, P.; Paredes, A.; Michinel Alvarez, H.; Zaragoza Fernandez, S. Use of statistical correlation for energy management in office premises adopting techniques of the industry 4.0. *DYNA* **2018**, *93*, 602–607. [CrossRef]

57. ASHRAE Ventilation and Acceptable Indoor Air Quality in Residential Buildings. Available online: https://www.ashrae.org/technical-resources/resources (accessed on 5 November 2020).
58. American Society for Testing and Materials. *Standard Guide for Using Indoor Carbon Dioxide Concentrations to Evaluate Indoor Air Quality and Ventilation*; ASTM D6245-12; American Society for Testing and Materials: West Conshohocken, PA, USA, 2012.
59. Lee, Y.S.; Guerin, D.A. Indoor environmental quality differences between office types in LEED-certified buildings in the US. *Build. Environ.* **2010**, *45*, 1104–1112. [CrossRef]
60. Ng, L.C.; Wen, J. Estimating building airflow using CO 2 measurements from a distributed sensor network. *HVAC&R Res.* **2011**, *17*, 344–365. [CrossRef]
61. Robledo-Fava, R.; Hernández-Luna, M.C.; Fernández-de-Córdoba, P.; Michinel, H.; Zaragoza, S.; Castillo-Guzman, A.; Selvas-Aguilar, R. Analysis of the Influence Subjective Human Parameters in the Calculation of Thermal Comfort and Energy Consumption of Buildings. *Energies* **2019**, *12*, 1531. [CrossRef]

applied sciences

MDPI

Article

Assessing Urban Accessibility in Monterrey, Mexico: A Transferable Approach to Evaluate Access to Main Destinations at the Metropolitan and Local Levels

Ana Luisa Gaxiola-Beltrán [1], Jorge Narezo-Balzaretti [1], Mauricio Adolfo Ramírez-Moreno [1,*], Blas Luis Pérez-Henríquez [2], Ricardo Ambrocio Ramírez-Mendoza [1], Daniel Krajzewicz [3] and Jorge de-Jesús Lozoya-Santos [1]

[1] Tecnologico de Monterrey, School of Engineering and Sciences, Eugenio Garza Sada 2501 Sur, Monterrey 64849, Mexico; a01113679@itesm.mx (A.L.G.-B.); jnarezo7@gmail.com (J.N.-B.); ricardo.ramirez@tec.mx (R.A.R.-M.); jorge.lozoya@tec.mx (J.d.-J.L.-S.)
[2] Mexico Clean Economy 2050, Stanford University, Stanford, CA 94305, USA; blph@stanford.edu
[3] German Aerospace Center, Institute of Transport Research, Rudower Chaussee 7, 12489 Berlin, Germany; daniel.krajzewicz@dlr.de
* Correspondence: mauricio.ramirezm@tec.mx

Citation: Gaxiola-Beltrán, A.L.; Narezo-Balzaretti, J.; Ramírez-Moreno, M.A.; Pérez-Henríquez, B.L.; Ramírez-Mendoza, R.A.; Krajzewicz, D.; Lozoya-Santos, J.d.-J. Assessing Urban Accessibility in Monterrey, Mexico: A Transferable Approach to Evaluate Access to Main Destinations at the Metropolitan and Local Levels. *Appl. Sci.* **2021**, *11*, 7519. https://doi.org/10.3390/app11167519

Academic Editor: Vicente Julian

Received: 30 June 2021
Accepted: 6 August 2021
Published: 17 August 2021

Abstract: Cities demand urgent transformations in order to become more affordable, livable, sustainable, walkable and comfortable spaces. Hence, important changes have to be made in the way cities are understood, diagnosed and planned. The current paper puts urban accessibility into the centre of the public policy and planning agenda, as a transferable approach to transform cities into better living environments. To do so, a practical example of the City of Monterrey, Mexico, is presented at two planning scales: the metropolitan and local level. Both scales of analysis measure accessibility to main destinations using walking and cycling as the main transport modes. The results demonstrate that the levels of accessibility at the metropolitan level are divergent, depending on the desired destination, as well as on the planning processes (both formal and informal) from different areas of the city. At the local level, the Distrito Tec Area is diagnosed in terms of accessibility to assess to what extent it can be considered a part of a 15 minutes city. The results show that Distrito Tec lacks the desired parameters of accessibility to all destinations for being a 15 minutes city. Nevertheless, there is a considerable increase in accessibility levels when cycling is used as the main travelling mode. The current research project serves as an initial approach to understand the accessibility challenges of the city at different planning levels, by proving useful and disaggregated data. Finally, it concludes providing general recommendations to be considered in planning processes aimed to improve accessibility and sustainability.

Keywords: urban accessibility; Distrito Tec; Monterrey Metropolitan Zone (MMZ); Urban Basic Geostatistical Areas (AGEBs); 15 minutes city; UrMoAC

1. Introduction

Currently, planet Earth and all its inhabitants are living during the key moment to stop climate change. According to UN-Habitat [1], cities are responsible for more than 60% of the world's greenhouse emissions and are thus an important part of the problem. Nevertheless, the authors of the present paper believe that cities are also the solution, but only if immediate and effective actions are taken within them.

Urban planners and decision-makers have been exploring new ways to reconfigure and build cities, with the objective of making more livable, safe, affordable, environmentally friendly and sustainable urban realms. However, the rate of change is not fast enough to meet the needs to stop climate change. Certain cities are advancing faster than others; therefore, the current project has decided to analyse to what extent the City of Monterrey,

Mexico, is transforming to become a more accessible environment for its inhabitants by using sustainable modes of transport to provide access to its main destinations.

Monterrey is one of the largest cities in Mexico, with an area of 324.8 km^2 and 1,142,994 inhabitants according to the 2020 census of the National Statistics and Geography Institute (INEGI, by its acronym in Spanish). This city shares its urban environment and transportation infrastructure with eighteen municipalities, which comprise the Monterrey Metropolitan Zone (MMZ): Abasolo, Apodaca, Cadereyta Jiménez, El Carmen, Ciénega de Flores, García, San Pedro Garza García, General Escobedo, General Zuazua, Guadalupe, Juárez, Monterrey, Pesquería, Salinas Victoria, San Nicolás de los Garza, Hidalgo, Santa Catarina and Santiago. According to INEGI, the population of the MMZ is 5,341,177 inhabitants within a 7657 km^2 area.

Even though Monterrey is one of the wealthiest cities in Mexico, due to its industrial workforce and productivity, it faces several environmental and urban mobility challenges [2]. Monterrey has been previously identified as the second-most-polluted city in Latin America and holds the ninth place amongst most polluted cities in the world [3]. This pollution problem is produced by carbon and co-pollutant emissions from both traffic and industry. The city's urban mobility statistics have shown an increasing trend in the number of vehicles, while the use of public transportation has been decreasing [2]. Public transportation in the MMZ lacks security, is expensive and is often described as timely inefficient, making private vehicles the preferred choice for inhabitants [3].

The weather in Monterrey can also affect the citizens' willingness to use certain modes of transportation, such as walking and bicycles. Temperatures in the MMZ can reach extremely high values surpassing 35 °C during the summer months [4,5].

According to the Sustainable Urban Mobility Program for the Monterrey Metropolitan Area 2020 (PIMUS, by its acronym in Spanish), the mode share (the percentage of travellers using a particular type of transportation) is 46% for cars, 20% for public transport, 19% for walking, 6% for transport network companies (TNCs) and taxis, 5% for institutional transportation, 2% for school transport, 0.8% for bicycles and 0.3% others. Furthermore, the travel purposes are 44% work, 18% education, 14% shopping, 12% accompany or picking up someone, 4% recreation, 3% health and 5% other [6].

The elevated number of vehicles in the city, together with an inefficient urban design, results in traffic congestions, which at the same time contribute to an increase in carbon and co-pollutant emissions due to the repetitive motor starts and stops [7]. This worsens Monterrey's pollution. Citizens from the MMZ also suffer from accessibility disparity when it comes to their mobility options, which is related to socio-economic factors. The higher the income, the more options the citizens have for travelling within the city and to access certain opportunities. This disparity is more evident in such a vehicle-oriented urban morphology [3].

There have been different efforts and projects to improve urban accessibility within the MMZ; an example is Distrito Tec, which is a project led by the Instituto Tecnológico de Monterrey (ITESM) university in collaboration with the community in the vicinity of the campus and local authorities. It aims to improve the urban environment near the university campus for citizens and campus users. This project is one of the City Improvement Districts (CIDs) implemented in Monterrey, emerging as proposals of urban regeneration to increase citizen engagement at strategic points of the MMZ. Figure 1 shows the facilities of the ITESM Monterrey Campus, allotments and green areas within Distrito Tec.

This particular CID is an area of 452 hectares located at the south of Monterrey City. It has a population of 26,333 inhabitants and comprises 24 surrounding neighbourhoods of the university [8]. Distrito Tec is of special interest to investors and local property owners due to the high demand of services (food, lodging, leisure, entertainment, among others) required by the community of students, professors and collaborators at Tecnológico de Monterrey. This dynamic environment results from the proximity and integration of this polygon to the university, and it is an interesting small-scale representation of an urban environment [9].

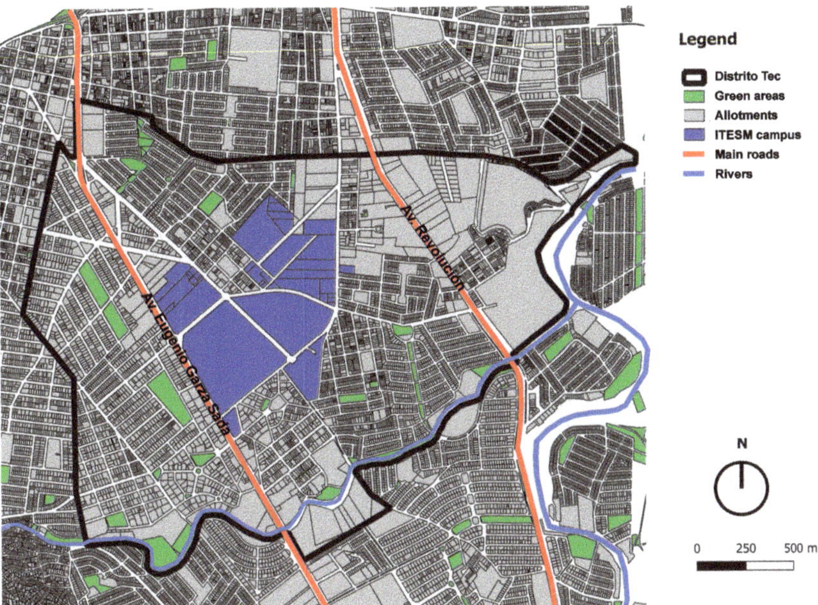

Figure 1. Urban representation of Distrito Tec, in the City of Monterrey.

In response to the current increasing mobility problems in the MMZ, the present study proposes the analysis of urban accessibility using the Urban Accessibility Computer (UrMoAC) software developed by the German Aerospace Centre (DLR) [10]. The analysis was performed at two scales: (i) the metropolitan level (entire MMZ) and (ii) the local level, using the Distrito Tec Area (neighbourhoods in the vicinity of the Tecnológico de Monterrey university campus). Both scales have the objective to measure accessibility to destinations that most people frequently visit: schools, main employment centres, supermarkets and hospitals. Nevertheless, each scale of analysis has a specific scope and methodology to evaluate accessibility, as will be explained in Section 2.

It is important to mention that the whole social, economic and, to a certain extent, political structure of the city behaves as a metropolitan area, despite the actual political divisions. Therefore, it is key to comprehend the urban needs at the different scales to develop public policies and interventions that respond to specific issues.

In contrast to traditional urban planning theories and procedures, taking an approach from urban accessibility has been demonstrated to have an incredible potential to better understand the systemic and complex nature of cities [11]. According to [12], accessibility can be defined as the "extent to which the land use-transport system enables (groups of) individuals or goods to reach activities or destinations by means of (combination of) transport mode(s)". Furthermore, it has been argued that accessibility consists of four components: (I) transport, (II) land use, (III) temporal and (IV) individual [12]. Accessibility concentrates on studying and evaluating how people access, or not, the different opportunities (destinations) of the urban realm, by taking into account the distributions of activity locations and the available transport modes within a given area.

It is important to mention that there are certain variables that affect the peoples' behaviour when accessing opportunities in the city, which go beyond the availability of activity locations and transport alternatives. These variables can be related to social preference (e.g., a family prefers one school over another), demographic (group ages of the population in a specific area), entitlement to health services (whether a person has the right to receive attention at a specific hospital or not), or level of service of the opportunities (this has to do with opening hours, capacity and type of service provided), among others.

It is crucial to fully analyse these variables before suggesting or making any interventions in the land use or mobility network of a given area, as they drastically affect the level of accessibility for the local population. Nevertheless, to do so goes beyond the scope of work of the current research project, which aims to provide a preliminary diagnosis of urban accessibility at the metropolitan and local level and relate it, only for the local level, to the 15 minutes city planning approach.

"La ville du quart d'heure" or 15 minutes city is an urban concept developed by Carlos Moreno where he imagines a city where every urban dweller can access her/his daily necessities within a maximum of 15 min of travel time by foot or bicycle [13]. The 15 minutes city implies an urban shift from car-oriented cities to proximity-based cities, upon the idea that "quality of urban life is inversely proportional to the amount of time invested in transportation" [13].

In this sense, the 15 minutes city concept emphasizes urban planning at the local (neighbourhood) level and concentrates on promoting accessibility rather than mobility. The focus is on diversifying land use to guarantee that every part of the city has enough green space, housing, public services, recreation areas, and jobs, at the local level, instead of developing more or higher capacity transport networks. Micromobility plays a key role in 15 minutes cities, as it promotes people's ability to access all local opportunities by walking or by using a bicycle. Hence, there is a special interest in creating open streets that foster activity in the public space and make walking and cycling safe and comfortable [13].

As a result of recent planning trends such as the 15 minutes city and a transition of transport modes from vehicles to active mobility (such as walking, cycling, etc.), accessibility has taken a spotlight in the planning paradigm by encouraging ideas such as mobilising people rather than motorised vehicles; creating access, not mobility; and thinking first at the local level [14].

To analyse to what extent Mexican cities are prepared to become accessible or 15 minutes cities and have a safe, comfortable, and realistic transition to active mobility, it is necessary to understand the transport modes and the availability of activity locations, as well as cultural, economic, political and social factors that may promote or oppose such transformation. Hence, the use of urban accessibility measures represents an adequate approach to comprehend travel times and distances and their implications in a social, economic, and environmental dimension. The results represent a key input to detect accessibility issues and develop solutions for them. They also are usable knowledge that can be easily transmitted to the local population to socialise and cocreate interventions, projects, and public policies that can transform the local environment into a safer, more livable, affordable, accessible and sustainable place.

Finally, it must be said that large Latin American cities have specific socio-economic and spatial characteristics that make them very different from cities of similar sizes in other parts of the world [15]. Recognising and working with these specific characteristics has been an important part of the development of the project, to suggest and reach results that are relevant and tailor-made for the nature and context of the MMZ and Distrito Tec.

2. Materials and Methods

2.1. Software

Urban Mobility Accessibility Computer (UrMoAC) is an open-source tool developed by the German Aerospace Centre (DLR) [10] to compute accessibility measures within a geographical area, which can be aggregated for variable areas. UrMoAC can calculate, among others, the minimum time and distance required to reach the nearest specified destination (places such as hospitals, schools, parks, industries, etc.) using a specific transport mode or a combination of them (walking, cycling, motorised vehicles, and public transport), while taking into consideration the real road network and its constraints (speed limits, directions, and modes of transport). UrMoAC is a command-line tool written in the Java programming language, which reads its inputs from a PostgreSQL/PostGIS database.

UrMoAC was selected as the most suitable tool to perform the required accessibility analysis since it provides accurate and disaggregated results based on real distances and close-to-reality transportation behaviours. Furthermore, the open-source QGis software was used to visualise the information obtained from UrMoAC [16].

2.2. Data Processing

The current project was based on two scales of analysis. The first is at the metropolitan level, which aims to highlight and demonstrate the diverging levels of accessibility in the MMZ and relate them to relevant urban planning paradigms and socio-economic indicators such as the marginalisation levels. The second is at the local level, considering the Distrito Tec Area. Here, the objective was to analyse the accessibility levels to main employment centres, schools, supermarkets, and hospitals, travelling by foot or bicycle, to determine whether the area can be, or not, considered as a 15 minutes city.

For the current analysis, UrMoAC was programmed to start the measurement from urban blocks (origins) to several different destinations (hospitals, schools, supermarkets, and main employments centres). It is important to mention that there are socio-economic factors, such as social preference and willingness to pay, that can affect the decision of people to access one destination over another. Specifically, higher-income social groups tend to prefer private hospitals and schools to fulfil their health and education needs. In contrast, public services usually have access limitations related to the capacity of the building, level of service, entitlement, opening hours, working days, demand, among others. Including all these variables in the analysis goes beyond the scope of work from the current research project, as it aims to provide a preliminary approach to urban accessibility based on the supply of land uses and transport networks at different scales. In this sense, only public hospitals and schools were taken into account, embracing the idea that public establishments are available for everyone despite their income.

The transport modes used for the computation were either walking (with an average speed of 3.6 km/h) or bicycle (average speed of 12 km/h) to analyse to what extent the MMZ is ready to shift from car-oriented mobility to generating accessibility through micromobility and proximity-based land use planning. The aforementioned average speeds for each transport mode are the ones proposed by the UrMoAC tool [10]. Public Transport was not considered as it requires General Transit Feed Specification (GTFS) data files to run the accessibility computation, which are not currently available for the areas of study analysed in this work.

As mentioned before, the computation was designed to start every measurement from urban blocks. Depending on the measurement desired, the destinations would vary considering either: the closest available, a specific number of destinations that can be reached (e.g., the five nearest hospitals), or with a 15 min travel time constraint to see how many destinations are available within that travel time. The reasons for generating different measurements were to demonstrate UrMoAC's capacity to compute diverse and relevant results, to integrate social and capacity factors into the measurement (e.g., one school alone cannot fulfil the entire education demand of a given district; therefore, more than one has to be included) and to analyse the results from different approaches that provide useful insights about the accessibility of the study area.

For the results to be legible, understandable, and useful, the individual results from each block were aggregated into Basic Geostatistical Areas (AGEBs, by its acronym in Spanish) and then mapped. The AGEBs are the basic aggregation areas used by the National Institute of Statistics and Geography (INEGI, by its acronym in Spanish) census; therefore, they are an unit of analysis widely used for mapping purposes in Mexico.

For the realisation of this project, the following data were gathered: a compendium of the urban geostatistical cartography of Nuevo León (Mexico) state (from 2016, as it is the most recent version), the urban block subdivision of the 18 municipalities that form the MMZ, the National Statistical Directory of Economic Units 2019 (DENUE, by its acronym in Spanish), and the OpenStreetMap road network from 2021.

Table 1 shows the databases of the variables (municipalities, blocks, AGEBs, DENUE, and roads), the institutions, and the URLs for all the gathered data in this study.

Table 1. Databases and URLs of the used resources in this study.

Variable	Database	Institution	URL (All Data Was Accessed on 20 February 2021)
Municipalities	INEGI	INEGI	https://www.inegi.org.mx/app/biblioteca/ficha.html?upc=702825218867
Blocks	INEGI	INEGI	https://www.inegi.org.mx/app/biblioteca/ficha.html?upc=702825218867
AGEBs	INEGI	INEGI	https://www.inegi.org.mx/app/biblioteca/ficha.html?upc=702825218867
DENUE	DENUE	INEGI	https://www.inegi.org.mx/app/descarga/default.html
Roads	OSM	OSMF	https://www.openstreetmap.org/#map=10/24.6488/-100.5263

The DENUE is elaborated by INEGI and offers a large compendium of all the economic activities registered within the Mexican territory, which can be categorised according to the North American Industry Classification System (NAICS). The NAICS is a standard created to allow high comparability in business statistics among North American countries (Canada, USA, and Mexico). For this study, the categories displayed in Table 2 were analysed.

Table 2. Categories of the analysed economical activities, according to the NAICS.

Code	Category	What Does It Include?
611112	Public sector preschools	Economic units of the public sector which are mainly dedicated to providing preschool education including indigenous, and community types.
611122	Public sector primary schools	Economic units of the public sector which are mainly dedicated to providing primary education including indigenous, community, and adult types.
611132	Public sector general secondary	Economic units of the public sector which are mainly dedicated to providing secondary education including tele-secondary (televised lessons), community, and adult types.
611142	Public sector technical secondary	Economic units of the public sector which are mainly dedicated to providing technical secondary education including indigenous types.
611152	Public sector terminal technical	Economic units of the public sector which are mainly dedicated to providing technical high school education of a terminal nature.
611162	Public sector high schools	Economic units of the public sector which are mainly dedicated to providing general or technical baccalaureate education, of a preparatory nature, including tele-baccalaureate (televised lessons).
622112	General public sector hospitals	Economic units of the public sector which are mainly dedicated to providing medical services for a wide range of diseases among children, women, the elderly, or patients in general. These economic units have facilities for the hospitalisation of patients and are known as general, paediatric, geriatric, and women's disease hospitals.
622212	Public sector addiction medical services and psychiatric hospitals	Economic units of the public sector which are mainly dedicated to providing treatment to patients who require hospitalisation due to psychological disorders and addiction.
622312	Public sector hospitals of services for other medical specialties	Economic units of the public sector which are dedicated primarily to the care of specific diseases or the condition of an apparatus or system. These economic units have facilities for the hospitalisation of patients and are known as oncology, gynaecological-obstetric, pneumonia, and cardiology hospitals.
462111	Retail trade supermarkets	Economic units (supermarkets) which are mainly dedicated to the retail trade of a wide variety of products, such as food, clothing, cleaning supplies and household items, organised into sections or specialised exhibition areas that facilitate direct access by the public to the goods.

For the metropolitan-level analysis, the following accessibility measures were performed: number of employments centres with 51 or more employees accessible within a 15 min travel time, the distance and time to the nearest public school (including all levels from preschool to high school), and nearest hospital (including codes 622112, 622212 and 622312).

For the Distrito Tec's analysis, the computations were: number of employment centres with 51 or more employees accessible within a 15 min travel time, the number of retail trade supermarkets available within a 15 min travel time, and the time required to access the five nearest schools or hospitals; here, an individual study was conducted for each category mentioned in Table 2.

The aforementioned measurements were evaluated under the following (see Table 3) criteria, which were determined in compliance with each transport mode's average speed (according to UrMoAC) and the cumulative opportunity accessibility metric (where more opportunities in the surroundings relate to higher accessibility levels) [17]. It is important to highlight the complexity of a standardised rating scale for the number of destinations reachable within a specific time or distance, when comparing different services, as socio-economic and demographic features must be considered to develop an accurate measurement. Therefore, the current research project performed a preliminary approach and diagnosis by evaluating the number of destinations available within a 15 min travel time either by bicycle or walking. Comparisons of accessibility across different services were not evaluated.

Table 3. Standardised rating scale.

Scale (Accessibility)	Time (min)	Distance Cycling (m)	Distance Walking (m)
Low	>30	>6000	>1800
Medium	15–30	3000–6000	900–1800
High	0–15	0–3000	0–900

2.3. Study Area

2.3.1. Monterrey Metropolitan Area

With 74 metropolitan areas in 2015, as declared by the National Population Council (CONAPO, by its acronym in Spanish), the conformation of large metropolitan areas has been one of the main urbanisation trends in Mexico since the 1950s Century [18]. The creation of metropolitan areas has been fuelled by the centralisation of economic activities, such as industries and services, in large- and, to a lower extent, middle-sized cities. These cities grew demographically from rural to urban migration, led by people who were looking for better employment opportunities and living conditions. Yet, the metropolitan areas have failed to achieve an urban development that can cope, in a sustainable manner, with the social, economic, and environmental demands [18].

The marginalisation index is an indicator used by CONAPO that serves to demonstrate how metropolitan areas have divergent conditions and how planning paradigms and processes have failed to create a city where all inhabitants have the same conditions and access to opportunities [19]. It relates to a lack of social opportunities and the absence of the capacity to generate them, as well as to deprivation and inaccessibility of basic goods and services. This indicator is based on 10 different socio-economic factors [19]:

1. Percentage of the population within 6 and 14 y who do not attend school;
2. Percentage of the population 15 y or older who have not completed their basic education;
3. Percentage of the population without entitlement to access the public health system;
4. Percentage of dead children from women between 15 and 49 y;
5. Percentage of particular occupied households without running water;
6. Percentage of particular occupied households without drainage connected to the public network or a septic tank;
7. Percentage of particular occupied households without a toilet connected to running water;
8. Percentage of particular occupied households with dirt floors;

9. Percentage of particular occupied households with some level of overcrowding;
10. Percentage of particular occupied households without a fridge.

Figure 2 shows the marginalisation levels in the MMZ according to the CONAPO [20], as well as the location of Distrito Tec within the MMZ. It can be observed that most of the AGEBs in the MMZ lie within the low (22.91%) and very low (41.42%) marginalisation levels, with few classified as medium (20.76%), high (4%), or very high (1.63%). There are 9.19% of the AGEBs that are not categorised. It is worth mentioning that the regions with higher marginalisation levels are located in the outskirts of the MMZ, usually places with low levels of accessibility.

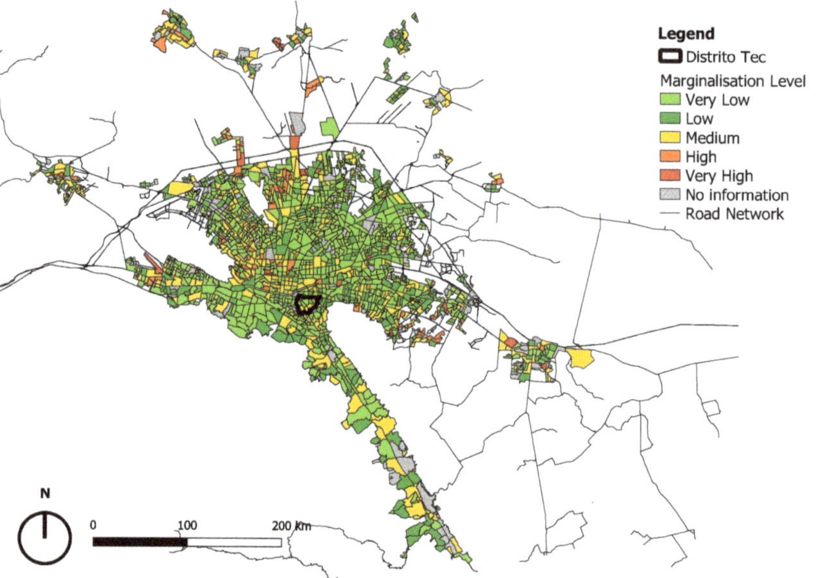

Figure 2. Marginalisation levels in the Monterrey Metropolitan Zone (MMZ).

Clearly, the MMZ is not an exception to the challenges that Mexican metropolitan areas face. There exists room for improvement in terms of creating central and well-located social housing, diversifying economic clusters throughout the city to promote decentralisation, and rethinking urban planning concerning land uses to guarantee enough supply to satisfy local needs.

Furthermore, changing demographics have to be taken into account. Figure 3 displays the population distribution in the MMZ as in Monterrey Municipality. The MMZ population has a pyramidal shape where most of its inhabitants are 40 y or younger. Even though the distribution of the Monterrey Municipality is similar to the one of the MMZ, there is a faster shift towards a more rectangular shape, meaning that the city is facing a stationary growth [21]. This population trend will result in a decreasing demand for infrastructure related to children (such as schools and kindergartens) and an increase in demand in infrastructure related to elderly persons (such as hospitals and retirement/nursing homes).

The analysis at the metropolitan level has the objective to demonstrate the performance of the MMZ in terms of accessibility to different location types. The results will allow urban planners and policymakers to identify underperforming areas in relationship to specific variables. This information can then be translated into a hierarchy-based intervention plan, starting with the areas with lower levels of accessibility. Hence, the analysis at the metropolitan level does not relate to the 15 minutes city concept, as a broad analysis of each area of the city would be required and the addition of more variables (destinations) will be needed. Nevertheless, some of the results can provide useful insights and findings to determine whether certain areas of the city meet the 15 min travel time parameters for a specific variable.

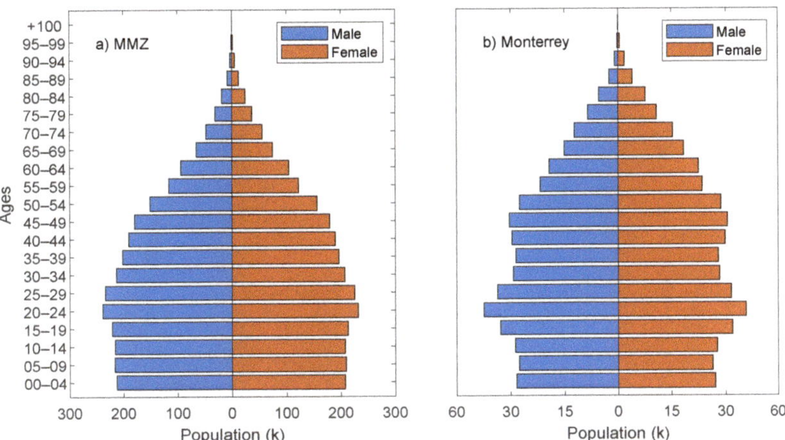

Figure 3. Population distribution (thousands) by ages 0–100+ in the MMZ (**a**) and Monterrey (**b**).

The current research project departed from analysing accessibility at a metropolitan scale, to understand and assess the socio-spatial relationships that drive the city. Hence, the two major destinations that promote travel were studied: main employment centres and public schools. Additionally, a variable for public hospitals was added, as it was considered that access to health is key to promoting better living conditions.

2.3.2. Distrito Tec Area

By considering a smaller scale of analysis, the level of complexity of the area reduces, allowing incorporating new variables and destinations within the scope of work of the current research project. Therefore, the local area analysis, considering Distrito Tec (See Figure 1) as the study area, can take a preliminary approach to analyse to what extent Distrito Tec meets the requirements of accessibility to different destinations, using walking or cycling as the transport modes, to be considered a 15 minutes city.

It is important to mention that Distrito Tec is located within a central area of the city. Hence, in comparison to other areas, it is considered privileged in terms of its surroundings (see Figure 2). Based on this, it is expected that the levels of accessibility within such an area will be higher than the ones of suburban areas.

The neighbourhoods surrounding the Distrito Tec Area are heterogeneous and affect the levels of accessibility of the area. The same counts for the mobility patterns. Therefore, it is important to give a brief characterisation of each one of them.

According to Figure 4, the northern area has very low to medium marginalisation levels and is mainly residential with single-family zoning. The commercial activities are located on the main avenues and offer some additional services such as community schools.

The eastern area has faced important changes since the 1950s, as it used to have a brick factory that was dismantled to allow several urban renewal processes to happen. There is a contrast between the socio-economic levels of the population. In the neighbourhoods in the vicinity of the brick factory, a low-income worker population predominates; in contrast, on the opposite side of the river, there are high-income gated residential communities. The land use variety is low, with most being residential with single-family zoning. The small numbers of commercial places and services tend to be located around the main avenues where vast shopping centres have been built (e.g., Nuevo Sur shopping mall).

The western area is the oldest, dating to the 1920s. It is characterised by being a middle-income residential area with single-family zoning. Through time, a heterogeneous land use mixture has emerged in the area, having many commercial activities and services located on the main avenues.

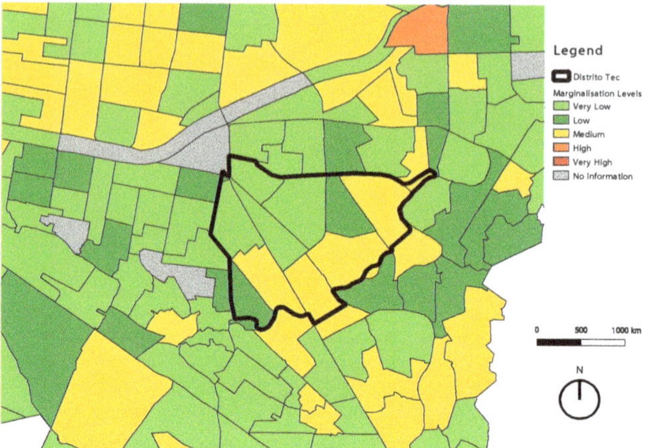

Figure 4. Marginalisation levels in Distrito Tec.

The polygon of Distrito Tec was designed based on the neighbourhoods' boundaries, which are located in the vicinity of the university campus. However, the AGEBs do not fully correspond to such limits; thus, some AGEBs will not be completely encompassed within the polygon. For the current analysis, 11 AGEBs were considered as part of the Distrito Tec Area, as presented in Figure 4.

The first set of variables to be analysed is access to public kindergartens and schools. For this section, a specific accessibility measure was computed for each education level to understand how the local people, depending on their age, are able of accessing education places. Considering that the closest destination might not be able to address the entire demand of the AGEB, as well as social preference factors, the accessibility measures were evaluated for the five closest locations (kindergartens, primary, secondary and high schools).

3. Analysis and Results

3.1. Monterrey Metropolitan Area

Figure 5 shows the average distance from every AGEB to the closest public schools using bicycles as the transport mode. It can be observed that most of the AGEBs have a public school in a range of less than 1000 m. This is due to the high density of schools (especially primary and secondary) that most areas of the MMZ have. Building schools has been one of the priorities of the authorities, as there is a high demand for them given that Mexican demographics maintain a pyramid structure. This means that there is a considerable amount of people younger than 30 y, as observed in Figure 3a,b, which shows the age distribution of the population of the MMZ and Monterrey, respectively, according to the INEGI 2020 census.

Interestingly, many suburban areas that are not even physically integrated into the city have high accessibility levels (e.g., some northern and southwest regions). This phenomenon reduces the travel dependence from suburban areas to central areas to access education.

The south and southeast of the city have the lowest access levels, which relates to the fact that both are high-income areas where residents tend to prefer private schools and that some of these areas were designed and built as gated communities with only residential land use.

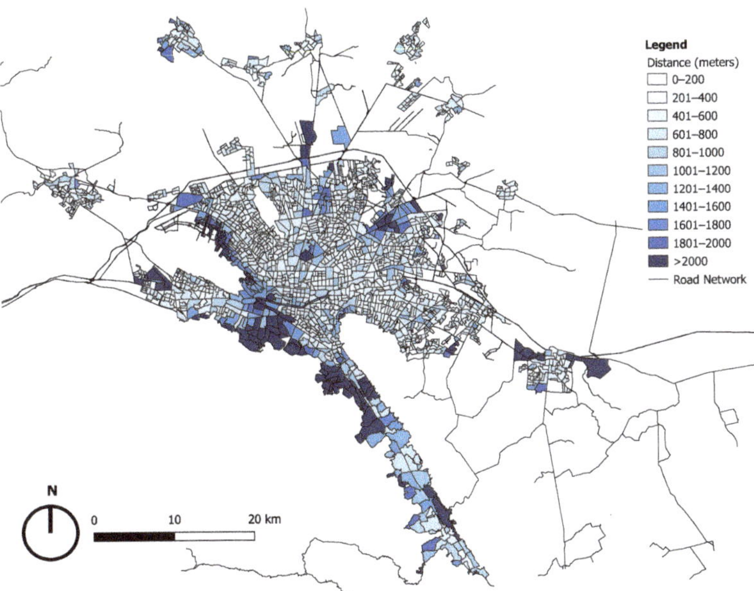

Figure 5. Average distance to the closest public school travelling by bicycle.

Figure 6 compares the performance of the transport modes (bike and walking) in accessing the closest public school at the metropolitan level. The first important thing to highlight is that despite the transport mode used, 75% of the AGEBs can access a school in less than 10 min. This is related to the high density of schools at the metropolitan level, as observed in Figure 5. By comparing the transport modes, it becomes evident that cycling achieves considerably higher levels of accessibility: while it takes 9.7 min for 75% of the AGEBs to reach the closest school by walking, for the same percentage, it only takes 3.5 min by cycling.

It is important to mention that Figure 5 is just a preliminary approach to understand accessibility to schools; this variable will be disaggregated in the following section, to analyse how access varies depending on each school's education level.

Figure 7 presents the average travel time by AGEB to the closest public hospital by foot. The map shows a considerable divergence of accessibility between AGEBs, with many going beyond the 30 min travel time, mainly due to the low density or nonexistence of hospitals, predominantly in suburban areas.

Central areas show high access to hospitals. This can be attributed to the historical conformation of the city, which started at the centre and gradually sprawled. Therefore, central areas have existed for longer periods, allowing the authorities to implement through time the necessary health infrastructures in these areas. In contrast, many suburban areas are relatively new, and some have lacked formal planning procedures (people built homes without having any official permit from the government or public administration), while others have been designed following urban paradigms that purposely create isolated and monofunctional areas, such as gated communities.

The southeast of the map represents an excellent example of how planned urban communities can lack access criteria in their planning procedures and how these problems are not exclusive to lower-income areas or highly marginalised ones that have been built without any formal planning. In this sense, what some people think could be a solution (urban planning) can trigger a problem.

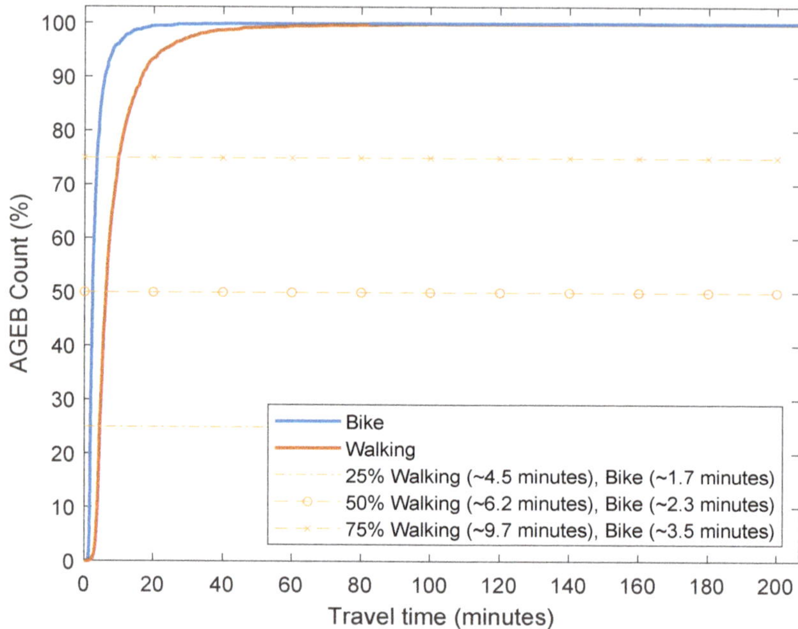

Figure 6. Comparison of the performance of transport modes to access the closest public school at the metropolitan level.

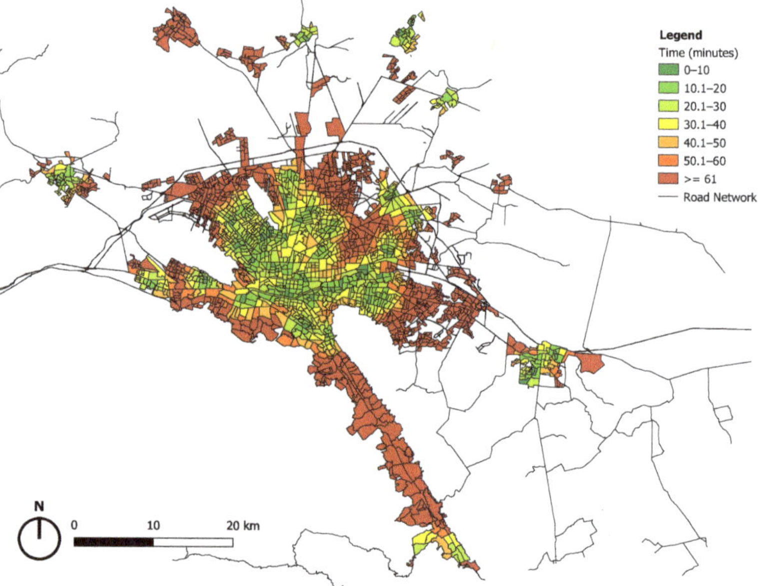

Figure 7. Average travel time (minutes) to the closest public hospital by foot.

The COVID-19 pandemic has demonstrated how important it is to have a robust and accessible healthcare system [22]. Therefore, poor access to health services should be unacceptable, as observed in many AGEBs of the MMZ. The MMZ needs to urgently tackle this problem by developing new public health centres in any area of the city that does not meet the desirable access parameters.

Figure 8 compares the performance of both transport modes (walking and bicycle) to access the closest public hospital. For this specific variable, the differences between one mode and the other are more evident. This is related to the lower number of destinations available at the metropolitan level (in contrast with Figure 6, the number of hospitals is evidently lower than that of schools). Hence, faster transport modes, such as bicycles, will outperform walking by a considerable margin: for 75% of the AGEBs to reach the closest public hospital travelling by foot, it takes 46.8 min; in contrast, it only takes 14.7 min by bicycle.

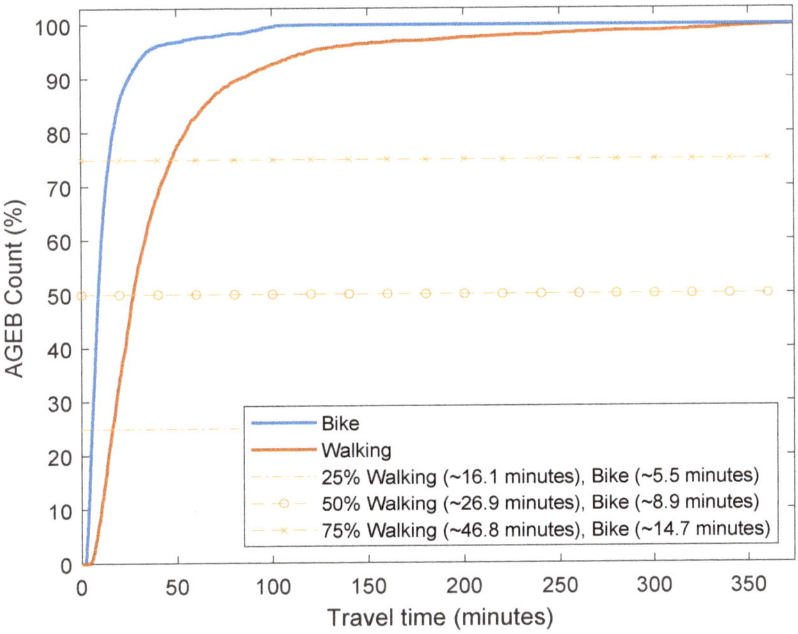

Figure 8. Comparison of the performance of transport modes to access the closest public hospital at the metropolitan level.

Figure 9 shows the number of economic activities that employ 51 persons or more (main employment centres) that can be accessed within a 15 min travel time by bicycle. It can be appreciated that most of the AGEBs with higher levels of accessibility (+73 destinations) are located in the central areas of the city, with a gradual decline towards the outskirts. At first sight, such a result might be perceived as unusual, considering that the MMZ is an industrial city and that most industries are located in suburban areas. Nevertheless, industries usually have very large complexes that occupy vast areas of land and make them separate from one another, reducing access to them in short periods. In contrast, even though they are smaller in size and number of employees, the economic activities that concentrate in central areas of the city (such as commerce and services) require smaller areas of land compared to industries and are located closer to each other.

The reduced number of main employment centres at the northern, western, eastern, and southeast AGEBs implies that residents in those areas have to travel long distances and periods of time to central areas to work, relying heavily on motorised modes of transport. This traffic has high economic, environmental and social costs for the city and its residents, and can be tackled by creating new clusters of economic activities in areas with low access and by improving transport networks, especially public transport. It is important to isolate and analyse each particular case, as some of these areas lack main employment centres given that, historically, they were independent of the city, but the sprawl has reached them and forced them to integrate with its economy. Some other areas were lacking main employment centres on purpose, as many of the AGEBs located to the southeast of the MMZ were designed as high-income gated communities (especially golf clubs) with only residential land use. As all the residents from these areas require accessing the opportunities located within central areas, they generate demand for public infrastructure that is extremely expensive and that should not exist if planning policies were to assure land use diversity and accessibility parameters before authorising building permits.

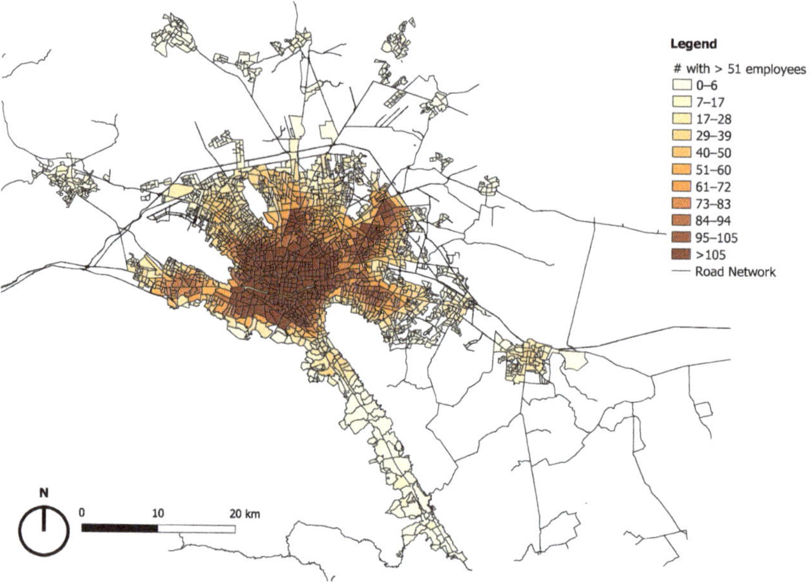

Figure 9. Number of economic activities that employ ≥51 persons within a 15 min travel time by bicycle.

The three variables previously presented demonstrate how complex urban areas truly are and how inaccurate generalisations can be. The present exercise thus serves as a preliminary approach to demonstrate that every city faces many different and particular challenges. Consequently, any solution, in order to generate a beneficial change, must be based on a full understanding of the specific desired conditions in the city.

The current research demonstrates that using accessibility measures is extremely useful for identifying specific needs from different areas of the city and that despite the accessibility measure chosen, the data obtained are relevant. The information generated can be used as a departure point for prioritising interventions and public policies depending on the performance of each area on a given variable. By doing so, much time and many resources can be saved.

This first section of analysis concludes that the levels of accessibility at the metropolitan level are divergent and that they drastically vary, even in the same location, depending on the measured variable. The results are a relevant input to obtain a general diagnosis of the state of the MMZ in terms of accessibility. These results can also be compared to different socio-economic variables, such as marginalisation rates, to better understand the social, economic, political, and environmental factors that drive the city. This information represents a solid departure point to then analyse what happens at smaller scales, without ignoring that each area of the city is bound to metropolitan interrelationships.

3.2. Distrito Tec Area

Figure 10 shows the average time by AGEB to access the five closest public kindergartens. As most children going to kindergartens are not able to cycle, the accessibility measure was computed for travelling by foot. Even though the number of kindergartens within the polygon is low (four), the surrounding areas at the north, west, and southwest have considerably high densities. In contrast, the areas in the east and the southeast have very low densities.

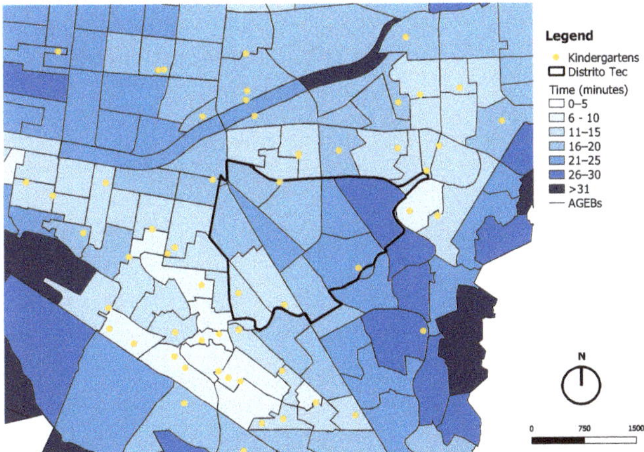

Figure 10. Average time to the 5 closest public kindergartens travelling by foot.

Three of the eleven AGEBs within Distrito Tec meet the constraint to access the closest five kindergartens in a maximum of 15 min, with all three AGEBs in the 11–15 min class. The rest of the AGEBs have considerably higher access times with four AGEBs in the 16–20 min class, three in the 21–25 min class, and one in the 26–30 min class. The lowest accessibility levels are represented in the AGEBs located to the southeast of the polygon. This is due to the relatively low density of destinations in the surrounding areas, especially to the east.

As most AGEBs within the polygon do not meet the 15 min travel time criteria, it can be argued that in terms of access to kindergartens by foot, the Distrito Tec Area cannot be considered a 15 minutes city. Nevertheless, additional information should be collected to know how many children in the Distrito Tec Area are within the age range to go to kindergartens. In the case that this number is low, the computation could be made to the closest kindergarten, as this one could host all the local demand. Such a scenario would considerably increase the levels of accessibility.

Figure 11 presents the average time by AGEB to access the five closest public primary schools by foot. Again, the density of destinations within the Distrito Tec Area is low, with only three schools. Yet, the surrounding areas at the north, northeast, west, and southwest have considerably high densities.

Only three out of eleven AGEBs from the Distrito Tec Area meet travel times of a less than or equal to 15 min. These AGEBs are located on two opposite corners of the polygon (northeast and southwest) and can meet the 15 min parameters due to the high number of destinations in the outer nearby areas. In contrast, six AGEBs require a travel time of 16–20 min to reach the closest five destinations. The higher travel time is due to the low number of destinations in central areas of the Distrito Tec Area. Finally, two AGEBs reach the 21–25 min class.

Given the previous results, it can be said that in terms of access to public primary schools, the Distrito Tec does not meet the requirements to be considered a 15 minutes city using only foot as the travel mode. However, by using other modes of transportation such as bikes, the travel times would drastically decrease and the accessibility level would increase.

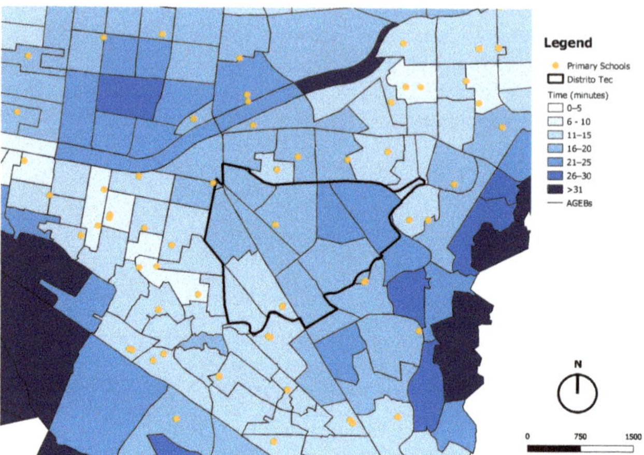

Figure 11. Average time to the 5 closest public primary schools travelling by foot.

As the age groups that attend secondary and high schools usually are able to use bicycles as their main travel mode, the following computations were performed considering bicycle as the transport mode. It is important to bear in mind that the average speed of a person travelling by bike is three-times higher than travelling by foot. Therefore, a much higher number of destinations can be accessed within the same time.

Figure 12 presents the average travel time by AGEB to reach the five closest public secondary schools travelling by bike. Even though the density of destinations in the Distrito Tec Area is extremely low (two) and remains low in the outer areas, the accessibility levels are rather high. All AGEBs can access the closest five destinations in a maximum of 10 min of travel time.

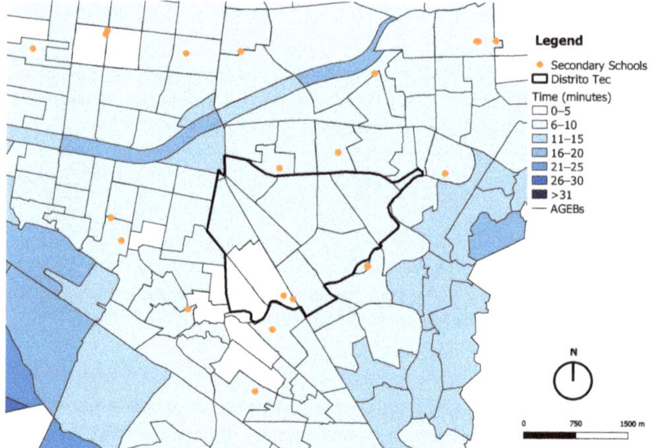

Figure 12. Average time to the 5 closest public secondary schools travelling by bicycle.

From the 11 AGEBs, two are in the lowest class of only 0–5 min of travel time. These AGEBs are located at the southwest border of the polygon, where the only two destinations inside the polygon are located and there are three others in the vicinity. In this sense, it can be said that using bicycles as the main travel mode, the Distrito Tec Area can be considered a 15 minutes city.

However, when the computation was run using foot as the travel mode, the accessibility levels drastically dropped (See Figure 13). As the share of trips done by bike is very low for the MMZ (only 0.8%), the current research project assumed that most of the population attending secondary schools uses other transport modes. In this sense, it is important to highlight the benefits, in terms of accessibility, that fostering the use of bicycles as the main travel mode would bring. One key factor to promote the use of bicycles is to create the safety conditions that users require; thus, implementing cycling lanes and open streets would be mandatory.

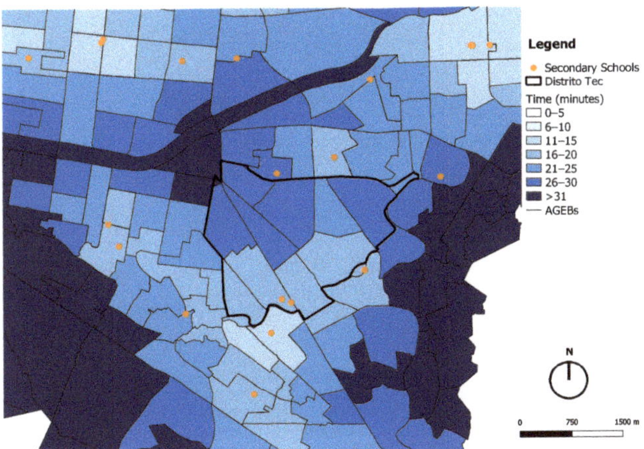

Figure 13. Average time to the 5 closest public secondary schools travelling by foot.

Figure 14 shows the average time by AGEB to reach the five closest public high schools by bike. The density of destinations is extremely low, not only in Distrito Tec (one) but in all the surrounding areas. Hence, the levels of accessibility are significantly lower compared to other variables. Only five out of eleven AGEBs in Distrito Tec can access the five closest destinations in a maximum travel time of 15 min. The other six are in the class above of a 16–20 min travel time.

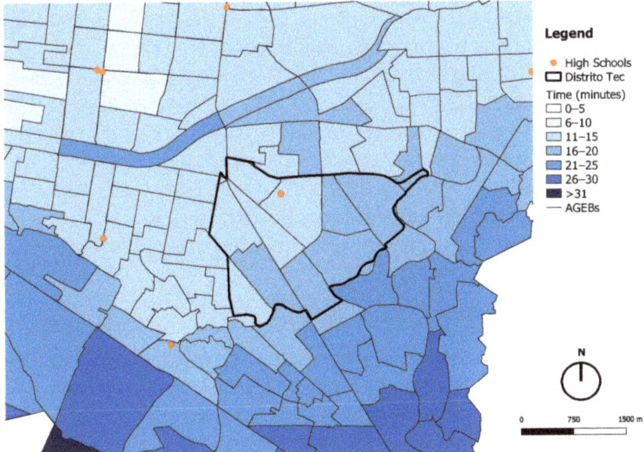

Figure 14. Average time to the 5 closest public high schools travelling by bicycle.

The surrounding areas at the south and east of Distrito Tec have an absolute lack of destinations. As a result, the population living there has to travel through Distrito Tec and other areas to reach a public high school. This phenomenon creates traffic in all the surrounding areas and other negative effects.

Again, a computation considering walking as the main travel mode to reach public high schools was included (see Figure 15). The accessibility results are worrying, as all the AGEBs within Distrito Tec, and the surrounding areas, demand a $\geq$31 min travel time. Hence, despite which travel mode is taken, it can be argued that the Distrito Tec Area cannot be considered a 15 minutes city regarding access to public high schools.

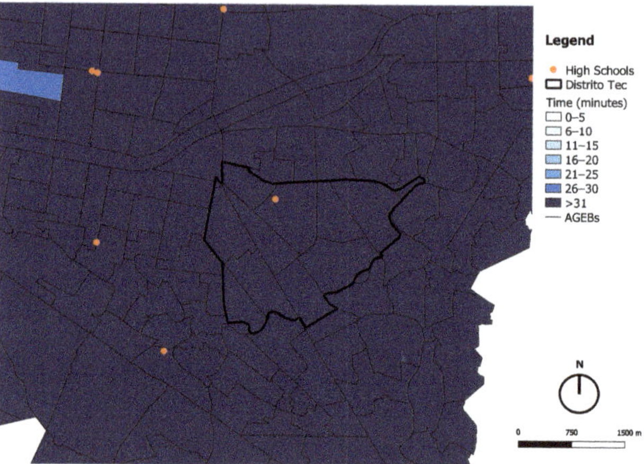

Figure 15. Average time to the 5 closest public high schools travelling by foot.

As seen along with the previous examples, education has many additional variables that have to be taken into consideration such as student capacity, local population age groups, etc., in order to entirely assess how accessibility behaves in the local area. Nevertheless, the accessibility measures are an excellent departing point for understanding the impact of transport modes and services' density on accessibility patterns.

The next variable is access to commercial activities, specifically to supermarkets. Due to a lack of public markets, as seen in many other Mexican cities, Monterrey's population does most of their grocery shopping at supermarkets. Hence, supermarkets are one of the most frequented destinations among the local population, and access to them is key to guaranteeing satisfactory quality of life.

According to the 2020 INEGI census, the average number of inhabitants per household in the MMZ is 3.18 people. This indicator is relevant given that people tend to buy groceries not for one, but for three persons when they go to the supermarket. For that reason, the bought items can be difficult or heavy to carry if the individual travels by foot or bicycle, so many people prefer to use motorised vehicles such as cars or public transport. Nevertheless, if accessibility to supermarkets is high, people would be encouraged to visit supermarkets more often than once per week and buy fewer items per visit so they could avoid the need to use a motorised vehicle.

Access to health is probably one of the most important things for an urban dweller, and this has been exemplified throughout the entire COVID-19 pandemic [22]. However, hospitals are very expensive infrastructures to build and run; hence, they usually have a metropolitan or regional radius of incidence in their planning processes (see Table 4). This section analyses what effects these planning processes have at the local level by measuring the average travel time by foot and bicycle to the closest public general hospital.

Table 4. Urban equipment regulatory system, recommended service radius. Source: Secretaría de Desarrollo Social (SEDESOL), 1999.

Type of Hospital	Regional	Urban
General Hospital (SSA)	60 km (or max. 2 h)	population centre
General Hospital (IMSS)	30 km to 200 km (or 30 min to 5 h)	1 h to the population centre
Urban Health Centre (SSA)	5 to 15 km (or 30 min)	1 km (or max. 30 minutes)
Medical Unit (ISSSTE)	30 to 60 min	30 min
Medical Unit (IMSS)	15 km (or 20 min)	5 km (or 10 min)
Health Clinic (ISSSTE)	only local scope	30 min
Hospital Clinic (ISSSTE)	2 h max.	population centre
Regional Hospital (ISSTE)	3 to 4 h	population centre

Figure 16 presents the average travel time by AGEB to the closest public general hospital by foot. The map shows that the number of destinations is extremely low, and within the Distrito Tec Area, there are no public general hospitals whatsoever. The surrounding areas of the polygon show a lack of destinations, with the only exceptions at the northeast area with two available hospitals and the southwest area with also two destinations.

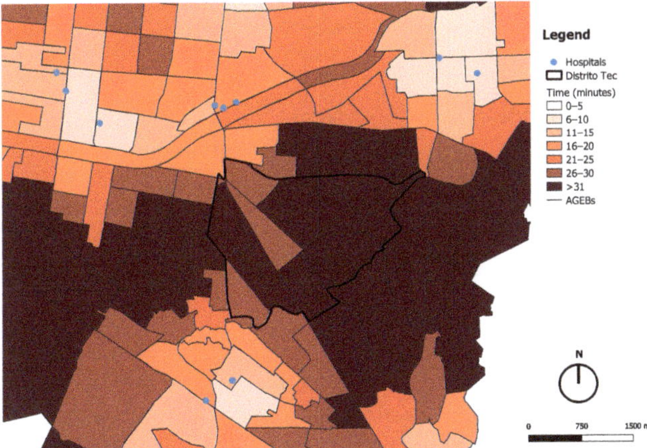

Figure 16. Average time to the closest public general hospital travelling by foot.

As a result of the lack of destinations available within Distrito Tec and in most surrounding areas, eight out of eleven AGEBs of the Distrito Tec's polygon require over a 31 min travel time to reach the closest destination. From the three AGEBs left, a travel time between 26 and 30 min is needed to reach the closest public general hospital. These results highlight that the location of hospitals follows metropolitan and regional planning processes, where access at the local scale is not relevant. Additionally, as seen in Figure 16, most of the hospitals tend to cluster. This has to do with the complementarity that one can give to others. A consequence of these localisation patterns is that travel distances from all neighbourhoods that do not have a nearby hospital are considerably high and demand motorised transport modes.

From the previous results, it can be said that Distrito Tec Area is far from becoming a 15 minutes city, if people travel only by foot.

Figure 17 shows the average travel time by AGEB to the closest public general hospital travelling by bike. In comparison to Figure 16, the travel times, both for Distrito Tec and the surrounding areas, drastically decrease. Nonetheless, many AGEBs still are beyond the 15 min parameter. In the Distrito Tec Area, six AGEBs require a 6–10 min travel time, three an 11–15 min travel time and two a 16–20 min travel time.

As nine out of eleven AGEBs meet the parameters to access a public hospital in under 15 min of travel time, the Distrito Tec Area could be considered a 15 minutes city.

Here, it becomes evident that transport modes dramatically influence access levels and that the location and number of destinations are not the only relevant factors in terms of accessibility.

There are two important lessons from the previous results. The first has to do with the importance that infrastructure at different scales has. One possible solution to improve the access to public general hospitals would be to increase the number of small clinics or medical centres throughout the city. By doing this, people could solve most of their basic health issues at local destinations and travel to large hospitals only when they have complications. The second lesson is that if public general hospitals need to have a metropolitan and regional scale and cluster in specific areas, they need to ensure high-quality transit connections. Building roads for cars is not the solution.

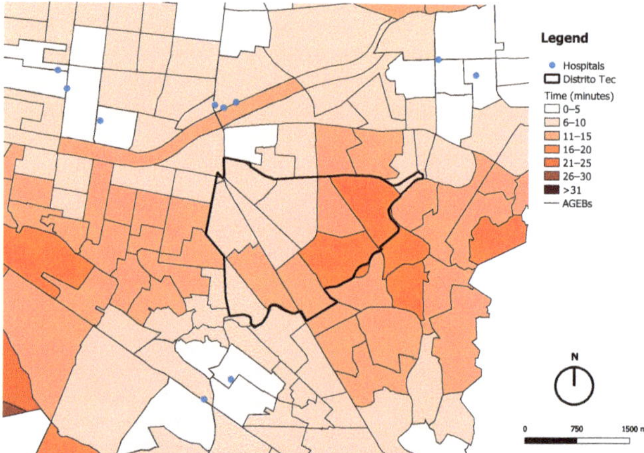

Figure 17. Average time to the closest public general hospital travelling by bicycle.

As 44% of the daily trips from the MMZ are travel to work, employment is the driving force of mobility. Thus, creating employment opportunities for the population at the local level is key to reducing the population's travel distances and times. By this, negative effects from traffic, such as accidents, pollution and low quality of life, would decrease.

It is important to mention that employment exists in a vast variety of ways. For example, there are many small- to middle-sized economic activities (such as grocery shops or dry cleaners) that are located throughout the entire Metropolitan Area; however, each one of them employs a relatively low number of persons (less than 50). In contrast, large economic activities tend to be less common but employ a larger amount of people ($\geq$51). These activities in the MMZ are usually industries (most located in suburban areas) and company headquarters or offices (most located in central areas).

As seen before, at the metropolitan level, the main employment centres (activities that employ $\geq$51 persons) are located in central areas of the city. As Distrito Tec is considered a privileged location in terms of centrality, it is expected that the number of main economic activities available at the local level will be high.

Figure 18 presents the number of economic activities that employ $\geq$51 persons (main employment centres) that can be reached from each AGEB within a 15 min travel time by foot. Clearly, most of the main employment centres are located on the two main avenues of Distrito Tec: Av. Eugenio Garza Sada and Av. Revolución.

Even though there is a high number of economic activities, the access levels are relatively low. The 11 AGEBs that compose the Distrito Tec Area are in the same access class of 6–17 destinations within a 15 min travel time by foot. From the surrounding areas, the northeast, east, southeast southwest, and west have very low access levels, all in the class of 0–6 destinations. This has to do with the low number of economic activities located in these areas. Once again, the population living in those areas has to travel to other parts of the city to work, causing traffic, as well as other negative effects.

People highly value the possibility to walk to their job; therefore, it is important to decentralise economic activities and create social housing in central areas that permit the local population to easily access their jobs. Figure 19 shows the number of economic activities that employ ≥51 persons that can be reached by AGEB using a bicycle. As soon as one looks at the map, it is evident that the access levels from the entire area drastically increase. Ten out of eleven AGEBs of the Distrito Tec Area can reach ≥105 destinations and the remaining one 95–105. The surrounding areas also present very high access levels, excluding the east. The eastern area is mainly residential with a gated community urban structure. It is also physically separated from the rest of the city by the river, with a small number of bridges that allow crossing to it. These two factors generate longer-distance trips to access job opportunities, which also results in longer travel times and motorised vehicle dependence.

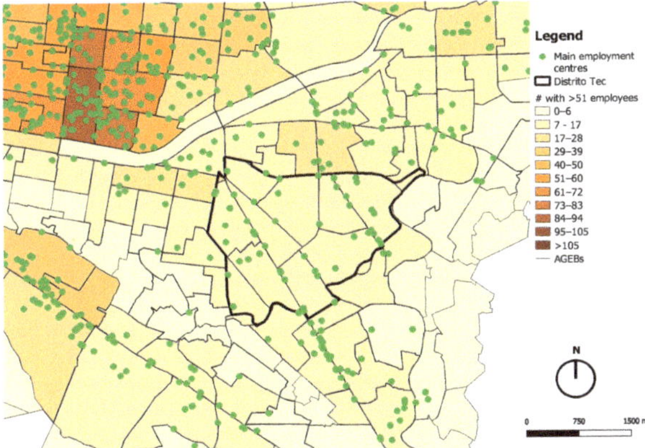

Figure 18. Number of main employment centres reachable within 15 min travelling by foot.

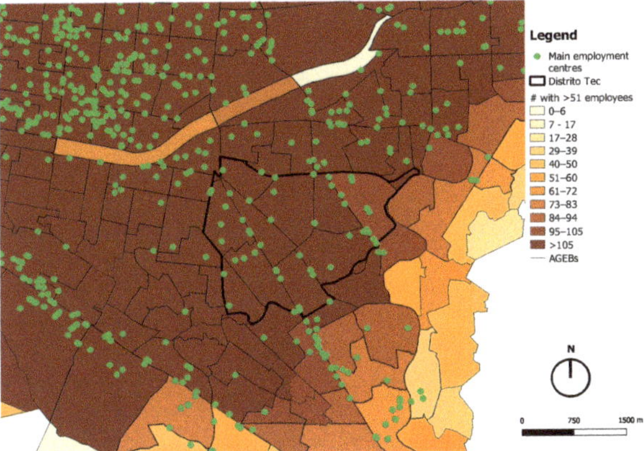

Figure 19. Number of main employment centres reachable within 15 min travelling by bicycle.

From the results presented in Figures 18 and 19, it can be said that in terms of access to main employment centres, the Distrito Tec Area can be considered a 15 minutes city. Nonetheless, additional information should be gathered to understand the specific type of economic activities located in the area and if the local population has the qualification, interests, and requirements to take those jobs.

Figure 20 presents the number of supermarkets that can be reached by foot within a travel time of a maximum of 15 min. There are five supermarkets within the Distrito Tec polygon; in the surrounding areas to the west and northeast, there is a high density. In contrast, to the east and southwest, there are no destinations available. To the south, the density is low with all supermarkets concentrating on Eugenio Garza Sada Avenue. The location of supermarkets has negative effects on the mobility of the Distrito Tec Area, as the population living in the eastern neighbourhoods has to travel to or through Distrito Tec to access a supermarket.

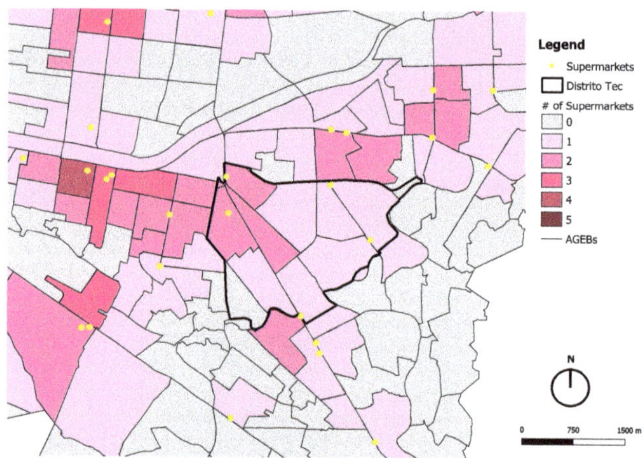

Figure 20. Number of supermarkets reachable within 15 min of travel by foot.

From the 11 AGEBs that conform the Distrito Tec Area, 2 cannot access a single supermarket, 6 can access 1 destination, and 3 can access 2 destinations. On that account, most of the AGEBs meet the parameters of the 15 minutes city using foot as the travel mode.

Figure 21 shows the number of supermarkets that can be reached by AGEB within a travel time of 15 min using a bicycle. Immediately, one can appreciate that the number of destinations becomes very high for pretty much the entire area, which is related to the fact that cycling speed is three-times higher than walking.

By bike, all of the AGEBs within Distrito Tec show high accessibility levels, only one AGEB being in the 31–50 destinations class, one AGEB in the 51–69 class, five AGEBs in the 70–93 class, and three AGEBs in the 94–195 class. Such results demonstrate that the Distrito Tec Area comfortably meets the parameters to be considered a 15 minutes city in terms of access to commercial activities (supermarkets). Additionally, the evidence collected is key to understanding the massive potential that cycling has in the city in terms of accessibility. By using this transport mode, the local population can drastically increase their access to commercial activities without relying on motorised vehicles. Once again, this should promote local authorities to create adequate conditions to make cycling safe and comfortable for the local population.

All the previous variables serve as a preliminary approach to understanding to what extent Distrito Tec is ready to become a 15 minutes city in terms of its road network and its destinations' availability. Nevertheless, as mentioned repeatedly, further variables and information have to be taken into consideration to provide a much more accurate analysis of the accessibility and the population's needs in the local area.

The results presented highlight the importance of thinking locally and not just at a metropolitan or regional scale when infrastructure and urban equipment are planned. They also demonstrate that micromobility, especially via bicycles, has a massive potential to increase local accessibility without, necessarily, having to increase the number of amenities in a given area.

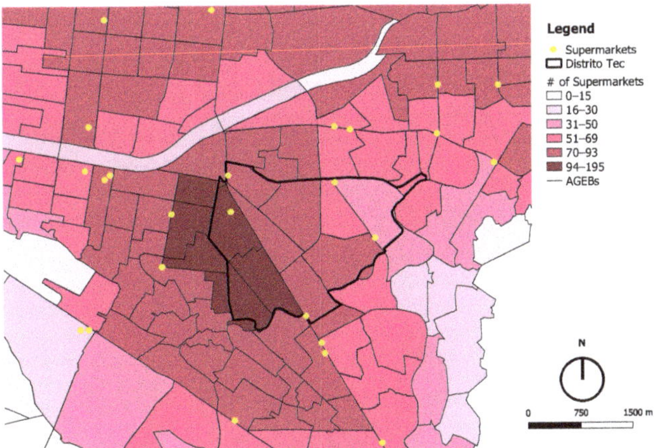

Figure 21. Number of supermarkets reachable within 15 min of travel by bicycle.

4. Discussion

The previous section illustrates that the urban planning processes, as well as the social, political, economic, and environmental factors that have structured the MMZ have failed to fulfil the accessibility conditions that the population requires. Hence, it becomes clear that a change of approach in planning is needed, where high-capacity transit and car-oriented mobility paradigms need to be left behind to adopt an accessibility-based process.

In order to move away from motorised vehicle dependence, not just mobility has to be restructured, but also the land use mix throughout entire cities. The localisation of opportunities in the city dictates the mobility patterns of the population; therefore, if cities want to concentrate on moving people rather than vehicles and creating access instead of traffic, a systemic change to the urban structure is required.

The previous results demonstrated that accessibility can provide extremely disaggregated data based on specific variables. This information is useful for having a starting point for planning processes by identifying priority areas in which to intervene and improve. Nevertheless, additional variables, especially concerning qualitative data (entitlement rights, a specific level of hospital attention, etc.) and social preference, have to be taken into account to obtain a more close-to-reality accessibility model.

After assessing and analysing the accessibility levels of the MMZ and of the Distrito Tec Area, the following recommendations regarding planning processes can be made:

1. **Rethink the planning of the city starting at the local level.** The results demonstrate that the MMZ is extremely heterogeneous and that not all of the population of urban areas has the same needs. Hence, before thinking about what interventions at a large scale have to be made, it is important to understand what is happening at the microlevel. To do so, speaking to citizens and opening citizen participation spaces are crucial to extract relevant information about what are the most important needs. If this process is replicated throughout the entire city, many of the metropolitan issues will be solved without having to perform metropolitan interventions. This is a win-win procedure, as the authorities spend less of their budgets and do so more wisely and local people obtain tailor-made solutions to their problems;

2. **Embrace land use mixture.** People need more than their house to carry out their daily lives. Therefore, each neighbourhood should have enough land use mixture to allow the local population to access commercial activities, recreation, education, employment, and health. If people can find all these opportunities in the vicinity of their household, mobility and all its negative effects would dramatically reduce. Furthermore, the short distances that people would require to travel would encourage the use of sustainable transport modes such as walking or cycling. The authorities

have a key role in influencing the structure of the city through building permits. Thus, they should not provide any permits to create residential areas unless they ensure that the needed land use mix will be available for the local population to address their daily needs;

3. **Promote the use of bicycles and other micromobility vehicles.** The results demonstrate that using a bike drastically increases the number of destinations that a person can reach in a given time. Therefore, the improvements that have to be made in terms of mobility and transport networks in the MMZ and at the local level should concentrate on increasing the quality and coverage of public transport for metropolitan journeys and creating open streets with safe and comfortable cycling lanes for short and local trips. By doing so, the number of destinations (opportunities) does not necessarily have to be increased, but only the ease of reaching them. If the MMZ wants to reduce the extreme use of motorised vehicles, the first step to take is to reduce the public space designated to them by creating a safe built environment for other transport users;

4. **Accessibility is a social and gender policy.** Usually, the groups with the lowest incomes live in areas of the city where accessibility levels are low and marginalisation levels are high; this is not an exception for the MMZ. Therefore, increasing accessibility in such areas will bring massive benefits to such populations as their transportation expenditures (for the specific case of women, this is known as the "Pink Tax") will decrease, as well as their travel distances and times, hence improving their quality of life and sense of belonging to their local community. Additionally, women in Mexican cities usually have more complex travel patterns than men, as they tend to visit many different destinations in one day (trip-chaining), carry heavy things, and accompany other persons (mobility of care) in their trips [23]. Consequently, making all these destinations easier and less expensive to access will bring important improvements to the travel behaviours and patterns of women. Finally, women argue that they usually feel unsafe when using public transport [23]; thus, using bikes and walking as their main transport modes will make them feel safer in their daily trips. In addition, further research should be conducted to assess the changing needs of an ageing society and accessible mobility in MMZ;

5. **Decentralise the city.** As in many other cities in the world, most of the main employment centres of the city are located in central areas. This localisation pattern results in a phenomenon where, during the morning, the origin of most trips to go to work is from suburban areas, and the destination is the centre of the city. In the afternoon, most people travel back home, and the mobility pattern is inverted, starting in the centre and ending in suburban areas. Such a way of structuring the city creates traffic, as well as extremely long average distances and travel times, which result in high economic, social, and environmental costs. Hence, it is important to apply planning processes or economic activities (especially commerce and services) that encourage the localisation of such activities in different parts of the city, creating subcentres and reducing the need for metropolitan mobility.;

6. **Increase the density of the city.** With only 698 people per square kilometre, the MMZ is not a dense city. Large sprawling cities with low densities go against accessibility, as there is not enough critical mass (demand) to implement some opportunities throughout the entire city. This generates a dependency on middle to large-distance motorised travel. Considering the nature of the MMZ, it is recommended that the city no longer sprawl. To do so, it has to generate all-income-level vertical housing in central areas. The authorities play a key role in doing so by putting in action economic incentives for developers (such as tax waivers or reductions) to build social vertical housing in central areas and by eradicating any permits to develop large monofunctional suburban residential gated communities. This is an ambitious plan for a city like the MMZ, as it has historically grown horizontally and with most of its population preferring single-family housing schemes rather than apartments.

Nevertheless, the transformation of its urban structure to a denser area will bring massive social, economic, and environmental benefits to the local population in the short, middle, and long term;

7. **Downsize urban infrastructure and equipment.** It has been demonstrated that high-capacity and large-scale destinations such as public general hospitals or shopping centres attract many persons who live in other districts. Therefore, having a larger number of small-scale opportunities dispersed along the city would allow locals to access their daily needs in their neighbourhood. This would drastically reduce middle to long distance mobility, as well as the use of motorised vehicles. It is important to mention that certain destinations, such as specialised hospitals, will still have a metropolitan radius of influence. For these kinds of opportunities, sustainable and diverse transport alternatives have to be guaranteed;

8. **Use of technology.** One of the major lessons that the COVID-19 pandemic has left is the massive potential that technology has to reshape education, commerce, and employment. This can generate important changes in the urban realm. By encouraging the use of technology in schools, commerce, and employment centres (e-learning, e-commerce, and home offices), mobility can drastically reduce the need to go to a specific location every day. Lowering the number of travelling people reduces the overloading of public transport, as well as the traffic levels, bringing important social, economic and environmental benefits. Additionally, spaces that are no longer needed, such as some office buildings, can be reconverted (taking advantage of their usual central location) to other uses such as commerce, housing, or schools. Finally, it must be said that using technology brings considerable economic benefits to employers, the government, and society, as mobility is expensive for all.

As part of the further research that can be performed to complement the current investigation, it will be important to incorporate more socio-economic and demographic factors that help provide a more accurate understanding of the qualitative differences that also affect the degree of accessibility in a given city. By doing so, the relationship between marginalisation levels and other social variables such as income, age, or gender with the level of accessibility will become clearer, providing useful data for developing public policies and urban interventions. Considering new accessibility measures to other destinations as green areas, universities and minor employment centres and more local commerce such as pharmacies or small shops can provide very useful data of the local level configuration of the city in different areas. These additional measures can also help to develop specific public policies based on the topic being analysed; for example, accessibility to green areas.

Even though the concept of the 15 minutes city focuses on transport modes such as walking and cycling, it would be relevant to run the accessibility measures using other transport modes, especially public transport and private vehicles. For this to be done, it is important to generate the necessary data, such as GTFS for public transport and a database that provides the average speed for private vehicles per road per time of the day.

The previous paragraph encloses a compilation of ideas that could complement and expand the current scope of work. It is important to consider that multiple variables play a dynamic role in the analysis of accessibility in a city, and by understanding this, it becomes clear that the possibilities to analyse accessibility from different perspectives are vast.

5. Conclusions

The urban realm is in constant and rapid transformation. Every day, new social, economic, and environmental needs arise, and cities have to evolve to satisfactorily overcome the challenges. However, this is not an easy task, as the number of variables and elements that conform the great system known as the city are uncountable.

Hence, dealing with complexity has become one of the biggest challenges for policy-makers and urban planners. The fast rate of change in cities demands tailor-made solutions for each specific context, in very short periods of time, that respond to short-, medium- and long-range time frames. Traditional planning has failed to do so, and this has to do, to a certain extent, with the approach taken. Humans like to think that a new invention can be the answer to all our problems, and the car was considered so. During the Twentieth Century, the car-dominated all urban planning processes, as it was seen as the ultimate alternative for mobilising goods and people within and around urban areas. Nevertheless, car-oriented planning was soon demonstrated to be unsustainable due to the extremely negative social and environmental effects that it has.

Taking more holistic, people-centric, and humble approaches to plan can be a more effective and equitable way to deal with complexity. Major urban planning and infrastructure interventions,on the other hand, in the urban realm take a long time to mature, and they can only be assessed after completion. Therefore, iterating planning processes can provide useful information about the rate of success of a given intervention to deal with a specific issue and allow the possibility to adjust the solution before being fully implemented. To do so, a huge amount of data has to be gathered and processed. On that account, technology plays a key role in urban planning, by providing high-quality open data, such as those collected from INEGI, and software, such as UrMoAC, with which urban planners can model at a high level of accuracy the state of a city. This allows a clear understanding of the issues of different parts of a city, permitting the elaboration of a hierarchy-based intervention.

Additionally, cities have to have a comparable and, ideally, standardised methodology to assess urban accessibility. By doing so, benchmarking, as well as the evaluation of public policies and urban interventions can be directly compared between cities of the same country, region, or even at a worldwide scale. This allows planners to argue to what extent a change is being made and at what rate. The current research project adds to that by using a transferable approach that can be replicated in different cities, making comparison of the results possible.

The results from the current research project demonstrate that taking an approach to urban planning based on accessibility provides many benefits in terms of sustainable planning. Accessibility measures permit obtaining highly disaggregated data of the performance of every part of the city regarding a specific variable, highlighting the areas with higher intervention needs and providing results that contrast how different transport modes perform for accessing a specific destination. Thus, using accessibility as the pillar of urban planning will allow cities into transform to more inclusive, environmentally friendly, and comfortable places to live.

Most cities in the world are still far from becoming 15 minutes cities, especially when whole metropolitan areas are considered, and the MMZ is not an exception. However, having this travel time frame as an objective should encourage policymakers, citizens and the private sector to transform cities from the local to the metropolitan level. It is also important to understand that in order to truly transform cities into better living environments, not only central areas need intervention, but the whole urban area. Everyone plays a key role in such a transformation, as it is necessary to agree that the current state of cities is going against sustainability and that an urgent urban, social, economic, political and environmental transformation is needed.

Author Contributions: Conceptualisation, J.d.-J.L.-S. and R.A.R.-M.; methodology, A.L.G.-B., J.N.-B. and M.A.R.-M.; software, A.L.G.-B., J.N.-B., M.A.R.-M. and D.K.; validation, J.N.-B. and D.K.; formal analysis, A.L.G.-B., J.N.-B. and M.A.R.-M.; investigation, A.L.G.-B., J.N.-B. and M.A.R.-M.; resources, R.A.R.-M. and J.d.-J.L.-S.; data curation, A.L.G.-B., J.N.-B. and M.A.R.-M.; writing—original draft preparation, A.L.G.-B., J.N.-B. and M.A.R.-M.; writing—review and editing, D.K., B.L.P.-H. and J.d.-J.L.-S.; visualisation, A.L.G.-B., J.N.-B., M.A.R.-M. and J.d.-J.L.-S.; supervision, R.A.R.-M., B.L.P.-H., D.K. and J.d.-J.L.-S.; project administration, M.A.R.-M., R.A.R.-M. and J.d.-J.L.-S.; funding acquisition, R.A.R.-M. and J.d.-J.L.-S. All authors read and agreed to the published version of the manuscript.

Funding: This research received funding by the CampusCity Initiative of the Tecnologico de Monterrey. CampusCity Initiative is funded by Fundación FEMSA. Fundación FEMSA had no role in study design, data collection and analysis, decision to publish, or preparation of the manuscript.

Institutional Review Board Statement: Not applicable.

Informed Consent Statement: Not applicable.

Data Availability Statement: The data used in this study were obtained from public databases and can be found in Table 1.

Acknowledgments: Special thanks to the Deutsches Zentrum für Luft-Raumfahrt (DLR) for making UrMoAC an open-source software and sharing relevant information about its use with our research team. The authors would also like to thank the Distrito Tec team, from the Instituto Tecnológico de Monterrey, for giving relevant information and feedback throughout the development of the entire project. We acknowledge as well the support of the California—Global Energy, Water & Infrastructure Innovation Initiative at Stanford University and the expedient proofreading by Josh Dimon

Conflicts of Interest: The authors declare no conflict of interest. The founders had no role in the design of the study; in the collection, analyses, or interpretation of data; in the writing of the manuscript; nor in the decision to publish the results.

References

1. Nations, U. Cities and Polution. Available online: https://www.un.org/en/climatechange/climate-solutions/cities-pollution#:~:text=According%20to%20UN%20Habitat%2C%20cities,cent%20of%20the%20Earth%27s%20surface (accessed on 28 May 2021).
2. Leo, A.; Morillón, D.; Silva, R. Review and analisys of urban mobility strategies in Mexico. *Case Stud. Transp. Policy* **2017**, *5*, 299–305. [CrossRef]
3. Chavez-Rodriguez, L.; Lomas, R.T.; Curry, L. Environmental justice at the intersection: Exclusion patterns in urban mobility narratives and decision making in Monterrey, Mexico. *DIE ERDE J. Geogr. Soc. Berl.* **2020**, *151*, 116–128. [CrossRef]
4. Montalvo-Urquizo, J.; Villarreal-Marroquín, M.G.; Hernández-Castillo, J.J.; Hernández-González, H.E. MWTP: Monterrey Weather, Traffic and Pollution Database for Geospatial Analysis. *arXiv* **2017**, arXiv:1703.04526.
5. Sisto, N.P.; Ramírez, A.I.; Aguilar-Barajas, I.; Magaña-Rueda, V. Climate threats, water supply vulnerability and the risk of a water crisis in the Monterrey Metropolitan Area (Northeastern Mexico). *Phys. Chem. Earth* **2016**, *91*, 2–9. [CrossRef]
6. de Nuevo León, G. Programa Integral de Movilidad Urbana Sustentable (PIMUS-ZMM). 2010. Available online: https://www.nl.gob.mx/publicaciones/documento-ejecutivo-pimus (accessed on 19 March 2021).
7. Hancke, G.P.; de Silva, B.C.; Hancke, G.P. The role of advanced sensing in smart cities. *Sensors* **2013**, *13*, 393–425. [CrossRef] [PubMed]
8. Tec, D. Distrito Tec. 2021. Available online: http://distritotec.itesm.mx/categorias/colonias/ (accessed on 19 March 2021).
9. Rezende, D.A.; Bustani, C.E.G. The urban districts and tourist improvement in the Strategic Digital City: The DistritoTec in Monterrey, Mexico. *TURyDES Rev. Tur. Desarro. Local* **2018**, *11*, unpaginated.
10. Krajzewicz, D.; Heinrichs, D.; Cyganski, R. Intermodal Contour Accessibility Measures Computation Using the 'UrMo Accessibility Computer'. *Int. J. Adv. Syst. Meas.* **2017**, *10*, 111–123.
11. Mora, H.; Gilart-Iglesias, V.; Pérez-Del Hoyo, R.; Andújar-Montoya, M.D. A comprehensive system for monitoring urban accessibility in smart cities. *Sensors* **2017**, *17*, 1834. [CrossRef] [PubMed]
12. Geurs, K.; Ritsema van Eck, J.R. *Accessibility Measures: Review and Applications. Evaluation of Accessibility Impacts of Land-Use Transportation Scenarios, and Related Social and Economic Impact*; Technical Report 90715; Netherlands Environmental Assessment Agency: The Hague, The Netherlands, 2001. Available online: https://www.pbl.nl/en/publications/Accessibility_measures_review_and_applications (accessed on 19 March 2021).
13. Moreno, C.; Allam, Z.; Chabaud, D.; Gall, C.; Pratlong, F. Introducing the "15-minutes City": Sustainability, Resilience and Place Identity in Future Post-Pandemic Cities. *Smart Cities* **2021**, *4*, 6. [CrossRef]
14. Oswald Beiler, M.R.; Phillips, B. Prioritizing Pedestrian Corridors Using Walkability Performance Metrics and Decision Analysis. *J. Urban Plan. Dev.* **2016**, *142*, 04015009. [CrossRef]
15. Krajzewicz, D.; Nieland, S.; Balzaretti, J.; Heinrichs, D. Assessing Sustainable Development Goals: A Transferable Approach Using Contour Accessibility Measures at the Example of Berlin and Mexico City. In Proceedings of the MOVICI-MOYCOT 2018: Joint Conference for Urban Mobility in the Smart City, Medellin, Colombia, 18–20 April 2018, pp. 1–7. doi:10.1049/ic.2018.0010 [CrossRef]
16. Park, S.; Nielsen, A.; Bailey, R.T.; Trolle, D.; Bieger, K. A QGIS-based graphical user interface for application and evaluation of SWAT-MODFLOW models. *Environ. Model. Softw.* **2019**, *111*, 493–497. [CrossRef]
17. Kwan, M.P. Space-time and integral measures of individual accessibility: A comparative analysis using a point-based framework. *Geogr. Anal.* **1998**, *30*, 191–216. [CrossRef]
18. López-Alonso, M. Growth with Inequality: Living Standards in Mexico, 1850–1950. *J. Lat. Am. Stud.* **2007**, *39*, 81–105. [CrossRef]
19. CONAPO. Anexo C Metodología de Estimación del Índice de Marginación Urbana. 2010. Available online: http://www.conapo.gob.mx/work/models/CONAPO/Resource/862/4/images/06_C_AGEB.pdf (accessed on 28 May 2021).

20. CONAPO. Datos Abiertos del Índice de Marginación. Available online: http://www.conapo.gob.mx/es/CONAPO/Datos_Abiertos_del_Indice_de_Marginacion/ (accessed on 28 May 2021).
21. Saroha, J. Types and Significance of Population Pyramids. *World Wide J. Multidiscip. Res. Dev.* **2018**, *4*, 59–69.
22. Bai, X.; Nagendra, H.; Shi, P.; Liu, H. Cities: Build networks and share plans to emerge stronger from COVID-19. *Nature* **2020**, *584*, 517–520. [CrossRef] [PubMed]
23. Soto-Villagrán, P. *Análisis de la Movilidad, Accesibilidad y Seguridad de las Mujeres en tres Centros de Transferencia Modal (CETRAM) de la Ciudad de México*; Technical Report IDB-TN-1780; Banco Interamericano de Desarrollo: Mexico City, Mexico, 2019. [CrossRef]

MDPI
St. Alban-Anlage 66
4052 Basel
Switzerland
Tel. +41 61 683 77 34
Fax +41 61 302 89 18
www.mdpi.com

Applied Sciences Editorial Office
E-mail: applsci@mdpi.com
www.mdpi.com/journal/applsci